Reproductive Biology of South American Vertebrates

William C. Hamlett
Editor

Reproductive Biology of South American Vertebrates

With 102 Illustrations

Springer-Verlag
New York Berlin Heidelberg London Paris
Tokyo Hong Kong Barcelona Budapest

William C. Hamlett
Indiana University School of Medicine
South Bend Center for Medical Education
University of Notre Dame
Notre Dame, Indiana 46556, USA

Frontmatter illustration by William C. Hamlett, 1992.

Library of Congress Cataloging-in-Publication Data
Reproductive biology of South American vertebrates/edited by William C. Hamlett.
p. cm.
Includes bibliographical references and index.
ISBN-13:978-1-4612-7701-9 e-ISBN-13:978-1-4612-2866-0
DOI: 10.1007/978-1-4612-2866-0

1. Vertebrates—South America—Reproduction. I. Hamlett, William C.
QP251.R44447 1992
596'.016'098—dc20 91-36123
Printed on acid-free paper.

Softcover reprint of the hardcover 1st edition 1992

Production managed by Karen Phillips; manufacturing supervised by Jacqui Ashri.
Typeset by Asco Trade Typesetting Ltd., Hong Kong.

9 8 7 6 5 4 3 2 1

Preface

South America is one of the few areas of the world that includes vast areas of as yet undisturbed natural habitats. These areas are home to as many as half the world's species of plants and animals. Many of the animals inhabiting these areas are of direct and immediate economic importance, while others are of unknown potential value. With economic development of these areas, there is a coincident disruption of natural habitats that have a direct impact on the reproductive capabilities of the affected species. Reproduction is central to the survival of all species and, regardless of the environmental stresses imposed on them, scientific attention must be focused on reproductive biology as a way to deal with these pressures. It is vitally important to gather as much research data as possible on the reproductive biology of the species on this continent in order to provide governmental agencies and scientists with the most accurate information on which to base decisions regarding development. This volume is intended to draw attention to these pressing matters by gathering a wide representation of scientists actively engaged in reproduction research relating to South American vertebrates. It is intended that this volume will serve as a resource for individuals and organizations interested in reproductive biology and species survival. An additional benefit is that economically important species, as well as potentially important ones such as sharks, will be discussed. Hence, this volume will contribute information on the reproductive biology of animals that are rare or endangered, thereby aiding conservation efforts as well as providing invaluable biological data. An effort has been made to have all the major vertebrate groups represented including Chondrichthyes, Osteichthyes, Amphibia, Reptilia, Aves, Marsupilia, and Mammalia.

This volume was written to have an important impact not only on the synthesis of available information, but also in the setting of research priorities for some years to come. It is hoped that this volume will generate a sense of community among scientists in the diverse field of vertebrate reproductive biology and that workers will organize individual workshops focused on particular vertebrate groups to thoroughly debate a topic and arrive at novel investigative approaches.

The world's largest tropical rain forest is in South America and has been assaulted in recent years by a drive for economic development. This and the consequent deforestation has and will continue to have an important impact on the plants and animals resident there. In 1984, paving of Highway BR-364 connecting Cuiaba, the capital of Mato Grosso, and Porto Velho, the capital of Rondonia, Brazil, was completed. As a consequence, Rondonia's population has doubled in the 1980's and more than 20 percent of the rain forest has been destroyed. The concern lies not in the development of the area so much as the uncontrolled way in which this development is handled. Now 30 to 40 percent of the lands in the Northwest Region have been placed in reserves. Research efforts by reproductive biologists in these preserves and in the jungle is critically needed to insure adequate resource information on which to best plan development and coordinate efforts that will mutually benefit man and the animal life there.

The proposed 500 mile road, financed by Japan, linking existing highways in western Brazil and Peru to the Pacific coast will accelerate deforestation of the rain forest by providing a shorter supply route for timber destined for Asia. This route would open vast tracts of mostly undisturbed rain forest. Amid the growing international debate relating to the impact of human migration, development, and deforestation, there emerges a critical need to collect and analyze data on the reproductive biology of the impacted species, many of which are almost completely unknown. The impact of damage to South America's equatorial rain forest will play a role in the determination of world climate and directly affect at least half of the earth's animal and plant species that live there. Researchers estimate that 15 to 20 percent of the Amazon's primary forest has been cleared in the last two decades and that, if 40 percent is deforested, the remainder may suffer irreparable damage.

A government study conducted in Brazil has concluded that in less than ten years the largest species of commercial fish in the Amazon River has declined by 50 percent. This is due primarily to the impact of the soaring population in the Amazon basin and its tributaries and predatory fishing practices. The human population has grown from 11 million to 19 million in 15 years. There is a growing shortage of fish used for food. In the Amazon between Tabatinga and Itacoatiara, the harvest of the tambaqui, an important food fish, fell from nine tons in 1979 to half that in 1986. Over the same period, harvests of the economically important pirarucu fell from 1,140 to 380 tons.

Other factors adversely impacting the fish population are that this region is the world's largest hunting grounds for ornamental aquarium fish. In 1974, Amazonia exported six million live aquarium fish and, by 1985, the number had risen to 18 million. There are no organized efforts to farm ornamental species and the stock levels are unknown. Nonetheless, the supplies are diminishing.

Past and proposed hydroelectric dams have devastated areas of the

Amazon at the expense of its animal life. Some proposals would require new hydroelectric dams to include fish farms and restock the rivers and reservoirs. It is critical to understand the reproductive biology of these species in order to make the proper and most efficient decisions.

Aside from the damage to the rain forests, other populations of animals face increasing pressures, among these are marine fish and mammals. Japan's large harvests of fish and whales have had a direct impact on stocks. Other species with emerging commercial importance include sharks. These animals are migratory; hence, they are susceptible to pressure from environmental pollution all along their range. They have a long gestation period, up to two years in some species, and give birth to a relatively few number of young. Therefore, they have a large reproductive investment in a reduced number of offspring. More and more pounds of shark meat are being harvested and a renewed interest in the shark leather industry is emerging. A serious threat is the practice of finning, where the fishermen cut off the shark's fins and discard the carcass in the water. The fins are valued in Asia for shark fin soup. More than 35 million pounds of shark, three times the annual domestic catch, are wasted by finning in the United States. These animals are, consequently, very susceptible to over fishing. This practice could decimate species in a relatively short time. It is essential to gather information on their reproductive biology, including migration patterns, litter size, and sex ratios of adults and young.

Planning for protection of the environment involves a broad consideration of issues such as global warming, stratospheric ozone depletion, urban ozone, wetlands and tropical forest destruction, and hazardous waste cleanup. Successful strategies to deal with these diverse problems must come from societies that develop the scientific and engineering manpower to address environmental problems and also expand public education as to the role of the individual in confronting these issues. Environmental problems are not solely the responsibility of either industrialized countries or emerging ones. Likewise, solutions must come from all sectors. We, as reproductive biologists, hope this volume will make a contribution towards our common goal of assuring the continued prosperity and longevity of South America.

William C. Hamlett, Ph.D.

Acknowledgments

I would like to express my sincere gratitude to my colleagues for their contributions to this volume and for their patience during the editorial process. On behalf of all the contributors I wish to thank Prof. Mauricio Moscovici, President of the XIII Federative International Congress of Anatomy, for facilitating the original workshop on which this volume is based. I wish to acknowledge the cooperation of Drs. Harold Stolberg and David Keeland, Division of International Programs, National Science Foundation, in supporting the original concept of a workshop devoted to reproductive biology of South American vertebrates. The financial support of the National Science Foundation made the workshop possible. Thanks are due the editorial department at Springer-Verlag New York, for their invaluable assistance in completing this endeavor. A special thanks goes to all the anonymous reviewers whose careful consideration of the manuscripts contributed to the success of this venture. Finally, my warmest appreciation goes to my wife, Martha, and our children, Hayes, Kathleen, and Alex, for providing the motivation for me to complete this project.

Contents

Eutheria

Contributors

Abbott, David H., Wisconsin Regional Primate Center and Department of Obstetrics and Gynecology, University of Wisconsin, Madison, Wisconsin 53715, USA

Bagnara, Joseph T., Department of Anatomy, University of Arizona, Tuscon, Arizona 85724, USA

Blackburn, Daniel G., Department of Biology, Life Sciences Center, Trinity College, Hartford, Connecticut 06106, USA

Caldwell, Janalee P., Department of Zoology and Oklahoma Museum of Natural History, University of Oklahoma, Norman, Oklahoma 73019, USA

Cikutovic, Marcos A., Departamento de Ciencias Biologicas, Facultad de Ciencias de la Salud, Universidad de Antofagasta, Antofagasta, Chile

Ferri, Antonio G., Sistema Integrado de Bobliotecas da Universidade de Sao Paolo, Cidade Universitaria, Armando de Slles Oliviera, Cx. Postal: 8191—CEP: 05508. Sao Paolo, SP, Brazil

Fitzpatrick, Lloyd C., Environmental Effects Research Group, Division of Environmental Sciences, Department of Biological Sciences, University of North Texas, Denton, Texas 76203, USA

Gruber, Samuel H., Bimini Biological Field Station, Division of Marine Biology and Fisheries, Rosentiel School of Marine and Atmospheric Science, University of Miami, Florida 33149-1098, USA

Guerra, Carlos G., Instituto de Investigaciones Oceanologicas, Universidad de Antofagasta, Casilla 170, Antofagasta, Chile

Hamlett, William C., Indiana University School of Medicine, South Bend Center for Medical Education, University of Notre Dame, Notre Dame, Indiana 46556, USA

Harder, John D., Department of Zoology, 1735 Neil Avenue, The Ohio State University, Columbus, Ohio 43210, USA

Hellgren, Eric C., Caesar Kleberg Wildlife Research Institute, Texas A&I University, Kingsville, Texas 78363, USA

Hess, David L., Department of Reproductive Physiology and Behavior, Oregon Regional Primate Research Center, Beaverton, Oregon 97006, USA

Lacher Jr., Thomas E., Archbold Tropical Research Center, Clemson University, Clemson, South Carolina 29643-1019, USA

Leyton, Víctor, Departamento de Morfología Experimental, Facultad de Medicina, Universidad de Chile, Casilla 2988, Correo Central, Santiago, Chile

Lochmiller, Robert L., Department of Zoology, Oklahoma State University, Stillwater, Oklahoma 74078, USA

Machado-Allison, Antonio, Instituto Zoologia Tropical, Universidad Central de Venezuela, Caracas, 1041-A, Venezuela

Marmontel, Miriam, University of Florida, School of Forest Resources and Conservation, Gaineseville, Florida 32601, USA

Menezes, Naercio A., Museu de Zoologia, Universidade de Sao Paolo, Av. Nazare 481, CEP 04263, Sao Paolo, SP, Brazil

Miglino, Maria A., Universidade de Sao Paolo, Fac. de Medicina Veterinaria e Zootecnia, Cidade Universitaria, Av. Lucio Martins Rodrigues—Travessa 4 Bloco 5, Departamento de Cirurgia e Obstetrica, 05508—Sao Paolo, SP, Brazil

Moore, H.D.M., Institute of Zoology, Zoological Society of London, Regent's Park, London NW1 4RY, United Kingdom

Odell, Daniel K., Sea World of Florida, Orlando, Florida 32821-8097, USA

Rasmussen, L.E.L., Department of Chemical and Biological Sciences, Oregon Graduate Institute, Beaverton, Oregon 97006, USA

Rastogi, Rakesh K., Facultá di Zoologia, Universitá di Napoli, 80134 Napoli, Italy

Rasweiler IV, John J., Department of Obstetrics and Gynecology, Cornell University Medical College, 1300 York Avenue, New York, New York 10021, USA

Reynolds III, John E., Eckerd College, St. Petersburg, Florida 33733, USA

Schwassmann, Horst O., Department of Zoology, 223 Bartram Hall, University of Florida, Gainesville, Florida 32611, USA

Tolson, Peter J., Toledo Zoological Society, 2700 Broadway, Toledo, Ohio 43609, USA

Valencia, José, Departamento de Ciencias Ecologicas, Facultad de Ciencias, Universidad de Chile, Casilla 653, Santiago, Chile

de M. Vazzoler, Anna Emilia A., Nucleo de Pesquisas em Limnologia, Ictiologia e Aquicultura, Departamento de Biologia, Universidade Estadual de Maringa, Av. Colombo 3690, CEP 87020, Maringa/Parana, Brazil

Vitt, Laurie J., Oklahoma Museum of Natural History and Department of Zoology, University of Oklahoma, Norman, Oklahoma 73019, USA

Wake, Marvalee H., Department of Integrative Biology and Museum of Zoology, University of California, Berkeley, California 94720, USA

Chondrichthyes

1

Modes of Reproduction in the Elasmobranchs of Brazil

William C. Hamlett, Antonio G. Ferri, and Maria A. Miglino

Introduction

The elasmobranch fishes include the sharks, skates, and stingrays. These cartilaginous fishes are primarily marine organisms, but some species enter brackish to fresh water estuaries, lagoons, and bays. Some sharks and stingrays, including the bull shark, *Carcharhinus leucas*, certain sawfish (Pristidae), potamotrygonid stingrays, and dasyatid stingrays, can inhabit purely freshwater rivers and lakes. The elasmobranch fauna of Brazil is rich and diverse and is of interest from both scientific and economic standpoints. Of the estimated 350 extant species of sharks world-wide, some 25% are of limited use in fisheries, 20% are of considerable use, and 7% represent major fisheries species (Compagno 1984). The approximately 470 species of batoids include the sawfishes, guitarfishes, skates, stingrays, and electric rays.

Elasmobranch reproductive modes are either oviparous or viviparous and have been further categorized as is shown in Table 1.1 (Hamlett 1989a).

Oviparous elasmobranchs include sharks of the Scyliorhinidae and the skates Rajidae. Category A for viviparous species includes species that rely primarily on yolk for growth and development, but may also be nourished by nutrient histotroph, "uterine milk," secreted by uterine villi. Included in this grouping are various dogfishes of the Squaliformes. Category B has species in which the maternal uterus develops hypertrophied uterine villi termed trophonemata. They produce a copious histotroph during the latter stages of embryonic development to nourish the embryos after their yolk stores have been consumed. Included in this category are members of the Torpedinidae, Dasyatidae, Myliobatidae, and Mobulidae. Category C includes the Odontaspididae, Alopiidae, and Lamnidae. Embryos have very small yolk stores that are depleted at an early age while the mother continues to ovulate during gestation. The embryos have precocious development of functional dentition and begin to sup on uterine eggs (oophagy). Some species practice intrauterine cannibalism by feeding on their siblings.

Table 1.1. Elasmobranch species of Brazil by reproductive mode

I. Oviparous
II. Viviparous
  A. Internal incubators with or without uterine villi
  B. Internal incubators with trophonemata
  C. Internal incubators with oophagy and intrauterine cannibalism
  D. Internal incubators with yolk sac placenta

Usually only one embryo per uterus survives and it achieves gigantic proportions. The newborn is a huge, practiced predator. Category D is restricted to the sharks Carcharhinidae and Sphyrnidae. Embryos are initially yolk dependent. As yolk sac reserves are depleted, the sac becomes modified as a functional placenta. The placenta forms an apposition with the uterine wall to produce a functional uteroplacental unit that carries out nutrient and respiratory functions.

Growth rates among elasmobranchs, in particular sharks, are slow when compared to bony fish. The majority of sharks have a life span of 20 to 30 years with some smaller species having relatively short life spans of 10 to 15 years. While the smaller species can mature in two to three years, most sharks do not mature until they are six or seven years or older. All elasmobranchs employ internal fertilization and more maternal energy is invested in producing fewer young when compared to teleosts.

Male sharks have paired intromittent organs called claspers which are modified pelvic fins. During copulation, one clasper is inserted into the female genital opening and sperm flows down a groove in the clasper from the genital pore. Paired siphon sacs fill with seawater which, during copulation, is squirted to flush sperm from the clasper into the female genital tract.

Mating generally precedes ovulation and sperm are stored for a variable period of time in the shell or nidamental gland. This is a specialized region of the upper oviduct. In some species, viable sperm can be stored for up to one year. The gestation period of placental sharks generally ranges from nine to eleven or twelve months. Many sharks require an additional year to recuperate before mating again. Lemon sharks require 13 to 15 years to mature and become sexually active (Gruber 1991). They have a 12-month gestation period and mate every other year. Usually 8 to 12 pups are born but first year mortality approaches 50%. Some smaller sharks, such as the Atlantic sharpnose *Rhizoprionodon terraenovae*, mature after two years but their reproductive potential is limited to three to four pups per season. The general life history pattern of elasmobranchs is characterized by slow growth, delayed maturity, low fecundity, and production of relatively few offspring that are born at a developmentally advanced stage.

The most prolific authority on the elasmobranchs of Brazil is Victor Sadowsky. He has worked for over thirty years documenting reproductive

Table 1.2. Representative elasmobranchs of Brazil by reproductive modes

	OVIPAROUS	
	(SHARKS)	
Shroederichthys tenuis	*Scyliorhinus haeckeli*	
	(SKATES)	
Raja agassizi	*Raja cyclophora*	
Raja platana	*Raja castelnaui*	*Psammobatis extenta*
Sympterygia acuta	*Sympterygia bonaparti*	
	VIVIPAROUS	
	Internal Incubators With or Without Uterine Villi	
	(SHARKS)	
Etmopterus gracilispinis		*Etmopterus pusillus*
Ginglymostoma cirratum		*Heptranchias perlo*
Hexanchus griseus		*Mustelus schmitti*
Notorynchus cepedianus		*Squaliolus laticaudus*
Squalus cubensis		*Squalus laticaudus*
Squatina argentina		
	(GUITARFISHES)	
Rhinobatos percellens		*Rhinobatos horkelii*
Zapteryx brevirostris		
	(SAWFISHES)	
Pristis pectinata		*Pristis perotteti*
	Internal Incubators with Trophonemata	
	(RAYS)	
Torpedo puelcha		*Narcine brasiliensis*
Discopyge tschudii		*Dasyatis sayi*
Dasyatis guttata		*Dasyatis americana*
Dasyatis centroura		*Gymnura altavela*
Gymnura micrura		*Aetobatus narinari*
Myliobatis freminvillei		*Rhinoptera bonasus*
Manta birostris		*Mobula hypostoma*
	(FRESHWATER RAYS)	
Potamotrygon motoro		*Potamotrygon circularis*
Elipesurus spinicauda		*Disceus thayeri*
	Internal Incubators with Oophagy and Intrauterine Cannibalism	
	(SHARKS)	
Alopias superciliosus		*Alopias vulpinus*
Carcharodon carcharias		*Cetorhinus maximus*
Eugomphodus taurus		*Odontaspis noronhai*
Isurus oxyrhincus		*Odontaspis taurus*
Lamna nasus		*Rhincodon typus*

(*continued*)

Table 1.2. (*Cont.*)

Internal Incubators With Yolk Sac Placenta	
(SHARKS)	
Carcharhinus acronotus	*Carcharhinus brevipinna*
Carcharhinus falciformis	*Carcharhinus isodon*
Carcharhinus leucas	*Carcharhinus limbatus*
Carcharhinus longimanus	*Carcharhinus obscurus*
Carcharhinus plumbeus	*Carcharhinus porosus*
Carcharhinus signatus	*Galeocerdo cuvier*
Isogomphodon oxyrhynchus	*Mustelus canis*
Mustelus higmani	*Negaprion brevirostris*
Prionace glauca	*Rhizoprionodon lalandei*
Rhizoprionodon porosus	*Sphyrna lewini*
Sphyrna media	*Sphyrna mokarran*
Sphyrna tiburo	*Sphyrna tudes*
Sphyrna zygaena	

From Figueiredo 1977

parameters, describing species, and establishing a shark museum in Cananeia. His work, primarily on the elasmobranchs of the littoral zone of Sao Paulo, has been extensive and we, as students of elasmobranchs, own a deep debt of gratitude to him. His works are far to extensive to enumerate here but a sample of the breadth of his work can be appreciated by a review of some of his contributions (Sadowsky 1965, 1967, 1970, 1971, 1973, 1974, 1976; Sadowsky et al. 1984, 1985).

Oviparous Species

The skates (Fig. 1) constitute a strictly oviparous group (Bigelow and Schroeder 1953). Fertilization is internal and eggcases are formed by the shell gland shortly thereafter. The cases are rectangular with straight, horny tendrils at each corner. They are laid in muddy or sandy areas and adhere to small pieces of rock, shell, or seaweed. The eggcases are deposited when the embryos are at a very early stage of development and the egg yolk is the sole source of organic material for development. Embryogenesis ensues and, at hatching, the embryo emerges through a transverse slit at one end of the eggcase. Embryonic development may take from three to eight months depending on the species and habitat.

Very little is known of the biology of the Brazilian freckled catshark, *Scyliorhinus haeckeli*. It inhabits coastal waters of Brazil from depths of 37 to 402 meters and deposits 6 to 7 cm long eggcases (Bigelow and Schroeder 1948; Springer and Sadowsky 1970). At present there is no fisheries activity for this species.

In oviparous species, yolk is passed up the ductus vitellointestinalis within the yolk stalk to the internal yolk sac, which is confluent with the

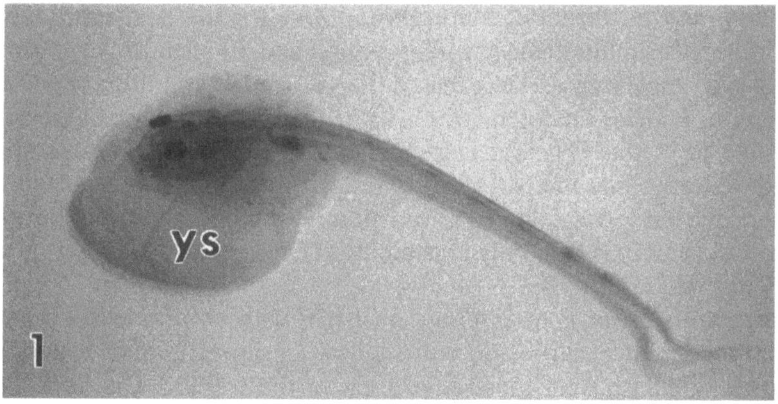

Figure 1.1. An oviparous skate embryo, *Raja erinacea*, is nourished by contents of its yolk sac (ys).

alimentary canal of the embryo. Yolk is also likely digested in the yolk syncytial-endoderm complex of the external yolk sac, as has been demonstrated in preimplantation embryos that develop a placenta (Hamlett and Wourms 1984; Hamlett et al. 1987).

Internal Incubators With or Without Uterine Villi

The uterine epithelium in some species may display a smooth contour or show varying degrees of uterine villi. The uterus produces histotroph or "uterine milk," the quantity and chemical composition of which varies from species to species. The major components are lipid and protein. Histotroph production may be minimal to extensive. In its simplest form the uterine epithelium is smooth, as in *Squalus acanthias*, or it may form rows of short villi, as in other squaloids. *Squatina* also shows the same simple uterine organization. If the dry weight of the egg and embryo are compared, these simple internal incubators show a 15 to 55% weight loss during gestation (Ranzi 1932).

Gilbert and Bevelander (1959) found that no nutrient transfer from mother to fetus was effected during gestation in the spiny dogfish, *Squalus acanthias*. Ultrastructural analysis of the gravid uterus and yolk sac by Jollie and Jollie (1967a, 1967b) led to the conclusion that the uterus functions only in respiration and osmoregulation. Yolk is digested in the endoderm of the external yolk sac, as well as being transported to the fetal alimentary tract.

Very little scientific information exists on the reproductive biology of the Brazilian guitarfish, *Rhinobatos horkelii*; however, recent studies (Leesa 1982; Leesa et al. 1986) provide some information. In April, in coastal waters 5 to 15 m in depth, females rapidly undergo parturition, copulation, ovulation, and fertilization. Maturation of ova for the next reproductive cycle proceeds simultaneously with gestation and all fertilized eggs are enclosed in a single eggcase in each of the two functional uteri. Following copulation, females inhabit depths greater than 50 m and the eggs remain in a state of "lethargy" until December. At this time, females return to coastal waters, while the embryos continue to develop. It is reported that the period from fertilization to parturition is 12 months but that development of the embryos lasts only four months (Leesa et al. 1986). Litter sizes are from 4 to 12.

The sawfish *Pristis* is an internal incubator with uterine villi. Largely a bottom feeder, it occurs widely in shallow, inshore, marine waters and estuarine, brackish, and freshwater areas in the tropics. The large tooth sawfish, *Pristis perotteti*, matures at a length of approximately three meters which takes about 10 years (Thorson 1982). An average of 7.3 young per litter are born in freshwater, following a gestation period of five months. The size at birth is 73 to 80 cm total length and breeding probably occurs every other year (Thorson 1976). Prior to parturition the saw is soft, leathery, and covered by a firm membranous sheath (Gudger 1951) that protects both the embryo and mother.

Development in the whale shark, *Rhincodon typus*, is still a matter of debate. Based on a large eggcase, 30 cm long, 14 cm wide, and 9 cm thick containing a near full-term embryo 36 cm in length found in the Gulf of Mexico, it was initially supposed that the mode of reproduction was oviparity (Baughman 1955). Wolfson (1983), however, has suggested that the Gulf of Mexico eggcase may have been aborted before term. The relative rarity of whale shark eggs, the thinness and lack of tendrils on the only known case, and the considerable yolk store suggests that the whale shark may be an internal incubator.

The nurse shark, *Ginglymostoma cirratum*, is an internal incubator with intrauterine development primarily supported by the large yolk deposits (Compagno 1984). The number of young is 21 to 28 per litter.

Internal Incubators With Trophonemata

All rays are initially nourished by yolk sac contents. As these stores are depleted during development, the maternal uterine epithelium hypertrophies to form spatulate to villiform secretory appendages termed trophonemata (Figs. 1.2, 1.3) (Wood-Mason and Alcock 1891). In the cownose ray, *Rhinoptera bonasus*, embryonic growth to 215 mm disc width is accomplished at the expense of yolk. Only the left oviduct is functional and a single embryo develops per reproductive cycle. Embryos are initially surrounded by an eggcase produced by the nidamental gland. As the embryo grows and escapes from the eggcase, trophonemata develop and produce nutrient histotroph. By the time the embryos are 257 mm disc

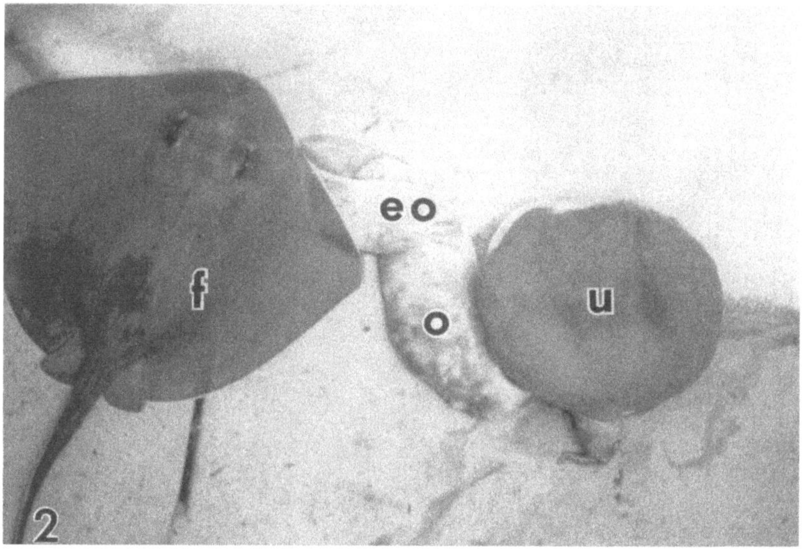

Figure 1.2. A term Southern stingray fetus (f), *Dasyatis americana*, is shown adjacent to the everted uterus (u) which contained it; eo = epigonal organ; o = ovary.

width, the trophonemata have reached their maximum length, 2 to 3 cm (Hamlett et al. 1985d). Growth to term, 405 mm disc width, is effected by ingestion and absorption of histotroph. Early in gestation, histotroph is thin and present in modest amounts. The uterine milk of *Dasyatis violacea* is 13.3% organic substance (Needham 1942). As development proceeds, both the quantity and viscosity of histotroph increases. Secretory epithelial acini composed of 8 to 10 cells elaborate the copious histotroph, and the stomach and spiral valve become engorged with the creamy fluid. In the cownose ray, the continuous supply of metabolites from trophonemata account for a 3000% increase in wet weight between the egg and newborn. In the Southern stingray, *D. americana* (Figs. 1.2, 1.3), the weight gain is 3750% (Hamlett 1990b).

The freshwater stingrays (Potamotrygonidae) are adapted for spending their entire lives in freshwater. The reproductive biology of *Potamotrygon circularis* and *P. motoro* has been the most well documented (Thorson et al. 1983). Both animals are found in South American river systems from the Amazon drainage of western Brazil and southern Colombia. Reproduction differs from the cownose ray in that both oviducts are functional and the mean litter size is 6.3 for *P. motoro*. No shell or eggcase is formed in *P. circularis* and the well vascularized trophonemata reach a maximum length of 20 mm. As with other rays, initial development is reliant on yolk stores while histotroph supports growth to term.

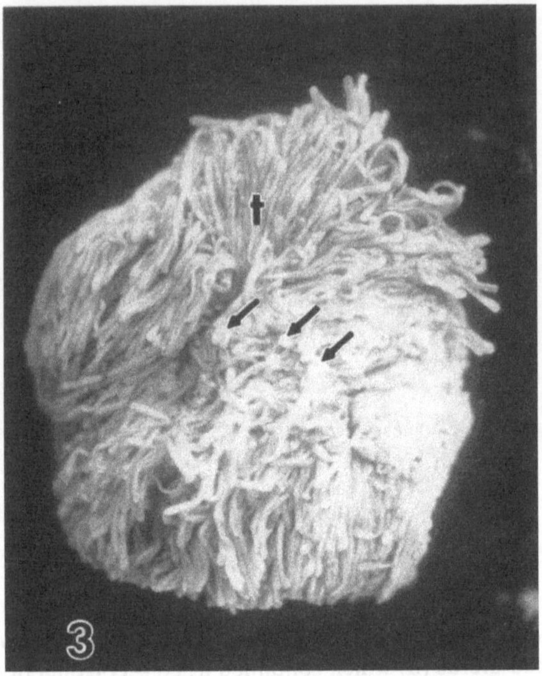

Figure 1.3. Secretory trophonemata (t) and histotroph (arrows) are visible in the everted uterus of the Southern stingray previously depicted in Figure 1.2.

Internal Incubators with Oophagy and Intrauterine Cannibalism

Among the elasmobranch fishes, oophagy has been reported in the families Odontaspidae (*Odontaspis taurus*, Springer 1948; Bass et al. 1975; *Pseudocarcharias kamoharai*, Fujita 1981; Gilmore et al. 1983), Lamnidae (*Lamna nasus*, Lohberger 1910; Shann 1911, 1923; Bigelow and Schroeder 1948), and Alopidae (*Alopias vulpinus*, Gubanov 1972; *A. pelagicus*, Otake and Mizue, 1981).

The reproductive biology of the sand tiger shark, *Odontaspis taurus*, has recently been studied in some detail (Gilmore et al. 1983; Hamlett 1983; Gilmore 1991). Embryonic egg eating (oophagy) produces a relatively large fetus at parturition. These large young have a marked size advantage over other newborn sharks. The sand tiger is the only species that is a

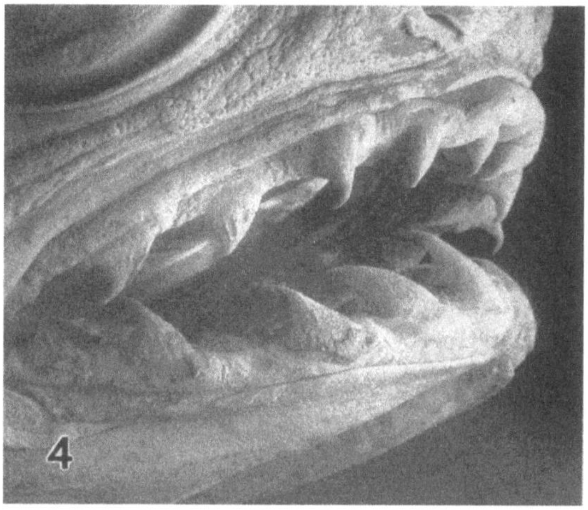

FIGURE 1.4. A scanning electron micrograph of a 60 mm total length sand tiger embryo, *Odontaspis taurus*, shows multiple rows of precociously erupted teeth (×120).

documented intrauterine cannibal (Gilmore et al. 1983). Among sand tiger populations in North Carolina and Florida, males and females congregate in specific mating grounds each year. After mating, males leave the area, presumably for deeper waters. Ovulation occurs every 24 hours and initial ovulations release two ova, one destined for each oviduct. Subsequent ovulations release two, three, and then up to 18 multiple ova. After two weeks, multiple eggs are released but not fertilized. Initially, embryos rely on yolk stores for growth. By the time the dominant embryos are 60 mm in total length (TL) (Fig. 1.4), they have developed large functional dentition. At this time, the first embryo to develop tears open its eggcase and gains access to the uterus. At 100 mm, TL the embryo begins to feed on egg capsules containing embryos. By 350 mm TL, all embryos have been consumed and the sole remaining embryo in each uterus continues to feed on eggcases containing unfertilized ova. A month prior to parturition ovulation ceases and, at birth, the 100 cm TL fetus resembles the adult (Gilmore 1991).

Like other lamnoid sharks, where reproduction is based on oophagy, it is believed that the great white shark, *Carcharodon carcharias*, also displays this mode of reproduction, although pregnant females have yet to be described adequately. Likewise, the basking shark, *Cetorhinus maximus*, is presumed to be oophagous. The shortfin mako, *Isurus oxyrinchus*, is also a oophagus, as is the longfin mako, *I. paucus*. The porbeagle, *Lamna nasus*, also employs this reproductive strategy.

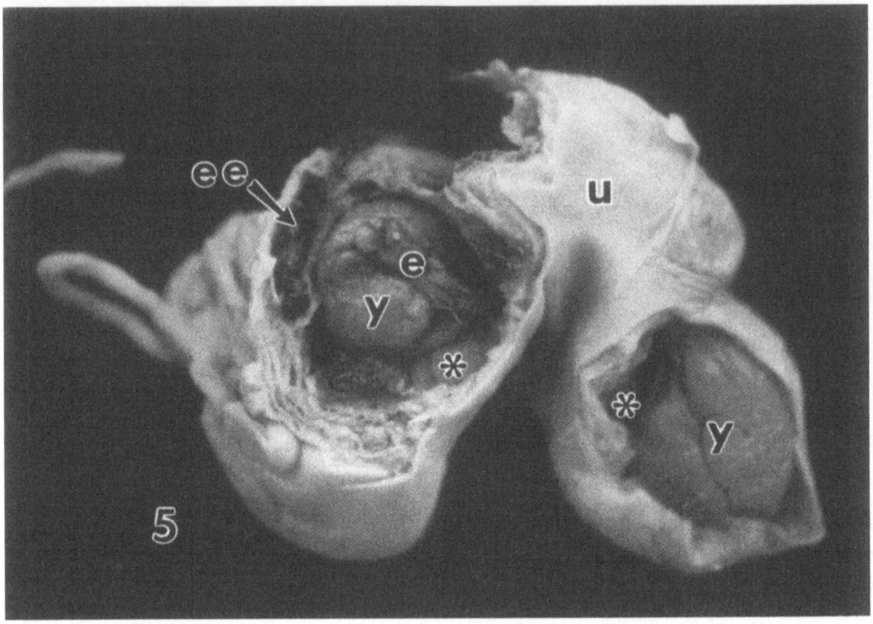

Figure 1.5. The thick gravid uterus (u) of the blacknose shark, *Carcharhinus acronotus*, contains preimplantation embryos (e) and their attached yolk sacs (y) in individual uterine compartment (asterisks). Excess egg envelope (ee) is stored in reservoir (arrow) continuous with each uterine compartment.

Internal Incubators with Yolk Sac Placenta

Yolk sac placentae develop in the requiem and hammerhead sharks. Following internal fertilization, dorsal and ventral uterine flaps grow to create separate uterine compartments to house each individual embryo (Fig. 1.5). The increased surface area presumably aids in respiratory and excretory functions. Each embryo is enclosed by a diaphanous egg envelope that is of sufficient size to accommodate the term fetus (Fig. 1.6). Initially, each embryo utilizes yolk stores in the yolk sac for growth and development. Yolk is digested in the yolk syncytial-endoderm-complex of the yolk sac and is also transported through the ductus vitellointestinalis in the yolk stalk to reach the gut of the embryo (Hamlett 1983, 1986, 1987, 1989b, 1990a, 1991). As the embryos are nourished by the placenta, they grow to fill the now thin, nearly transparent uterus (Fig. 1.7). During gestation the maternal liver produces yolk precursor materials to yolk a crop of eggs for the next ovulatory cycle, and it has been proposed that these same precursors are transported across the placenta to nourish the embryos (Hamlett et al. 1985a, 1985b, 1985c). As yolk stores are depleted, the flaccid yolk sac is not resorbed into the body wall but persists. It becomes regionally mod-

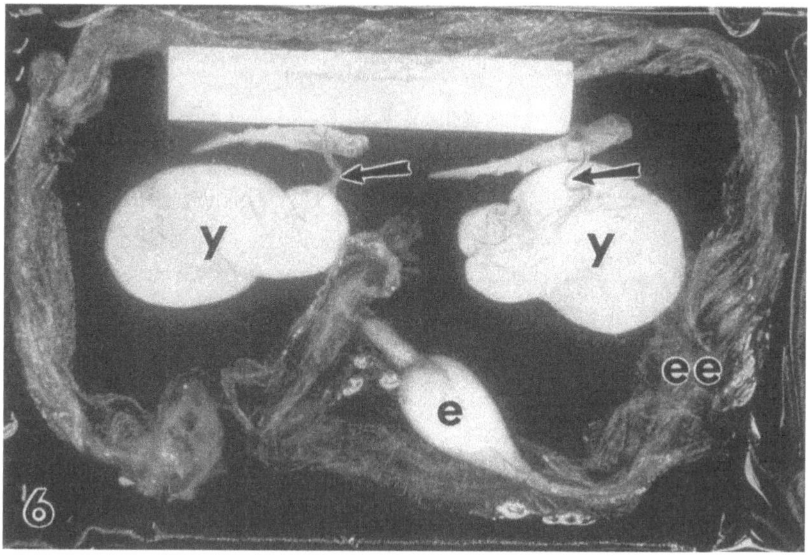

Figure 1.6. Two preimplantation lemon shark embryos are shown attached by their yolk stalks (arrows) to the yolk mass (y). An egg envelope (ee) that encloses each embryo is unfolded to show its full length; e = unfertilized egg.

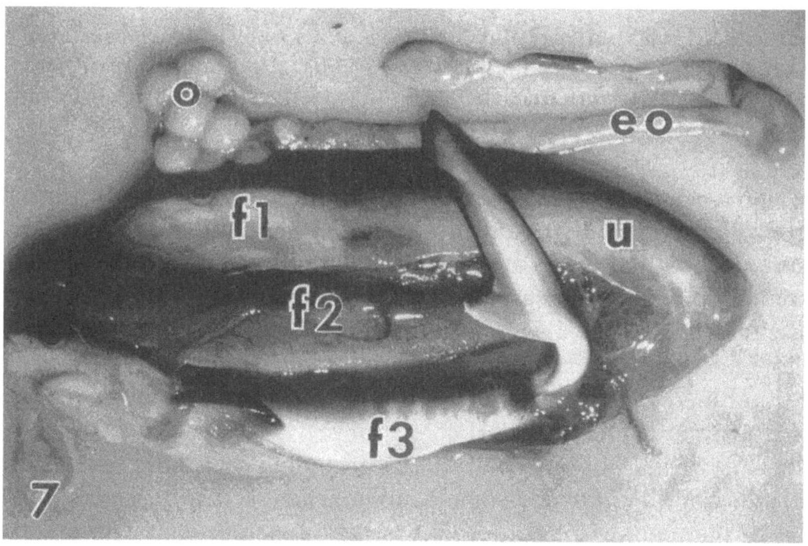

Figure 1.7. Three term Atlantic sharpnose fetuses (f1–f3), *Rhizoprionodon terraenovae*, occupy the thin, partially transparent uterus (u); o = ovary containing five ripe ova; eo = epigonal organ.

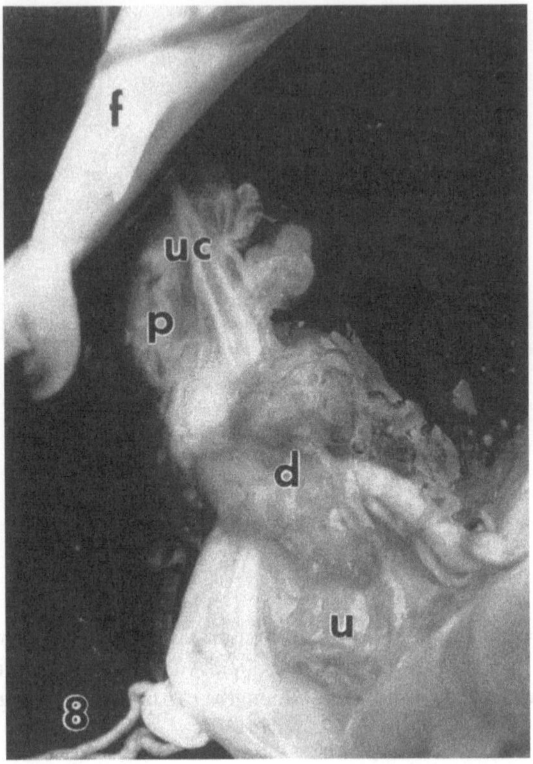

Figure 1.8. A full term blacknose fetus (f) is attached by its umbilical cord (uc) to the proximal portion (p) of the placenta. The distal portion (d) of the placenta interdigitates with uterine attachment sites (u) to form the uteroplacental complex.

ified into a functional placenta that is composed of a smooth, proximal portion, implicated in steroid production (Hamlett 1990a), and a distal, rugose portion that forms an apposition with the uterine wall. The uteroplacental complex nourishes the embryo to term (Fig. 1.8). The uteroplacental complex is also presumably involved in waste removal and osmoregulation. A gestation period of 9 to 12 months is typical of large placental sharks. Litter size ranges from three to six in the blacknose shark to 4 to 135 in the blue shark (Hamlett 1989a).

In some viviparous placental sharks, including sharpnose species of the genus *Rhizoprionodon* and some hammerheads, accessory vascular extensions of the umbilical cord develop termed appendiculae. These presumably are involved in nutrient absorption, and perhaps gaseous exchange, and osmoregulation.

The tiger shark, *Galeocerdo cuvieri*, is the only carcharhinid internal incubator that does not develop a yolk sac placenta, but it is uncertain whether internal incubation in the tiger shark is primitive or has been brought about by the evolutionary loss of a placenta (Compagno 1984).

Discussion

Elasmobranchs grow slowly to a relatively large size, mature late, reproduce every one to two years, and produce a small number of large offspring. For this type of animal to survive, the habitat they occupy must remain relatively stable and competition and predation must be relatively low. In recent years these requirements have been severely altered by the activity of man. Wetland industrialization and coastal development has, in some cases, produced oil spills, chemical pollution, and physical degradation. Sharks often give birth in specific inshore nursery grounds year after year and pups may spend up to several years of their most sensitive period of development in these waters. Most of the world's elasmobranch fishes are concentrated on the continental shelves and adjacent slopes. Perhaps 5% are oceanic while 50% occur on the shelves from the coast to 200 m. Approximately 5% occur in freshwater, about 35% from 200 m to 2000 m and deeper, and 5% in all three habitats (Compagno 1991).

Recreational and commercial shark fishing, employing mechanized fleets, pelagic longline, gillnets, and deep bottom-trawl, are landing more sharks annually for meat and fins. Commercial catches of shark in the United States increased from less than 500 tons in 1980 to 7144 tons in 1989. The catch figures for 1990 showed an estimated 20% decline which may reflect already dimishing populations. In Florida, 100,000 sharks were taken in 1989 exclusively for the Asian fin market (Gruber 1991). Expanding fisheries may, therefore, be threatening the survival of commercially familiar and other as yet undescribed and poorly known elasmobranch species. The ecological impact of removing such apex predators is incalculable.

References

Bass AJ, D'Aubrey JD, Kistnasamy N (1975) Sharks of the east coast of southern Africa. IV. The families Odontaspidae, Scapanorhynchidae, Isuridae, Cetorhinidae, Alopiidae, Orectolobidae, and Rhiniodontidae. Oceanogr. Res. Inst., Durban, South Africa. Invest. Rep. 39

Baughman JL (1955) The oviparity of the whale shark, *Rhincodon typus*, with records of this and other fishes in Texas waters. Copeia 1955 (1):54–55

Bigelow HB, Schroeder WC (1948) *Fishes of the Western North Atlantic*, Part I. Lancelets, Cyclostomes, and Sharks. Sears Foundation for Marine Research, New Haven, Connecticut

Bigelow HB, Schroeder WC (1953) *Fishes of the Western North Atlantic*, Part II. Sawfishes, Guitarfishes, Skates, and Rays. Sears Foundation for Marine Research, New Haven, Connecticut

Compagno LJV (1984) FAO Species catalogue. Vol. 4. Sharks of the World. An annotated and illustrated catalogue of shark species known to date. Part 2. Carcharhiniformes. FAO Fish. Synop. (125) Vol. 4, pt. 2

Compagno LJV (1991) The evolution and diversity of sharks. Underwater Nat. 19:15–22

Figueiredo JL (1977) Manual de Peixes Marinhos do Sudeste do Brasil. I. Introducao. Cacoes, raias e quimeras. Sao Paulo, Museu de Zoologia, USP. 104 pp

Fujita K (1981) Oviphagous embryos of the pseudocarchariid shark, *Pseudocarcharias kamoharai*, from the Central Pacific. Jap. J. Ichthyology 28:37–44

Gilbert PW, Bevelander G (1959) Permeability of uterine mucosa of gravid *Squalus acanthias* to phosphorus. Bull. Mt. Desert Island Biol. Lab. 68–69

Gilmore RG (1991) The reproductive biology of lamnoid sharks. Underwater Nat. 19:64–67

Gilmore RG, Dodrill JW, Linley PA (1983) Reproduction and embryonic development of the sand tiger shark, *Odontaspis taurus* (Rafinesque). Fishery Bull. 81:201–225

Gruber SH (1991) Life styles of the sharks. Underwater Nat. 7–14

Gubanov YP (1972) On the biology of the thresher shark, *Alopias vulpinus* (Bonnaterre), in the northwest Indian Ocean. J. Icthyol. 12(4):591–600

Gudger EW (1951) How difficult parturition in certain viviparous sharks and rays is overcome. J. Elisha Mitchell Sci. Soc. 67:56–86

Hamlett WC (1983) Maternal-fetal relations in elasmobranch fishes. Ph.D. Dissertation, Clemson University, Clemson, South Carolina

Hamlett WC (1986) Prenatal nutrient absorptive structures in selachians. In Uyeno T, Arai R, Taniuchi T, Matsuura K (eds.). *Indo-Pacific Fish Biology*. Ichthyol. Soc. Japan, Tokyo. pp. 333–344

Hamlett WC (1987) Comparative morphology of the elasmobranch placental barrier. Arch. Biol. (Brux.) 98:135–162

Hamlett WC (1989a) Reproductive strategies in elasmobranchs of the North Carolina shelf. In George RY, Hulbert AW (eds.). *North Carolina Coastal Oceanography Symposium*, NOAA-NURP Rept. 89–2. pp. 455–482

Hamlett WC (1989b) Evolution and morphogenesis of the placenta in sharks. J. Exp. Zool. Suppl. 2:35–52

Hamlett WC (1990a) Elasmobranch species as models for studies of placental viviparity and its endocrine regulation. J. Exp. Zool. Suppl. 4:129–131

Hamlett WC (1990b) Functional organization of uterine trophonemata and fetal alimentary structures in the Southern stingray. Amer. Zool. 30:39a

Hamlett WC (1991) From egg to placenta: Placental reproduction in sharks. Underwater Nat. 19:61–63

Hamlett WC, Schwartz FJ, DiDio LJA (1987) Subcellular organization of the yolk syncytial-endoderm complex in the preimplantation yolk sac of the shark, *Rhizoprionodon terraenovae*. Cell Tissue Res. 247:275–285

Hamlett WC, Wourms JP (1984) Ultrastructure of the preimplantation shark yolk sac placenta. Tissue & Cell 16:613–625

Hamlett WC, Wourms JP, Hudson JS (1985a) Ultrastructure of the full term shark yolk sac placenta. I. Morphology and cellular transport at the fetal attachment site. J. Ultrastruct. Res. 91:192–206

Hamlett WC, Wourms JP, Hudson JS (1985b) Ultrastructure of the full term shark yolk sac placenta. II. The smooth, proximal segment. J. Ultrastruct. Res. 91:207–220

Hamlett WC, Wourms JP, Hudson JS (1985c) Ultrastructure of the full term shark yolk sac placenta. III. The maternal attachment site. J. Ultrastruct. Res. 91:221–231

Hamlett WC, Wourms JP, Smith JW (1985d) Stingray placental analogues: Structure of trophonemata in *Rhinoptera bonasus*. J. Submicrosc. Cytol. 17:541–550

Jollie WP, Jollie LG (1967a) Electron microscopic observations on the yolk sac of the spiny dogfish, *Squalus acanthias*. J. Ultrastruct. Res. 18:102–126

Jollie WP, Jollie LG (1967b) Electron microscopic observations on accommodations to pregnancy in the uterus of the spiny dogfish, *Squalus acanthias*. J. Ultrastruct. Res. 20:161–178

Leesa RP (1982) Biologie et dynamique des populations de *Rhinobatos horkeli* du plateau continental. Brest, Franca, Universite de Bretagne Occidentale. These doctorat 3 eme cycle. 238 pp

Leesa R, Vooren CM, Lahayne J (1986) Desenvolvimento e ciclo sexual das femeas, migracoes e fecundidade da viola *Rhinobatos horkeli* (Muller & Henle, 1841) do Sul do Brasil. Atlantica, Rio Grande, 8:5–34

Lohberger J. (1910) Ueber Zwei Riesige Embryonen von *Lamna*. (Beitrage zur Naturgeschichte Ostasiens) Abh. Bayer. Akad. Wiss. 4 (Suppl. No. 2), pp. 1–45

Needham J (1942) Biochemistry and Morphogenesis. Cambridge University Press, Cambridge

Otake T, Mizue K (1981) Direct evidence for oophagy in thresher shark, *Alopias pelagicus*. Jap. J. Icthyology 28:171–172

Ranzi S (1932) Le Basi Fisio-morfologichedello sviluppo embrionale dei selaci-Parte I. Pub. Staz. Zool. Napoli 13:209–290

Sadowsky V (1965) The hammerhead sharks of the littoral zone of São Paulo, Brazil, with the description of a new species. Bull. Mar. Sci. Miami 15(1):1–12

Sadowsky V (1967) Selachier aus dem litoral von São Paulo, Brasilien. Beitr. Neotrop. Fauna 5(2):71–88

Sadowsky V (1970) First record of broad-snouted seven-gilled shark from Cananéia, coast of Brazil. Bolm. Inst. Oceanogr., São Paulo, 18(1):33–5

Sadowsky V (1971) Estudio economico sobre los elasmobrânquios de la zona litoral paulista. CARPAS. 5/D. Tec., 1. Rio de Janeiro, p. 1–11

Sadowsky V (1973) First record of a basking shark *Cetorhinus maximus* (Gunnerus, 1765) in the Brazilian Atlantic. Bolm. Inst. Oceanogr., São Paulo, 22:1–10

Sadowsky V (1974) Fauna dos peixes cartilaginosos (ELASMOBRANCHII) da região lagunar de Cananéia (SP). In: 26ª Reunião Anual da SBPC. Resumos–Suplemento de Ciência e Cultura 26(7):204

Sadowsky V (1976) Primeiro registro de ocorrência de peixe elétrico, *Torpedo puelcha* Lahille, 1928 (TORPEDINIDAE) nas águas do Brasil. In: 28ª Reunião Anual da SBPC. Resumos–Suplemento de Ciência e Cultura 28(7):421

Sadowsky V, Amorim AF, Arfelli CA (1984) Second occurrence of *Odontaspis noronhai* (Maul, 1955). B. Inst. Pesca 11(único):69–79

Sadowsky V, Amorim AF, Arfelli CA (1985) Observaçẽes em tubarão-anão, *Squaliolus laticaudus* proveniente do litoral sul do Brasil. In: XII Congresso Brasileiro de Zoologia, Resumos, Campinas, UNICAMP, 12:172

Shann EW (1911) A description of the advanced embryonic stage of *Lamna cornubica*. Fish. Bd. Scotland Ann. Rept. 28:73–79

Shann EW (1923) The embryonic development of the porbeagle shark, *Lamna cornubica*. Proc. Zool. Soc. (Lond.) 11:161–171

Springer S (1948) Oviphagous embryos of the sand shark, *Carcharias taurus*. Copeia 1948:153–157

Springer S, Sadowsky V (1970) Subspecies of the western Atlantic cat shark, *Scyliorhinus retifer*. Proc. Biol. Soc. Wash. 83:83–98

Thorson TB (1976) Observations on the reproduction of the sawfish, *Pristis perotteiti*, in Lake Nicaragua, with recommendations for its conservation. In Thorson TB, (ed.). *Investigations of the Ichthyofauna of Nicaraguan Lakes*. Univ. of Nebraska, Lincoln

Thorson TB (1982) Life history implications of a tagging study of the largetooth sawfish, *Pristis perotteti*, in the Lake Nicaragua-Rio San Juan system. Env. Biol. Fish. 7:207–228

Thorson TB, Langhammer JK, Oetinger MI (1983) Reproduction and development of the South American stingrays, *Potamotrygon circularis* and *P. motoro*. Env. Biol. Fish. 9:3–24

Wolfson FH (1983) Records of seven juveniles of the whale shark, *Rhiniodon typus*. J. Fish Biol. 22:647–655

Wood-Mason J, Alcock A (1891) On the uterine villiform papillae of *Pteroplatea micrura*, and their relation to the embryo. Proc. R. Soc. London 49:359–367

2

Serum Steroid Hormones During Reproduction in Elasmobranchs

L.E.L. Rasmussen, David L. Hess, and Samuel H. Gruber

Elasmobranchs comprise one of the oldest, distinct vertebrate lines in existence today. While they are a relatively small group, consisting of only about 800 living species, they are highly adapted and, because of their size, numbers, and ecological role, may be considered a biologically successful group. Much of that success can be traced to the particular reproductive strategies evolved by the elasmobranch over the past 350,000,000 years. According to Lund (in press), elasmobranchs of the Carboniferous period reproduced in a manner similar to sharks today. The key reproductive elements include internal fertilization, production of a few well-formed young, slow growth, late maturity, and longevity. These characteristics are interpreted by ecologists (McArthur and Wilson 1967) as an adaptation to a stable, rather crowded environment. In contrast, bony fish such as the cod (*Gadus callaris*) grow quickly to maturity, produce thousands of eggs, and live only a few seasons. Such a reproductive tactic is more amenable to a fluctuating environment. Clearly, mode of reproduction is one of the most important elements for successful survival for a species. However, the reproductive strategy that served the elasmobranchs so well since the Carboniferous has today placed them in jeopardy. As mentioned, low fecundity is interpreted as an adaptation to a stable, more-or-less predictable environment. But today the activities of man, especially mechanized commercial fishing, have rendered the shark's environment unstable and unpredictable. Now, man competes with sharks for the same food and indeed preys on sharks at an ever increasing rate. These man-made ecological changes are simply too pervasive and occur too rapidly for sharks to adapt. If the elasmobranchs are to survive, management plans must be implemented that account for their low reproduction rate. Thus, an understanding of the varied modes of elasmobranch reproduction might provide the very key to their survival. Such considerations are very relevant now in the coastal regions of South America, including Brazil.

Accordingly, given the critical role of steroids in regulating reproduction, this chapter focuses on the levels of steroid hormones detectable in the serum of elasmobranchs, with special emphasis on changes in these concentrations during maturation and during adult functions of the reproductive system. Concentrations of steroid hormones in the serum of oviparous and aplacental viviparous (ovoviviparous) elasmobranchs are compared with our recent data on the viviparous placental sharks (Rasmussen and Gruber 1990, Rasmussen and Gruber in press).

As background to this presentation we give a brief synopsis of (1) reproductive modes of the elasmobranchs, (2) the endocrinological relationships of the pituitary and hypothalamus, and (3) the sources and putative physiological effects of steroid hormones.

Reproductive Modes of the Elasmobranchs

Elasmobranch reproductive modes span the range from external development in horny eggcases (oviparity) to internal fetal development within the female (viviparity) and may include a combination of both (ovoiviparity or aplacental viviparity; Hamlett 1989b). Oviparity is the usual mode of reproduction in skates and in many shark families such as the Scyliorhinidae (catsharks) and Rhiniodontidae (the whale sharks; Wourms 1977).

Viviparity is also a common reproductive modality in elasmobranchs. The majority of the sharks and all of the rays follow an aplacental form known as ovoviviparity as exemplified by *Squalus acanthias*, the piked dogfish.[1] Most carcharhinoid sharks of the families Carcharhinidae (requiem sharks) and Sphyrinidae (hammerheads) follow the yolk sac placental model (Wourms 1981). Indeed, placental viviparity occurs only in these two shark families. (For more comprehensive reviews, see Teshima 1981; Hamlett 1987, 1989a, 1989b; Hamlett and Wourms 1984; Hamlett et al. 1985a, 1985b, 1985c, 1987; Wourms et al. 1988.)

The Carcharhinidae, evolving early in the Jurassic period (Schaeffer 1967; Schaeffer and Williams 1977), have a unique place in evolution since they apparently use steroid hormones to regulate various reproductive processes and have some reproductive adaptations unique to anamniotes, including the development of a functional placenta. In tropical waters and some temperate regions, including many coastal regions of South America, these sharks have special importance as they are the dominant sharks, often both in variety, abundance, and biomass (Compagno 1984).

Pituitary, Hypothalamus, and Endocrine Relationships

In reviewing work on the control and regulation of steroid secretion and synthesis, Wourms (1977) reported that the elasmobranchs have the full complement of endocrine organs associated with reproduction in verte-

---

[1] All scientific names for sharks are according to Compagno (1984). After the first citation, the accepted common name (Compagno 1984) is used in this text As no catalogue similar to Compagno's is presently available for skates and rays, a variety of sources were consulted.

brates, including testes, ovaries, interrenals (adrenals), hypothalamus, and pituitary. Generally in vertebrates, gonadal functions are regulated by the hypothalamic-pituitary axis with secretory activity of the hypothalamus regulating gonadotropin release by the anterior lobe of the pituitary. Anatomically, cells that are the principal source of the gonadotropic activity are localized in the ventral lobe of the pituitary (Holmes and Ball 1974; Sumpter et al. 1978; Dodd and Sumpter 1984). This lobe, unlike its counterparts in other vertebrates, is apparently not anatomically connected to the hypothalamus either by a vascular portal plexus in the median eminence or by direct innervation (Demski 1989, 1990). Rather, circumstantial evidence indicates that gonadotropin-releasing hormone (GnRH) from the hypothalamus exerts its regulatory functions on pituitary gonadotropins via the systemic circulation. The gonadotrophs, as in other vertebrates, are responsible for the seasonal and developmental changes in the production of sex steroids that drive reproductive organ function and behavioral changes associated with reproduction (Callard et al. 1989a, 1989b). A gonadotropic role in steroidogenesis for the ventral lobe was confirmed in in vitro studies using follicular and luteal components of the ovary (Klosterman and Callard 1986).

Hormonal Regulation of the Reproductive Tract

TISSUE SOURCES OF STEROIDS

Chieffi (1967) suggested that the corpora lutea in female elasmobranchs and the testes in males appeared to be sources of three steroid hormones: estradiol (17β-estradiol), testosterone, and progesterone. Estradiol and traces of progesterone have been demonstrated in ovarian extracts from *Squalus suckleyi* (= *acanthais*), the piked dogfish (Wotiz et al. 1960). These steroids have also been identified in *Torpedo marmorata*, the electric ray, *Scyliorhinus canicula*, the small spotted catshark, and the piked dogfish (Chieffi and Lupo di Prisco 1961; Simpson et al. 1963; Gottfried 1964).

In the small spotted catshark, testicular extracts contained testosterone, progesterone, and androstenedione (Chieffi and Lupo di Prisco 1961). In 1964, Simpson et al. demonstrated the synthesis of testosterone from [^{14}C]progesterone in the testis of the piked dogfish. In in vitro studies with the small spotted catshark, Kime (1978) demonstrated that testicular tissue incubated with [^{14}C]progesterone yielded testosterone as well as androstenedione. The possibility that Sertoli cells within the seminiferous tubules might be steroidogenic, in addition to their principal role in the control of spermatogenesis, was supported by the biochemical localization within this tissue of an essential steroidogenic enzyme, 3β-hydroxysteroid dehydrogenase, in both the piked dogfish and the small spotted catshark (Collenot and Ozon 1964; Simpson and Wardle 1967).

The production by the ovary of estradiol, testosterone, and small amounts of progesterone was demonstrated in vitro in the small spotted catshark (Dodd et al. 1983). The production rates of estradiol and testosterone in the wall of the ovarian follicle increased as the follicle matured. The production of progesterone by the luteal tissue of the ovary was also demonstrated in another oviparous species, *Raja erinacea*, the little skate (Fileti and Callard 1988).

The occurrence and distribution of steroidogenic enzymes was investigated in detail in the ovoviviparous piked dogfish by Lance and Callard (1969). Their study identified steroids in ovarian extracts and localized steroidogenic enzymes, principally in the granulosa cells of the follicular wall of the ovary. As studied in enzymatically isolated granulosa cells from the ovary, increasing amounts of testosterone and estradiol were apparently synthesized as the follicles enlarged during gestation (Tsang and Callard 1983; Tsang 1986). Both the theca and granulosa layer of the follicular wall of the ovary participated in steroidogensis; however, the granulosa layer alone was able to produce all three steroids—estradiol, testosterone, and progesterone. The demonstration that the amounts produced by the granulosa cells was not increased by co-incubation with the theca provided the first instance of vertebrate ovarian granulosa cells, producing significant amounts of all three steroids without the presence of theca cells. Callard et al. (1989a) suggested that the differences in steroidogenic capacity exhibited by the theca and granulosa elements in different vertebrates are related to their differing substrate supplies. The biochemical basis for the action of estradiol was recently demonstrated by the identification of a high affinity, low capacity receptor specific for 17β-estradiol in the oviduct of the little skate (Reese and Callard 1987).

Effects of Exogenous Steroids

An early review of elasmobranch reproductive endocrinology described the stimulatory effect of exogenous steroids on sexual development in several oviparous species and two viviparous rays (Chieffi 1967). Other studies of exogenous steroidal effects on the development of reproductive organs and secondary sexual characteristics and on plasma steroid levels include those of Gottfried (1964) and Jenkins and Dodd (1980). The effects of exogenous estradiol on such structures as the ovary and the oviduct have implicated estradiol in the development and maintenance of the reproductive tract of female elasmobranchs (Dodd and Goddard 1961). Exogenous progesterone decreased the time of retention of the encapsulated egg in the uterus of little skates and also affected oviposition times (Koob and Callard 1985; Callard et al. 1988). These effects of exogenous steroid hormones administered during the maturation and during the reproductive cycle support the pivotal nature of such steroids in normal reproductive functions.

Periodic Fluctuations of Steroid Hormones in Serum: Correlations to Sequential Reproductive Events

Correlations between fluctuations in serum levels of steroid hormones and changes occurring during reproduction and reproductive development have been determined for only two oviparous elasmobranchs, the little skate and the small spotted catshark, and for the piked dogfish utilizing internal incubation (Dodd 1983; Callard et al. 1988). The data from these species suggest that estradiol. testosterone, and progesterone play fundamental roles in the growth and development of the gonads and of the reproductive tracts and perhaps the maintenance of pregnancy.

OVARIAN EVENTS

Our understanding of the possible roles for estradiol in reproduction was aided particularly by two studies. Estradiol in small spotted catshark (Dodd and Goddard 1961) has been implicated in vitellogenesis and in the development and maintenance of secondary sexual characteristics, such as shell glands and oviducts. The correlation of estradiol levels in the serum and the sizes of ovarian follicles and the condition of the shell glands (Koob et al. 1986) implicates estradiol in the development and maintenance of the elasmobranch reproductive tract.

The role of testosterone in the female elasmobranch is less clear. It has been suggested (Dodd 1983) that testosterone functions at the level of the central nervous system to regulate reproductive behavior. Testosterone could be acting either as an androgen, indicated by the binding of androgens in certain areas of the brain of the piked dogfish (Jenkins 1978), or as an estrogen, indicated by the presence of aromatase and other androgen-converting enzymes in the elasmobranch brain (Callard et al. 1978).

Seasonal changes related to ovarian events in the concentrations of circulating steroids have been investigated in the small spotted catshark and the little skate. Both estradiol and testosterone were found in the plasma of the female small spotted catshark. The serum levels of these steroids fluctuated in a synchronous and cyclical way throughout the year and were especially elevated during recrudescence of the ovary (period when the mature oocytes are beginning to be released from the ovary), reaching the highest levels during the winter season of active egg-laying.

In the little skate, progesterone levels in the serum peaked 48 hours prior to ovulation. Subsequently, the levels declined rapidly to baseline before oviposition. These fluctuations imply that progesterone may be involved in the events of ovulation, encapsulation of the egg, and oviposition in the little skate (Callard et al. 1988, 1989a). Fileti and Callard (1988) have suggested that corpora lutea from previous cycles, as well as current pre-ovulatory follicles, contribute to the large peak in progesterone levels in the serum observed by Koob et al. (1986).

Pregnancy

Estradiol is probably necessary for the development and maintenance of the reproductive tract and is interrelated with the action of relaxin, a peptide hormone of the ovary (Koob et al. 1984; Bullesbach et al. 1987; Callard et al. 1988).

The lengthy gestation period of the piked dogfish (20 to 22 months) and the relatively high levels of progesterone throughout most of gestation suggest that progesterone may have a role in maintaining pregnancy (Tsang and Callard 1987a). The corpora lutea of the ovary in vitro synthesize progesterone and may be the principal source of this hormone (Klosterman and Callard 1986). Further proof of progesterone production and its cyclical nature was provided by Tsang and Callard (1987a, 1987b). They found maximum production of progesterone in minces of corpora lutea at about six-month post-ovulation (Stage B), with production declining thereafter until birth.

Levels of Steroid Hormone in the Serum

Until recently (1990) measurements of the concentrations of steroid hormones in the serum of elasmobranchs were restricted to two sharks, the piked dogfish and the small spotted catshark (Sumpter and Dodd 1979; Garnier et al. 1989), two skates, *Raja radiata*, the thorny skate (Fletcher et al. 1969) and the little skate (Koob et al. 1986), and the electric ray, *T. marmorata* (Lupo di Prisco et al. 1967). Now, concentrations of estradiol, progesterone, testosterone, dihydrotestosterone, and corticosterone have been measured in the serum of the following species of placental sharks: *Carcharhinus acronotus*, the blacknose shark; *C. leucas*, the bull shark; *C. limbatus*, the blacktip shark; *C. perezi*, the Caribbean reef shark; *C. plumbeus*, the sandbar shark; *Negaprion brevirostris*, the lemon shark; and *Rhizoprionodon porosus*, the Caribbean sharpnosed shark, all of the family Carcharhinidae. These steroids were also measured in two sphyrnid sharks, *Sphyrna mokarran*, the great hammerhead, and *S. lewini*, the scalloped hammerhead, and in the aplacental carcharhinid, *Galeocerdo cuvieri*, the tiger shark, and the oviparous nurse shark (Rasmussen and Gruber 1990).

Male Elasmobranchs

Immature Male Elasmobranchs

Although the levels of steroid hormones in the serum have not been measured in immature males of most elasmobranch species, Sumpter and Dodd (1979) have indicated that estradiol and testosterone were undetectable in the plasma of immature small spotted catsharks. We reported serum hormone levels for estradiol (E)[2], testosterone (T), dihydrotestosterone (D), and progesterone (P) for two species of immature male requiem sharks, the tiger shark and the lemon shark (Rasmussen and Gruber 1990, in press).

The relative changes of five steroid hormones in the male lemon shark throughout the course of maturation are shown in Figure 2.1 (top). Inspection of these data reveals that T, D, and P increased in a systematic way during maturation. These increases were respectively 20-, 7.5-, and 5-fold. Increases in serum progesterone levels occurred predominantly at the latter stages of maturation, whereas T and D increased steadily throughout growth and maturation. E levels were similar between immature and mature males, and immature females (Figure 2.1). Therefore, we suggest that the steroid hormones T, D, and P have important but as yet undefined roles in the maturation of the male lemon shark.

Adult Male Elasmobranchs

The principal steroid hormone studied in the serum of males has been testosterone. Idler and Truscott (1966) were the first to definitively isolate, and subsequently quantify, testosterone in elasmobranch serum. In both the thorny skate and *R. ocellata*, the winter skate, no seasonal differences could be observed as serum T ranges overlapped in summer and winter samples (Table 2.1). This conclusion is based on our recalculation of data given in Idler and Truscott (1966) and Fletcher et al. (1969; Table 2.1).

Garnier et al. (1989) studied the yearly patterns of four serum steroid hormones in the small spotted catshark. The variation observed in P levels was apparently not seasonal. However, testosterone levels rose seasonally during September, reaching a maximum of 21 ng/ml in February (at mid-egg laying period); by May levels had returned to 5 to 6 ng/ml, where they remained throughout the summer. Both serum D and testis T levels followed a similar pattern. Often the highest levels of hormones reported occurred at mating times. Most sharks studied segregate by sex and come together only at mating (Myrberg and Gruber 1974; Gruber and Myrberg 1977). In many other male vertebrates (Anderson 1969; Bubenik et al. 1983), the highest testosterone levels occur at the time of mating. However, in the male small spotted catshark, the highest T levels were reported before mating when males were grouped together (Garnier et al. 1989). Perhaps the high hormonal levels have influences on the behavioral patterns related to the establishment of dominance rankings prior to mating.

In male lemon sharks, the four hormones, T, D, P, and corticosterone (C) that we measured all increased, and different rates, with age and

[2] Abbreviations are used as indicated in parentheses; i.e., E = estradiol; T = testosterone; D = dihydrotestosterone; P = progesterone; C3= corticosterone.

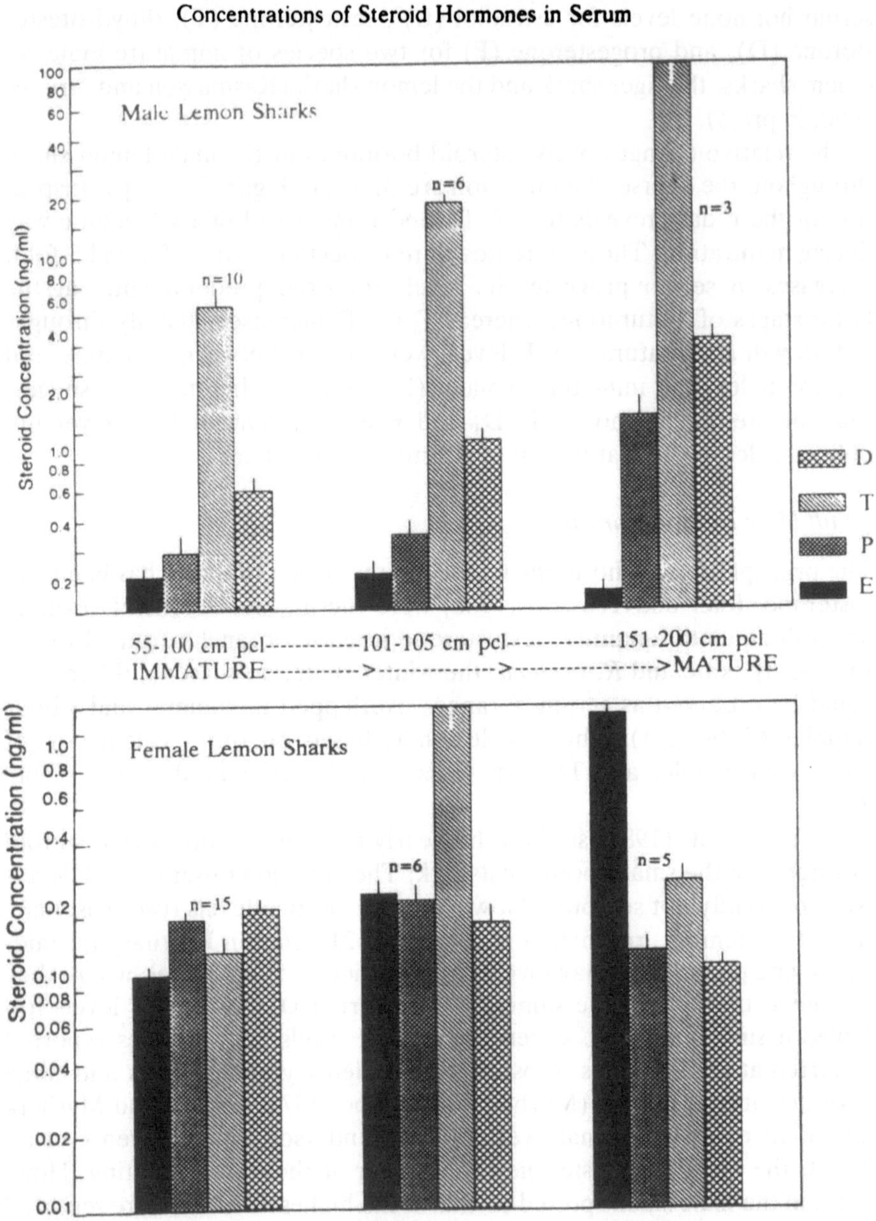

Figure 2.1. (Top) Mean concentrations of steroid hormones in the serum of male lemon sharks during growth and maturation. (Bottom) Mean levels of steroid hormones in the serum of female lemon sharks during growth and maturation. E = E-2 or 17β-estradiol; T = testosterone; D = DHT or dihydrotestosterone; P = progesterone.

Table 2.1. Concentrations of steroid hormones in the serum of adult male elasmobranchs.

Species (Season)	17β-Estadiol (pg/ml)	Progesterone (ng/ml)	Testosterone (ng/ml)	Dihydrotestosterone (ng/ml)	Reference
OVIPAROUS[d]					
Raja eglanteria (clear-nosed skate) n = 1	23	0.1	6.7[c]	4.4	Rasmussen and Hess Unpubl.
Raja radiata (thorny skate) n = 5 (July)			40–100[a]		Fletcher et al. 1969
Raja radiata n = 7 (Dec, Feb, August)			28–102[a]		Idler and Truscott 1966
Raja ocellata (winter skate) n = 3 (May)			22–208[a]		Idler and Truscott 1966
Scyliorhinus canicula (small spotted catshark)					
(May–June)	<100	0.1–0.9[a]	5–6[a]	*	Garnier et al. 1989
(Sept)	<100		12 ± 1.8[b]		
(Feb)	<100		21 ± 2.2[b]		
Ginglymostoma cirratum (nurse shark)					
(Fall)	42[c]	0.4	152.1[c]	1.3	Rasmussen Unpubl.
(Fall)	37	0.5	119.6	0.9	
(Fall)	9	0.06	8.1	1.0	
(Fall)	3	0.05	10.0	0.6	

(continued)

Table 2.1. (*Cont.*)

Species (Season)	17β-Estadiol (pg/ml)	Progesterone (ng/ml)	Testosterone (ng/ml)	Dihydrotestosterone (ng/ml)	Reference
VIVIPAROUS-Placental[d]					
Negaprion brevirostris (lemon shark)					
N = 5 (Spring)	10–100[a]	0.2–0.6	23–200	3–6.6	Rasmussen and Gruber In press
Carcharhinus acronotus (blacknosed shark)					
n = 2 (Spring)	20–51[a]	2.6–5.1	69–79	6–10	Rasmussen and Gruber 1990
Carcharhinus plumbeus (sandbar shark)					
n = 2 (May)	20–27[a]	0.5–0.7	40–73	2–4	Rasmussen Unpubl.
Carcharhinus leucas (bull shark)					
n = 3 (April)	9[c]	1.2	2.7	0.2[c]	Rasmussen and Gruber 1990
(May)	25	4.0	357.5	168.9	
(May)	52	7.3	171.6	15.7	
Carcharhinus perezi (Caribbean reef shark)					
n = 5 (April)	1–11[a]	0.03–0.2	0.9–3.1	0.1–1.0[a]	Rasmussen and Gruber 1990
	6 ± 2[b]	0.083 ± 0.04	1.63 ± 0.5	0.3 ± 0.2[b]	

Spyrna mokarran (great hammerhead)					
n = 2 (May)	21[c]	6.8	59.7	1.7	Rasmussen and Gruber 1990
(May)	13[c]	0.3	42.2	0.6	
Sphyrna lewini (scalloped hammerhead)					
n = 1 (May)	18[c]	3.7	40.4	9.1	Rasmussen and Gruber 1990
Carcharhinus limatus (blacktip shark)					
(Spring)	11[c]	2.5	25.1	1.6	Rasmussen and Gruber 1990
Rhizoprionodon porosus (Caribbean sharpnose)					
(May)	269[c]	2.9	41.0	4.7	Rasmussen and Gruber 1990
caudal vein	228[c]	1.9	13.0	1.0	Rasmussen, Hess, Gruber
testis artery	223[c]	3.1	17.0	1.3	This paper

[a] range
[b] mean
[c] individual values
[d] reproductive mode refers to female adaptations
* D levels follow same pattern at T; no concentrations were given

growth (Rasmussand and Gruber in press). Because of the limited sample size, we were not able to draw firm conclusions about seasonal influences on the serum steroid patterns although we suspect, as Garnier et al. (1989) suggested, that seasonal differences may well be significant in males. Nevertheless, we can state that both T and D were measurable in immature male lemon sharks and appeared to increase consistently during growth and maturation. Thus, androgens are apparently implicated in the normal process of sexual maturation. The stepwise 70-fold elevation of P between the subadult stage and adult stage provides even stronger evidence that this hormone may play an important role in the maturation of the male lemon sharks.

The high T levels in serum samples collected from adult male lemon sharks during the breeding season suggest a regulatory role in sexual behavior and reproductive function. The adult males of all species examined, including the male nurse sharks, had high T levels. Callard et al. (1978) suggested that high concentrations of T may represent precursors for the production of other as yet unidentified steroids. The presence of high levels of androgens may also be explained by a low metabolic clearance rate rather than a high production rate (Fletcher et al. 1969). However, we feel that the progressive and substantial increases in both T and D levels observed during growth are more indicative of a direct functional role for androgens in sexual maturation.

Although our data have been exclusively on samples form the caudal vein (thus representing circulating steroid hormones), we recently successfully obtained concurrent samples from the caudal and testicular veins of a living male Caribbean sharpnose shark. Analysis of these samples demonstraded two-fold higher P, somewhat higher T, and 1.5-fold higher D in the serum from a gonadal vessel than in that from the general circulation, whereas E levels were the same (last entry, Table 1). These data provide the first tangible hint that the testis may indeed be a source of P, T, and D in the male carcharhinid sharks.

Female Elasmobranchs

More data are available in the literature for female elasmobranchs, although most measurements have been made on immature animals. For clarity, we have divided this discussion of steroid hormones in the serum of female elasmobranchs into three sections: 1) levels during maturation; 2) changes in concentrations during the ovarian cycle; and 3) levels during pregnancy.

Levels of Steroid Hormones in Serum During Maturation

Studies of steroid hormones in serum during maturation are limited. E levels were measurable in aplacental elasmobranchs and reached ng/ml levels (Simpson et al. 1963; Craik 1979) (Table 2.1). Progesterone levels in immature females were undetectable or generally low, as seen in Table 2.2

TABLE 2.2. Steroid concentrations in the serum of adult female elasmobranchs.

Species (Season) (Stage)	17β-Estadiol (pg/ml)	Progesterone (ng/ml)	Testosterone (ng/ml)	Dihydrotestosterone (ng/ml)	Reference
OVIPAROUS					
Raja radiata (thorny skate) n = 7 (Feb)			0.2–6[a] 4.7[b]		Idler and Truscott 1966
Raja ocellata (winter skate) (May)			5.9[c]		Idler and Truscott 1966
Raja radiata n = 5 (Summer)			16–62[a] 38.03± 7.3[b]		Fletcher et al. 1969
Raja erinacea (little skate) n = 7					
[follicle <6mm]	<400	nd			
[follicle >18mm] (summer)	800–2000	nd 10.0[b]	4–36[a]		Koob et al. 1986
[encapsulation]	1000	12.0–36.0[a] 16.0[b]	4		
[egg retention]	500	2.0[b]	2		
[oviposition]	200	1.0[b]	1		
S. canicula	3,000–4,000[a]				Simpson et al. 1963
S. canicula (small spotted catshark) N = 25(March) [mature with yolk-filled oocytes]	20,000–82,000[a]				Craik 1979

(*continued*)

Table 2.2. (*Cont.*)

Species (Season) (Stage)	17β-Estadiol (pg/ml)	Progesterone (ng/ml)	Testosterone (ng/ml)	Dihydrotestosterone (ng/ml)	Reference
S. canicula					
(winter)	40,000		1		Sumpter and Dodd 1979
[peak egg-laying]					
(summer)	5,000		1		Sumpter and Dodd 1979
OVOVIVIPAROUS (Aplacental viviparous)					
Torpedo marmorata					
(electric ray)					
n = 8 [pre-gestation]	nd	73.0	15		Lupo di Prisco et al. 1967
n = 11 [gestation]	nd	nd	8		Lupo di Prisco et al. 1967
n = 8 [reproductively inactive]			24		Lupo di Prisco et al. 1967
Squalus acanthias					
(piked dogfish)					
[Stage A-recently ovulated]					Tsang and Callard 1987a
Oct.	250	2.5	0.2		
[Stage B 3.5–7cm embryo]					
6 mo.	500	4.3	0.3		
[Stage C-large yolk]	750	3.5	0.3		
[Stages A–C]	297–760[a]	2.6 ± 0.6[b]	0.3[b]		
[Stage D-near term-22 mo.]	781	0.4	0.8		
[Yolk almost gone]					
VIVIPAROUS-Placental					
Quiescent:					
Carcharhinus limbatus					
(blacktip)					
n = 3	0.04[b]	0.1[b]	0.2[b]	0.005	Rasmussen and Gruber 1990

Negaprion brevirostris (lemon shark) n = 3	1.2[b]	0.15[b]	0.3[b]	0.1	Rasmussen and Gruber (in press)
C. perezi (reef shark) n = 4	2–12[a]	0.01–0.07	0.003–0.02	nd	Rusmussen and Gruber 1990
Ovulatory State:					
Carcharhinus acronotus (blacknose) n = 4 [large ova]	381–1364[a]	0.06–0.1	0.05–0.8	0.01–0.3	Rasmussen and Gruber 1990
C. limbatus (blacktip) n = 1 [6 large ova in ovary]	370	4.9	0.9	0.2	Rusmussen and Gruber (in press)
Gestation:					
Rhizoprionodon porosus (sharpnose) n = 5 (Spring) [embryos, 1/2 term]	1,248–4,499[a]	0.014–0.08	0.02–0.06	nd–0.03	Rasmussen and Gruber 1990
N. brevirostris (lemon shark) n = 2 [near term]	400–600	0.2	0.02–0.2		Rusmussen and Gruber (in press)

nd = not detected
[a] range
[b] mean
[c] individual values

(Lupo di Prisco et al. 1967). In immature elasmobranchs, testosterone levels were in the ng/ml range (Idler and Truscott 1966; Lupo di Prisco et al. 1967). In mature little skates not actively ovipositing, wide ranges of E and T and no detectable P were reported (Koob et al. 1986; Table 2.2). These studies did not suggest any definitive roles for steroid hormones during maturation.

However, our recent study of maturation in carcharhinid sharks demonstrated that patterns for males and females followed different courses (Rasmussen and Gruber in press). In females, of the five hormones assayed, only E increased consistently during maturation (Fig. 2.1). The higher estradiol levels in the four to eight year old female lemon sharks as compared to the very young sharks (Fig. 2.1) support the concept that estradiol is important for maturation. In four to eight year old (101–105 cm pcl) female lemon sharks, the mean T levels were ten-fold higher than in the very young sharks (Fig. 2.1). These data may indicate a maturation role for T during the last stages of development.

Ovarian Cycles

Oviparous Elasmobranchs

In the small spotted catshark, high levels of E prior to ovulation (Craik 1979) suggest a functional role in yolk production or oocyte maturation (Table 2.2). More detailed studies by Koob et al. (1986) in the oviparous little skate indicated a role for estradiol in the maturation of the ovarian follicle where estradiol levels correlated directly with pre-ovulatory follicular size (Table 2.2). In the little skate, concurrently with the changing E levels, T predominated during the follicular phase of each ovulatory or egg-laying cycle and, similarly to E, declined about two-fold prior to encapsulation of the egg (Koob et al. 1986). We suggest that the high serum T levels measured by Fletcher et al. (1969) in the thorny skate resulted from the presence of samples collected from a few actively ovipositing females in the study population. For the oviparous skate, E and T are the predominant steroids during the follicular phase of each spawning cycle, whereas progesterone may have a specific, shorter term role during ovulation and encapsulation, as suggested by its transitory 16-fold elevation just prior to encapsulation of the ovulated egg.

Sumpter and Dodd's 1979 study of steroid hormones in the serum of the small spotted catshark demonstrated ten-fold seasonal variations in E and T levels, with a striking correlation between the levels of E and T. This oviparous shark displayed a clear, synchronous annual cycle. Hormonal levels increased in the fall, rising with the recrudescence of the ovary, reached high levels, and remained elevated during the extended winter egg-laying period. In the spring, E levels declined to lower levels, reaching their lowest levels in the summer. In contrast, T levels were relatively constant throughout the year, at about 20% of the levels of E. In this oviparous elasmobranch, as in the little skate, estradiol and testosterone

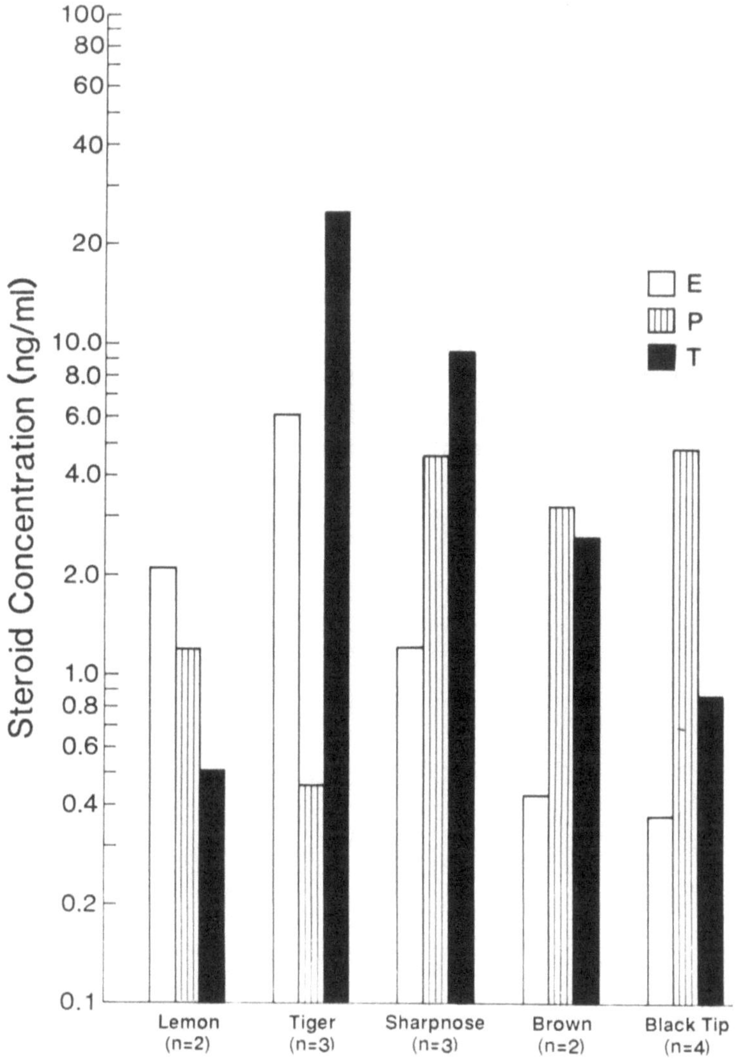

FIGURE 2.2. Mean concentrations of steroid hormones in the serum of five species of carcharhinid sharks during the periovulatory period. E = E-2 or 17β-estradiol; P = progesterone; T = testosterone (From Rasmussen and Gruber in press).

rose when the ovaries recrudesced and the percent of females with eggs increased.

Ovoviviparous (Internal Incubators) Elasmobranchs

Only two studies reported levels of serum steroid hormones in viviparous non-placental elasmobranchs during or near ovulation (Table 2.2). Just prior to gestation, the electric ray had very high P levels (Lupo di Prisco

1967), while the recently ovulated piked dogfish had moderately elevated progesterone. In the electric ray the T levels were moderately high but were low in the piked dogfish (Tsang and Callard 1987a). Tsang and Callard (1987a, 1987b) also observed that the cells of small follicles predominantly produced E, whereas the cells of larger follicles produced T. Measurements of steroid levels in the serum reflect the contribution of both the small E-contributing follicles and the larger T-contributing follicles.

Viviparous Elasmobranchs

During the periovulatory state (Fig. 2.2) and just prior to mating (Fig. 2.3) in the mature female lemon sharks, E levels were about two-fold higher compared to the concentrations during the reproductively quiescent period (Fig. 2.1). The E levels of placental carcharhinids generally appeared to be

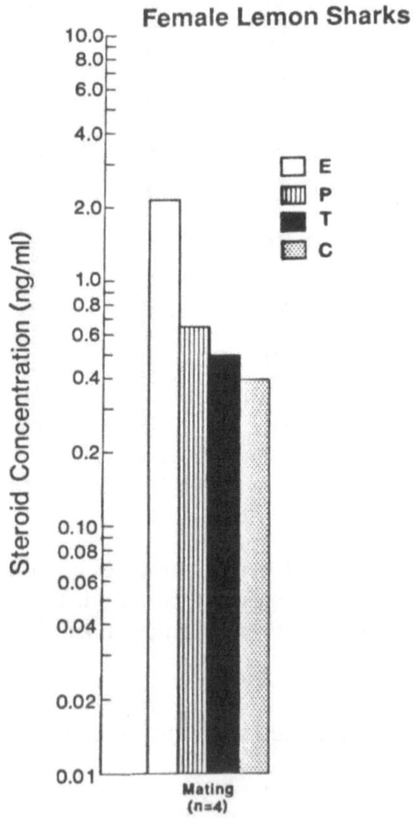

FIGURE 2.3. Mean levels of steroid hormones in the serum of adult female lemon sharks during the mating period. E = E-2 or 17β-estradiol; P = progesterone; T = testosterone; C = corticosterone (From Rasmussen and Gruber in press).

higher at ovulation than those of ovoviviparous piked dogfish sharks or electric rays (Table 2.2). However, E concentrations in oviparous sharks and skates were similar to those of the placental sharks (Table 2.2). Interestingly, the aplacental tiger shark has three-fold higher E levels than other requiem sharks (Table 2.2).

Among the placental sharks, reproductively quiescent Caribbean reef sharks and blacktip sharks had low T levels (Table 2.2; Rasmussen and Gruber 1990), whereas the reproductively active female lemon, sharpnose, blacktip, and sandbar (brown) sharks had several-fold higher T levels during the periovulatory period (Fig. 2.2). The apparent T elevation in the female lemon shark may not be significant because T levels were also relatively high during reproductively quiescent periods (Figs. 2.1, 2.2, 2.3). Again, the reproductively active aplacental tiger shark had 40-fold higher T levels than quiescent animals (Fig. 2.2), suggesting a pre-ovulatory role for testosterone.

Transitory elevations of P during the periovulatory period, as seen in the lemon, sharpnose, blacktip, and sandbar sharks (Fig. 2.2), suggest a possible pre-ovulatory initiation role for P (see also Rasmussen and Murra 1992).

Pregnancy

Aplacental Viviparous (Ovoviviparous) Elasmobranchs

Throughout gestation in the piked dogfish, E levels rose steadily to a three-fold elevation just prior to birth; however, absolute levels were relatively low compared to other sharks (Table 2.2) (Tsang and Callard 1987a). In contrast, P levels were relatively high during gestation but dropped substantially prior to birth. [A similar but less abrupt decline occurred during gestation in the electric ray (Lupo di Prisco et al. 1967).] In the piked dogfish, T levels gradually increased during gestation, from a low level to a three-fold elevation prior to birth (Tsang and Callard 1987a). In contrast, T levels in the electric ray during gestation were lower than pre- or post-gestation levels (Lupo di Prisco et al. 1967; Table 2.2).

Placental Viviparous Elasmobranchs

A different hormonal pattern is beginning to emerge for the placental sharks. In the sharpnose shark, mid-gestation data demonstrated high E levels and 15-fold lower P levels than in the piked dogfish (Table 2.2). At mid-gestation, T levels of the placental sharpnose shark were about 10-fold lower than T levels in the piked dogfish (Table 2.2).

Figure 2.4 shows that the levels of all five hormones (E, P, T, D, and C) did not differ significantly immediately before and after birth in the lemon shark. E levels were similar at term in both the lemon shark and the piked dogfish (Table 2.2). T levels in the piked dogfish increased to 750 pg/ml near term. An elevation in T levels did not occur in the lemon shark. P levels dropped four-fold prior to delivery in the piked dogfish (Table 2.2).

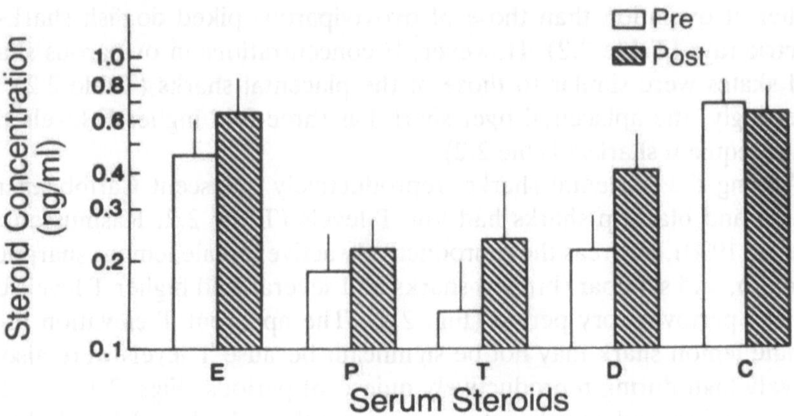

FIGURE 2.4. Mean concentrations of steroid hormones in the serum during the peri-panurient period in the lemon shark. E = E-2 or 17β-estradiol; P = progesterone; T = testosterone; D = DHT or dihydrotestosterone; C = corticosterone; n = 5 (From Rasmussen and Gruber in press).

This decline did not occur in the placental lemon shark. This lack of hormonal changes in the lemon shark in the period immediately prior to and after birth strongly supports the concept that the placenta is not producing large quantities of steroid hormones with roles in the birth process. Unlike most mammals as well as the ovoviviparous piked dogfish, the hormonal picture in this placental shark is static.

The foregoing data support the general concept that steroid hormones in the elasmobranchs, an ancient vertebrate group, play regulatory roles both in the process of maturation and during reproduction. The systematic manner in which testosterone, dihydrotestosterone, and progesterone change during male maturation and estradiol, testosterone, dihydrotestosterone, and progesterone change during female maturation suggests a continuum of regulatory mechanisms during maturation. Our studies reveal preliminary evidence of sophisticated hormonal cycles during the reproduction of viviparous placental sharks. In the lemon shark, our model species, estradiol and testosterone appear to play an important role, especially in pre- and peri-ovulatory events concerned with reproduction. It is possible that progesterone has specific time-related functions at critical reproductive stages. However, it is clear that much additional data are required to provide a complete framework necessary to understand elasmobranch reproductive processes.

ACKNOWLEDGEMENTS

This research was supported in part by NIH grants RR 00163 and HD-18185 and in part by a grant from NSF-OCE, 88 43425 to Samuel H. Gruber and a grant from Earthwatch. Our sincere appreciation is expressed to

Mr. Colin Higgs, Assistant Director, Department of Fisheries, Commonwealth of Bahamas, and the people of Bimini, Bahamas; to Michelle Pyle of Kawasaki Motors, U.S.A., for the use of a jet boat; Robert Attanazio Homizo and Ray Kurtz of Davie Marine for supplying skiffs; to Mr. Dan Schaad, Mercury Motors, Brunswick Corporation, for outboard motors; and to Honda power generators. We thank Mr. Andy Hudson of Earthwatch and Captain Morgan and the crew of the University of Miami's ORV Columbus/Iselin.

References

Anderson LL (1969) Sexual behavior and controlling mechanisms in domestic birds and mammals. In Cole HH, Cupps PT (eds.). Reproduction in Domestic Animals. Academic Press, New York. pp. 541–568

Bubenik GA, Bubenik AB, Schams D, Leatherland JF (1983) Circadian and circannual rhythms of LH, FSH, testosterone (T), prolactin, cortisol, T_3, and T_4, in plasma of mature, male white-tailed deer. Comp. Biochem. Physiol. 76A: 37–45

Bullesbach EE, Schwabe C, Callard IP (1987) Relaxin from an oviparous species, the skate (*Raja erinacea*). Biochem. Biophys. Res. Comm. 143:273–280

Brown CA, Gruber SH (1988) Age assessment of the lemon shark *Negaprion brevirostris* using tetracycline validated vertebral centra. Copeia 3:747–753

Callard G, Petro Z, Ryan KJ (1978) Phylogenetic distribution of aromatase and other androgen converting enzymes in the central nervous system. Endocrinol. 43:2283–2290

Callard IP, Klosterman L, Callard GV (1988) Reproductive Physiology. In Shuttlesworth TJ (ed.). Physiology of Elasmobranch Fishes. Springer-Verlag, Berlin pp. 277–317

Callard IP, Klosterman L, Sorbera LA, Fileti LA, Reese JC (1989a) Endocrine regulation of reproduction in elasmobranchs: Archetype for terrestrial vertebrates. J. Exp. Zool. Supplement 2:12–22

Callard G, Mak P, DuBois W, Cuevas ME (1989b) Regulation of spermatogenesis: The shark testis model. J. Exp. Zool. Supplement 2:23–34

Chieffi G (1967) The reproductive system of elasmobranchs: Development and endocrinological aspects. In Gilbert PW, Mathewson RF, Rall DP (eds.). Sharks, Skates and Rays. Johns Hopkins Press, Baltimore Maryland. pp. 553–580

Chieffi G, Lupo di Prisco C (1961) Identification of estradiol-17β, testosterone and its precursors from *Scyliorhinus stellaris* testis. Nature 190:169–170

Collenot G, Ozon R (1964) Mises en evidence biochimique et histochimique d'une Delta 5, 3-β hydroxysteroid deshydrogenase dans le testicule de *Scyliorhinus canicula*. Bull. Soc. Zool. France 89:577–587

Compagno LJV (1984) FAO species catalogue. Vol. 4. Sharks of the World. Parts 1 and 2. FAO Fisheries Synopsis No. 125. United Nations Development Programme, Rome, Italy

Craik JCA (1979) Simultaneous measurements of rates of vitellogenin synthesis and plasma levels of oestradiol in an elasmobranch. Gen. Comp. Endocr. 38: 264–66

Demski LS (1989) Pathways for GnRH control of elasmobranch reproductive physiology and behavior. J. Exp. Zool. Suppl. 2:4–11

Demski LS (1990) Elasmobranch reproductive biology: Implications for captive breeding. J. Aquaculture & Aquatic Sci. 5:84–95
Dodd JM (1983) Reproduction in cartilaginous fishes. In. Hoar WS, Randall DJ, Donaldson ME (eds.). Fish Physiology, Vol. 9. Academic Press, New York. pp. 31–95
Dodd JM, Goddard CK (1961) Some effects of oestradiol benzoate on the reproductive ducts of *Scyliorhinus canicula*. Proc. Zool. Soc. Lond. 137:325–332
Dodd JM, Dodd MHI, Duggan RT (1983) Control of reproduction in elasmobranch fishes. In Rankin JC, Pitcher TJ, Duggan RT (eds.). Control Processes in Fish Physiology. John Wiley & Sons, New York. pp. 221–285
Dodd JM, Sumpter JP (1984) Fishes. In Lamming GE (ed.). Marshall's Physiology of Reproduction, 4th edition. Churchill Livingston, Edinburgh. pp. 1–126
Fileti LA, Callard JP (1988) Corpus luteum function and regulation in the skate, *Raja erinacea*. Bull. MDIBL. 27:37–39
Fletcher GL, Hardy GC, Idler DK (1969) Production and metabolic clearance rates in sexually mature male and female skates (*Raja radiata*). Endocrinol. 100:1483–1495
Garnier DH, Coquil C, Chauvin J (1989) Seasonal variation of plasma and testicular sex steroid levels in a selechian (*Scyliorhinus canicula*). Gen. Comp. Endocr. 74(2):298
Gottfried H (1964) The occurrence and biological significance of steroids in lower vertebrates. A review. Steroids 3:219–241
Gruber SH, Myrberg AA (1977) Approaches to the study of the behavior of sharks. Am. Zool. 17:471–684
Hamlett WC (1987) Comparative morphology of the elasmobranch placental barrier. Arch. Biol. (Brux.) 98(2):135–162
Hamlett WC (1989a) Evolution and morphogenesis of the placenta in sharks. J; Exper. Zool. Suppl. 2:35–52
Hamlett WC (1989b) Reproductive strategies in elasmobranchs of the North Carolina shelf. In: Gorge RY, Hulbert AW (eds.). Carolina Coastal Oceanography. National Undersea Research Program Research Report 89-2. pp. 453–480
Hamlett WC, Schwartz FJ, Didio LJA (1987) Subcellular organization of the yolk syncytial-endoderm complex in the preimplantation yolk sac of the shark, *Rhizoprionodon terraenovae*. Cell Tissue Res. 247:275–285
Hamlett WC, Wourms JP (1984) Ultrastructure of the preimplantation shark yolk sac placenta. Tissue Cell 16:613–625
Hamlett WC, Wourms JP, Hudson JS (1985a) Ultrastructure of the full-term shark yolk sac placenta. I. Morphology and cellular transport at the fetal attachment site. J. Ultrastr. Res. 192–206
Hamlett WC, Wourms JP, Hudson JS (1985b) Ultrastructure of the full-terin shark yolk sac placenta. II. The smooth proximal segment. J. Ultrastr. Res. 207–220
Hamlett WC, Wourms JP, Hudson JS (1985c) Ultrastructure of the full-term shark yolk sac placenta. III. The maternal attachment site. J. Ultrastr. Res. 221–231
Holmes RL, Ball JN (1974) The pituitary gland. A comparative account. University Press, Cambridge
Idler D, Truscott B (1966) Identification and quantification of testosterone in peripheral plasma of skate. Gen. Comp. Endocr. 7:375–383
Jenkins N (1978) The endocrine control of reproduction in the dogfish (*Scyliorhinus canicula*) Ph.D. Thesis, University of Wales, U.K.

Jenkins N, Dodd JM (1980) Effects of synthetic mammalian gonadotropin releasing hormone and dogfish hypothalamic extracts on the levels of androgens and oestradiol in the circulation of the dogfish (*Scyliorhinus canicula*). J. Endrocrinol. 86:171–177

Kime DE (1978) Steroid biosynthesis by the testis of the dogfish *Scyliorhinus canicula*. Gen. Comp. Endocr. 34:6–17

Klosterman L, Callard IP (1986) Progesterone production by enzymatically dispersed cells from corpora lutea of the spiny dogfish, *Squalus acanthias*. Bull. Mt. Desert. Isl. Biol. Lab. 26:119–121

Koob TJ, Callard IP (1985) Progesterone treatment causes early oviposition in *Raja erinacea*. Bull. MDIBL. 25:138–139

Koob TJ, Tsang P, Callard IP (1986) Plasma estradiol, testosterone, progesterone levels during the ovulatory cycle of the skate (*Raja erinacea*). Biol. Reprod. 35:267–275

Lance V, Callard IP (1969) A histochemical study of ovarian function in the ovoviviparous elasmobranch *Squalus acanthias*. Gen. Comp. Endocr. 13:255–267

Lund R (in press) Observations on the relationships of the Chimaeroformes. In Pratt HL, Taniuchi T, Gruber SH (eds.). Elasmobranchs as Living Resources. NOAA Technical Report-NMFS, Seattle

Lupo di Prisco C, Vellano C, Chieffi G (1967) Steroid hormones in the plasma of the elasmobranch *Torpedo marmorata* at various stages of the sexual cycle. Gen. Comp. Endocr. 8:325–331

McArthur R, Wilson EO (1967) The Theory of Island Biogeography. Princeton University Press, Princeton, New Jersey 203 pp

Myrberg AA, Gruber SH (1974) The behavior of the bonnethead shark, *Sphyrna tiburo*. Copeia 3:358–374

Rasmussen LEL, Gruber SH (1990) Serum levels of circulating steroid hormones in free-ranging carcharhinoid sharks. In Wes Pratt H, Gruber SH, Taniuchi T (eds.). Elasmobranchs as Living Resources: Advances in Biology, Systematics and the Status of the Fisheries. NOAA Technical Report 90, NMFS, Seattle. pp. 145–155

Rasmussen LEL, Gruber SH (in press) Serum levels of reproductively-related circulating steroid hormones in the free-ranging lemon shark *Negaprion brevirostris*. Environ. Biol. Fish.

Rasmussen LEL, Murra FL (1992) Long-term studies of serum concentrations of reproductively-related steroid hormones in individual captive carcharhinids. Aus. J. Mar. Freshwater Res. 43.

Reese JC, Callard IP (1987) Receptors for estradiol-17-β in the oviduct of the skate, *Raja erinacea*. MDIBL 27:28–29

Schaeffer B (1967) Comments on Elasmobranch Evolution. In Gilbert PW, Mathewson RF, Rall DP (eds.). Sharks, Skates and Rays. Johns Hopkins Press, Baltimore, Maryland. pp. 3–37

Schaeffer B, Williams M (1977) Relationship of fossil and living elasmobranchs. Am. Zool. 17:293–302

Simpson TH, Wardle CS (1967) A seasonal cycle in the testis of the spurdog, *Squalus acanthias*, and the site of 3β-hydroxy-steroid dehydrogenase activity. J. Mar. Biol. Ass. U.K. 47:699–708

Simpson TH, Wright RS, Hunt SN (1963) Sex hormones in fish II. The oestrogens of *Scyliorhinus canicula*. J. Endocr. 26:499–507

Simpson TH, Wright RS, Hunt SV (1964) Steroid biosynthesis in the testis of the dogfish (*Squalus acanthias*). J. Endocr. 31:29–38

Sumpter JP, Dodd JM (1979) The annual reproductive cycle of the female lesser spotted dogfish, *Scyliorhinus caniculus L.*, and its endocrine control. J. Fish Biology 15:687–695

Sumpter JP, Follett BK, Jenkins N, Dodd JM (1978) Studies on the purification and properties of gonadotropin from ventral lobes of the pituitary gland of the dogfish (*Scyliorhinus canicula*). Gen. Comp. Endocr. 36:264–274

Teshima K (1981) Studies on the reproduction of Japanese smooth dogfishes, *Mustelus manazo* and *M. griseus*. J. Shimonoseki Univ. Fish. 29:113–199

Tsang P (1986) Endocrine correlates and regulation of follicular development and luteal function during the reproductive cycle of the viviparous, *Squalus acanthias*. Ph.D. Thesis, Boston University

Tsang P, Callard IP (1983) In vitro steroid production by ovarian granulosa cells of *Squalus acanthias*. Bull. MDIBL. 23:78–79

Tsang P, Callard IP (1987a) Morphological and endocrine correlates of the reproductive cycle of the aplacental dogfish *Squalus acanthias*. Gen. Comp. Endocrinol. 66:182–189

Tsang P, Callard IP (1987b) Luteal progesterone production and regulation in the viviparous dogfish *Squalus acanthias*. J. Exp. Zool. 241:377–382

Wotiz HH, Botticelli C, Hisaw FL, Olsen AG (1960) Estradiol-17β, estrone and progesterone in the ovaries of the dogfish (*Squalus suckleyi*). Proc. Nat. Aca. Sci. 46:580–585

Wourms J (1977) Reproduction and development in Chondrichthyan fishes. Amer. Zool. 17:379–410

Wourms J (1981) Viviparity: The maternal-fetal relationships in fishes. Am. Zool. 21:473–515

Wourms J, Grove B, Lombardi J (1988) The maternal-embryonic relationship in viviparous fishes. In Hoar WS, Randall DJ (eds.). Fish Physiology, XIB. Academic Press, New York. pp. 1–134

Ostheichthyes

3

Larval Ecology of Fish of the Orinoco Basin

Antonio Machado-Allison

Characteristics of the Orinoco River

The Orinoco River is one of the largest rivers of the world. During its course (2,060 km) it drains an area of basins approximately 9×10^5 km^2, situated in Colombia and Venezuela (Fig. 3.1). Its average discharge is estimated as 36,000 m^3/s (4–6 to 60.000 m^3/s). The main channel width varies from several meters in headwaters to 1,000 to 1,500 m in the middle to more or less 20 km at the mouth or delta. This distance is only exceeded by the Zaire in Africa and the Amazon in South America.

The Orinoco is considered a white water river with abundant sediment in suspension, carrying an average sediment load of about 200×10^6 tons/year. Due to its low inclination (4 to 5 m/100 km), and abundant rains (mean 1,500 mm), its waters are partially dammed by the sea. During the rainy season (June to November) this produces an extensive sedimentation in the deltaic region and numerous flooded lagoons on the margins, especially in the areas known as "Llanos" in Venezuela. During the dry season these flooded areas become isolated from the main channel and eventually dry out. The water temperature varies from 26°C to 32°C with an average of 28.3°C. The average transparency is about 0.8 m, increasing to 1.5 m in the lagoons. Dissolved oxygen varies from 90% in November to 65% in August and pH varies from 5.6 during rains to 7.5 in the dry season (Sanchez and Vasquez, 1986a).

The Orinoco river is a more or less continuous channel controlled in several places by precambrian rocks from the Guiana shield formation. There are few rapids in the middle region which are conspicuous during the dry season. The river bed is composed of a combination of coarse sand coming from the Guyana shield tributaries, plus fine sand. Due to the erosion of the Andes, clay is carried principally by the Guaviare (Colombia) and the Apure in Venezuela (Meade et al. 1983).

Due to its movement during the last million years, this river has formed an extensive arc which surrounds the mountain ridges of the Guyana shield (South) and the "Central" and "Oriental Llanos" (north and east). The

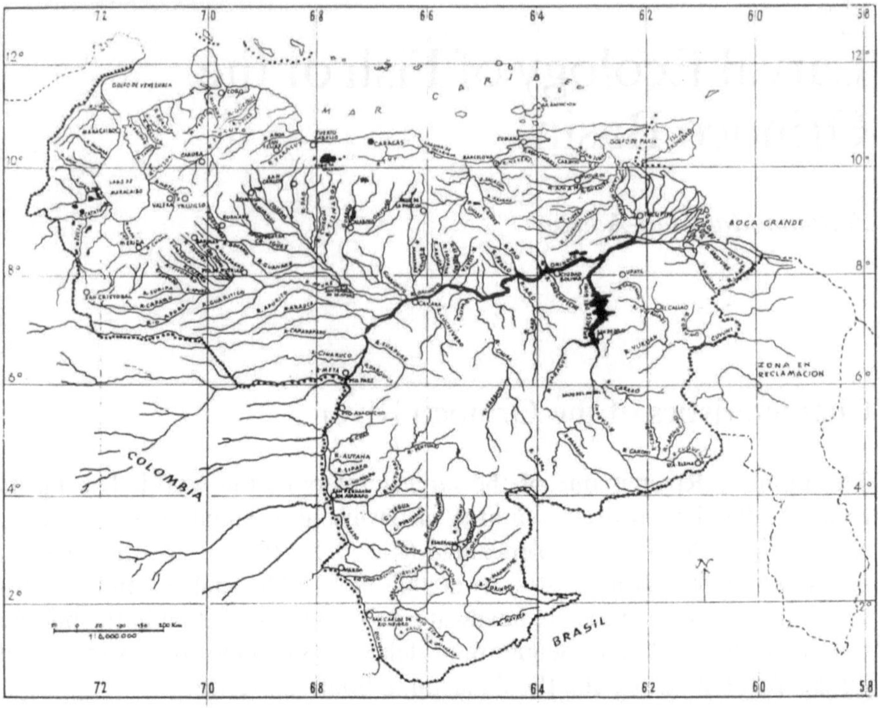

FIGURE 3.1. Map of Venezuela and Colombia (in part). Dark area depicts the Orinoco River Basin

movement has generated an extensive flood plain, highly complex and practically unknown in its abiotic and biotic constituents.

The marginal vegetation is closely related to, and limited by, the Mesa formation with a great number of savannas with an extensive gallery forest adapted to the direct influence of seasonal floods. During rains or floods a highly complex shore system is formed, with several lagoons and "caños" or "varzeas." In these areas a great diversity of aquatic life abounds (Colonello et al. 1986, Sanchez and Vasquez 1986b, Hamilton and Lewis 1987, Rodriguez and Lewis 1989). The aquatic environments are dominated by plants such as: *Eichhornia crassipes*, *Paspalum repens*, *Oxycarum cubense*, *Ludwigia helminthorriza*, *L. sediodes*, *Salvinia auriculata*, *Hymenachne amplexicaulis* and species of *Eleocharis*.

Areas with seasonal floodings and highly protected by a great quantity of aquatic plants (emergent and floating), are considered as "nursery areas" for numerous fishes. Larval stages of species such as *Colossoma macropomum*, *Piaractus brachypomus*, and *Prochilodus laticeps* penetrate the lagoons and flooded savannas. They feed on the abundant phyto- and zooplankton and larvae of aquatic insects. Numerous hiding places formed by the complex diversity of aquatic plants provides protection from predators. During the dry season the juveniles migrate back to the main channel where they complete their development (Machado-Allison 1982). Large adult characoids, such as *Brycon*, *Colossoma*, *Myloplus*, and *Piaractus*, enter the flooded forests and lagoons to feed on the great quantity of fruits and seeds (ichthyocory). As a result, they subsequently serve as seed dispersers.

Data on the diversity, relative abundance, life history and population dynamics of the fishes of the Orinoco are almost unknown. However, it has been suggested that the Orinoco Basin harbors nearly 1,000 species, the majority of which belong to the orders Characiformes, Siluriformes and Gymnotiformes (Mago-Leccia 1970, López et al. 1984, Machado-Allison 1987).

Life Histories of Fishes

Biologists have been intrigued for a long time by the great number of methods by which fish reproduce. There has been much discussion as to how life history patterns are products of natural selection, modulated by environmental changes. Few studies however, have attempted to obtain empirical data from natural systems. These results would further our knowledge of the reproductive ecology (reproduction, juvenile survival, clutch size, reproductive effort, adult survival, mean generation time, and intrinsic rate of increase) of the many species, which is necessary to improve programs of management and conservation.

Tropical freshwater species exhibit a diversity in morphological, physiological and ecological characteristics (Lowe-McConnell 1975, 1987). Despite few comparative studies on tropical fish reproduction to date (Welcomme 1969, Kramer 1978, De Silva et al. 1985), these diverse fish assemblages provide excellent systems for an evaluation of life history patterns (Winemiller 1989). The Orinoco drainage harbors one of the most diverse freshwater fish faunas in South America with about 1,000 known species, paradoxically derived from only a few basic stocks. More than 95% of the species belong to the orders Characiformes, Siluriformes (sensu Fink and Fink 1981), and the Cichlidae (Perciformes) (Mago 1970, Machado-Allison 1987). This situation makes it possible to plan ecological studies based on monophyletic groups with the purpose of establishing evolutionary trends within and among natural taxonomic groups.

Tropical, continental ichthyofauna, shows a great variety of types of reproduction. The "annual fishes" (Cyprinodontiformes: *Pterolebias* and *Rachovia*), spawn only once in their life within a single climatic cycle. The majority of the species that live in the Orinoco Basin included partial, multiple and total spawners. Viviparous species display a wide variety of parental care including constant parental care, nesting or mouth-carrying with partial parental care, or others that show no care at all. Table 3.1 summarizes today's knowledge of the representative types of reproduction.

Life history patterns, are the result of different selective pressures or factors imposed by the respective biotopes with the purpose to produce a maximum number of offspring able to reach sexual maturity. It has been suggested that flooded areas of large tropical rivers, such as the Orinoco play a critical role in the biological cycle of numerous fish species affecting aspects of reproduction, protection and feeding (Lowe-McConnel 1964, 1987, Welcomme 1979, Machado-Allison 1987).

In general, the Orinoco fish fauna, that exploit the flooded savannas possess only one reproductive season that is synchronized with the beginning of rains or flooding. This condition is similarly observed in other tropical areas of South America and Africa. Evidence supporting seasonal reproduction include:

1. Gonads mature just before the begining of the rainy season, insuring that the fish are reproductively mature when the rains and/or floods start;
2. Migrations ("ribasones") with reproductive purposes generally occur at the beginning of the flooding. Mature adults of prochilodontids and characoids species, leave the marginal lagoons, enter the main channel and travel several hundred Kilometers to spawn in headwaters of the main tributaries of the Orinoco. Example include *Prochilodus mariae*, *Colossoma macropomum*, and *Piaractus brachypomus*;
3. Ethology and appearance of secondary sexual characteristics in several species which include courtship, defense of territory, preparation of nesting sites and coupling, which are coincident with the begining of the rainy season. This is most evident in catfishes such as: ageneiosids, auchenipteriids, loricariids and callichthyids (Fig. 3.2);
4. Free swimming larvae of numerous species after the flooding.

The reproductive period in the Orinoco is influenced by factors such as: abundance of food (zoo- and phytoplankton, larval stages of aquatic insects, fruits and seeds), high variety of hiding places (aquatic plants, roots, holes in banks, and submerged logs), and physico-chemical changes in water-quality (transparency, pH, oxygen and temperature). However, if environmental conditions including water quality, and nutritional conditions are favorable throughout the year, numerous species could spawn several times yearly (Lowe-McConnell 1975, Machado-Allison 1987). This have been observed with the introduction of "piranhas" (*Serrasalmus* sp and *Pygocentrus* sp) in man-made dams or reservoirs in the Orinoco Basin.

We can divide the fishes into two groups: (1) those that perform all activities of the biological cycle within the flooded areas and "varzeas"; and (2) those that use these flooded areas only part of the time.

1) The majority of the species belonging to this group, possess ethological and physiological adaptations to cope with the drastic alterations that occur periodically during dry and wet seasons. During dry season, these adaptations manifest themselves as trophic, metabolic, and respiratory

TABLE 3.1. Summary of types of reproduction (fecundity, seasonality and parental care) in species of the Orinoco Basin.

Type & fecundity	Seasonality of reproduction	Movements & parental care	Examples
Total spawners; high fecundity	Very seasonal; synchronized with annual floods	Extensive migrations; no parental care	Colossoma, Mylossoma, Prochilodus, Curimata
Total spawners; low fecundity	Synchronized with floods	No movements or very restricted; generally nest construction & parental care.	Hoplosternum (floating), Hypostomus (holes), Loricaria (body), Ageneiosids (roots), Characids (roots)
	End of rainy season	Bury eggs in muddy bottoms.	Annual fishes, Cyprinodontiformes
Partial spawners	Synchronized with floods	Restricted movements; care of eggs in roots of aquatic plants or bottom	Characoids (plants), Serrasalmus (plants), Hoplias (bottom)
Partial spawners	Asynchronized with floods; permanent habitats	No movements; nesting behavior and complex parental care	Cichlidae such as Cichla, Astronotus, Aequidens
		viviparous	Poecilia, Potamotrygon

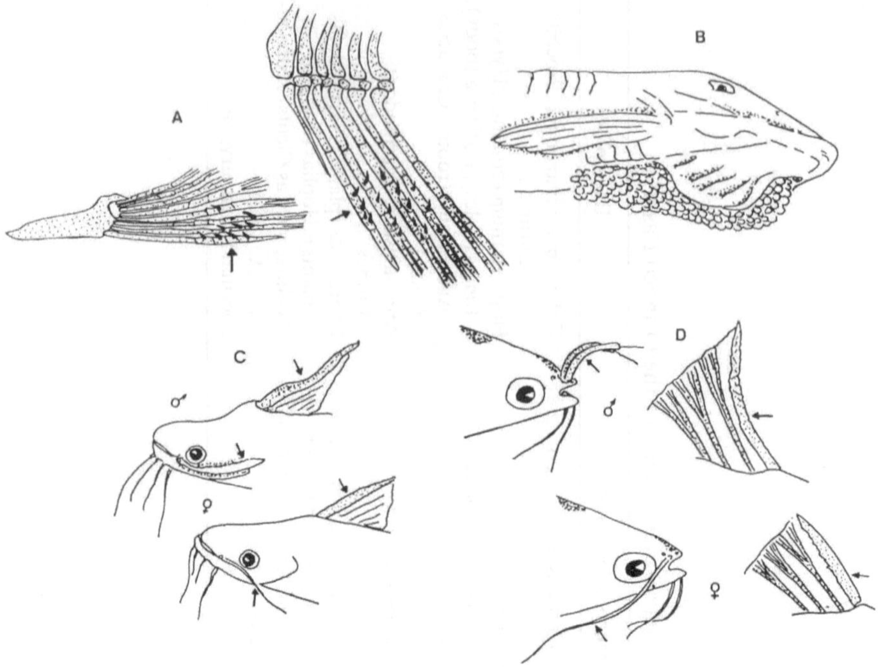

FIGURE 3.2. Sexual secondary characteristics in freshwater fish of the Orinoco Basin: (A) spines on the soft rays of male's pelvic and anal fins (several Characidae); (B) Upper lip development to carry and protect eggs in *Loricariichthys typus*; (C) and (D) sexual differences in the development of the first dorsal fin spine in Auchenipteriidae. (From:Machado-Allison and Lopez 1975; Mago-Leccia 1983, Machado-Allison 1987)

changes during dry season. A common reproductive trait is complex parental care for the offspring. These fish develop secondary sexual structures such as spines, tubercles, and lips as are observed in *Pseudoauchenipterus galeatus*, *Entomocorus gameroi*, *Epapterus blohmi*, and *Loricariichthys typus*; also have complicated courtship rituals, and construction of nests such that in *Hoplosternum littorale* and *Pterygopliicthys multiradiatus* (Figure 3.2, 3.3). The swimming larvae of these fish possess color patterns, such as bandings, that could deceive abundant potential predators in flooded areas (Figure 3.4). The larvae feed mainly on zooplankton and early stages of aquatic insects. Although there is no generalized studies dealing with larval behaviour of fishes in these areas, so far there is no evidence that early developmental stages of the species of this division form schools or exhibit other kinds of social behaviour.

2) These species of the second group generally spawn in open waters of the main channel. Their eggs and/or free-swimming larvae, carried by the

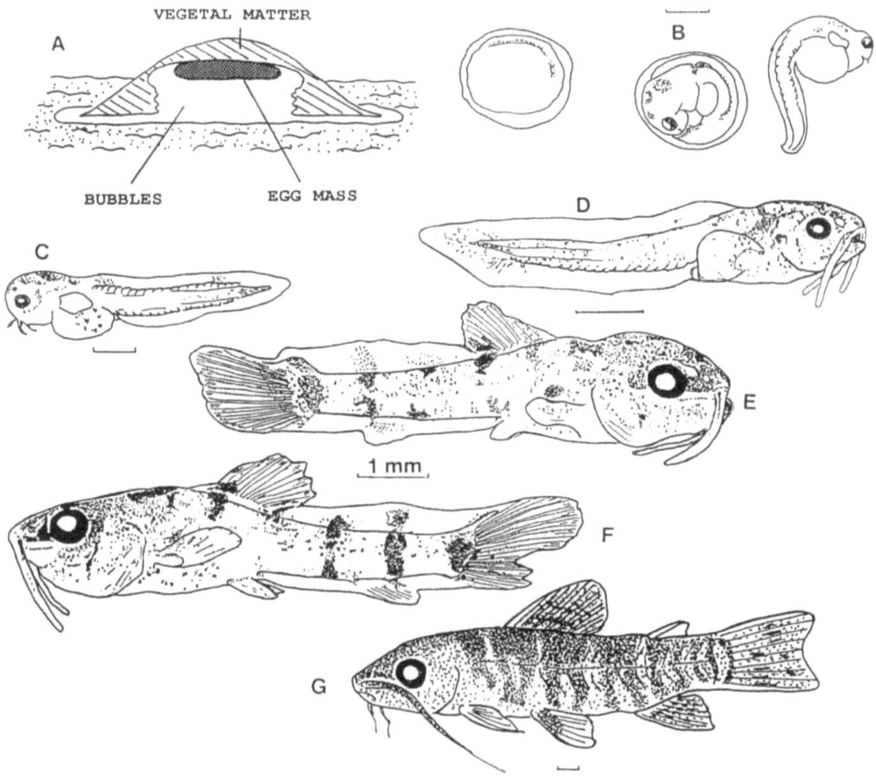

FIGURE 3.3. Developmental stages of *Hoplosternum littorale*: (A) Diagram of the floating nest indicating position of egg mass; (B) Eggs and embryo; (C) Yolk-sack larvae; (D) Early juvenile. (From: Machado and Zaret 1984; Machado-Allison, 1987)

water, enter the flooded savannas, where they start their development. Adults do not exhibit complex courtship behaviour nor parental care. Included in this group are a great number of species of great fishery value such as: *Brycon whitei*, *Colossoma macropomum*, *Mylossoma duriventre*, *Piaractus brachypomus*, *Brachyplatystoma filamentosum*, *B. juruense*, *B. vaillanti*, *Phractocephalus hemiliopterus*, *Pseudoplatystoma fasciatum*, and *P. tigrinum*.

An important aspect of the larval ecology of these groups is the "pelagic" behavior of early developmental stages. Almost all larvae have unpigmented or almost transparent bodies. During this stage of development they feed exclusively on phyto- and zooplankton (Table 3.2). After entering shallow waters on the margins of flooded areas, major changes in color patterns such as the development of bands, spots and other markings on the body and head regions occur (Figure 3.4). It was been suggested that the shifting in color patterns may have significance in social behavior,

TABLE 3.2. Comparative feeding habits and ontogenetic changes in fishes of the Orinoco Basin.

	Food resources									
Species	Plankton	Insects	Fruits/seeds	Detritous/mud	Algae	Fins	Scales	Fish	Decapods	Terrest. insects
Poecilia reticulata	Lar	Juv/Ad								Juv/Ad
Pyrrhulina filamentosa	Lar	Juv/Ad								Juv/Ad
Astyanax bimaculatus	Lar	Juv/Ad			Ad				Ad	Ad
Astyanax metae	Lar				Ad					
Apistogramma ramirezi	Lar	Juv/Ad			Juv/Ad				Ad	Ad
Moenkhausia lepidura	Lar	Juv/Ad			Ad					Ad
Gymnocorymbus thayeri	Lar	Juv/Ad			Ad					
Tatragonopterus argenteus	Lar	Juv/Ad			Ad					Ad
Loricariichthys typus	Lar			All*	All*					
Pseudohemiodon laticeps	Lar			All*	All*					
Pterigoplichthys multiradiatus	Lar			All*	All*					
Hypostomus plecostomus	Lar			All*	All*					
Hypophthalmus edentulus	All*									
Roeboides dayi	Lar	Juv/Ad			Juv/Ad		Juv/Ad			
Catoprion mento	Lar		Juv/Ad				All*	Juv		
Pygocentrus caribe	Lar	Lar/Juv			Ad	Juy		Ad	Ad	
Serrasalmus elongatus	Lar					Juv/Ad				
Serrasalmus irritans	Lar					Juv/Ad	Juv	Juv/Ad		
Serrasalmus medinai	Lar	Lar/Juv	Juv			Juv/Ad	Juv	Juv/Ad	Ad	
Serrasalmus rhombeus	Lar	Juv				Juv	Juv/Ad	Ad		
Pristobrycon striolatus	Lar	Juv	Ad			Juv	Juv/Ad	Ad	Ad	
Mylossoma aureum	Lar	Juv	Juv/Ad							
Mylossoma duriventre	Lar	Juv	Juv/Ad					Ad		
Colossoma macropomum	All*		Ad							
Piaractus brachypomus	Lar		Ad							
Triportheus angulatus	Lar	Juv/Ad	Ad							
Toracocharax stellatus	Lar	Juv/Ad								
Markiana geayi	Lar	Juv/Ad	Juv/Ad					All*		

Acestrorhynchus microlepis	?				All*	
Hydrolicus scomberoides	Lar				All*	
Raphiodontichthys vulpinus	Lar				All*	
Hoplias malabaricus	Lar					
Curimata cerasina	Lar			Juv/Ad		
Prochilodus mariae	Lar/Juv			Juv/Ad		
Hoplosternum littorale	Lar			Juv/Ad		
Schizodon sp	Lar	Juv/Ad	Juv/Ad		Ad	
Brachyplatystoma flavicans	Lar	Juv				
Rhabdolichops zareti	All*	All*				Ad
Adontosternarchus devenanzi		All*				Ad
Rhamphichthys reinhardti		All*				
Sternarchorhynchus mormyrops		All*				

Lar = larvae; Juv = juvenile; Ad = adult; All* = All developmental stages

Sources: Mago-Leccia and Zaret 1978; Joerger 1984; Nico and Taphorn 1985, 1988; Machado-Allison 1986, 1987; Machado and Garcia 1986; Lundberg et al 1987; Prejs and Prejs 1987; Marrero 1990.

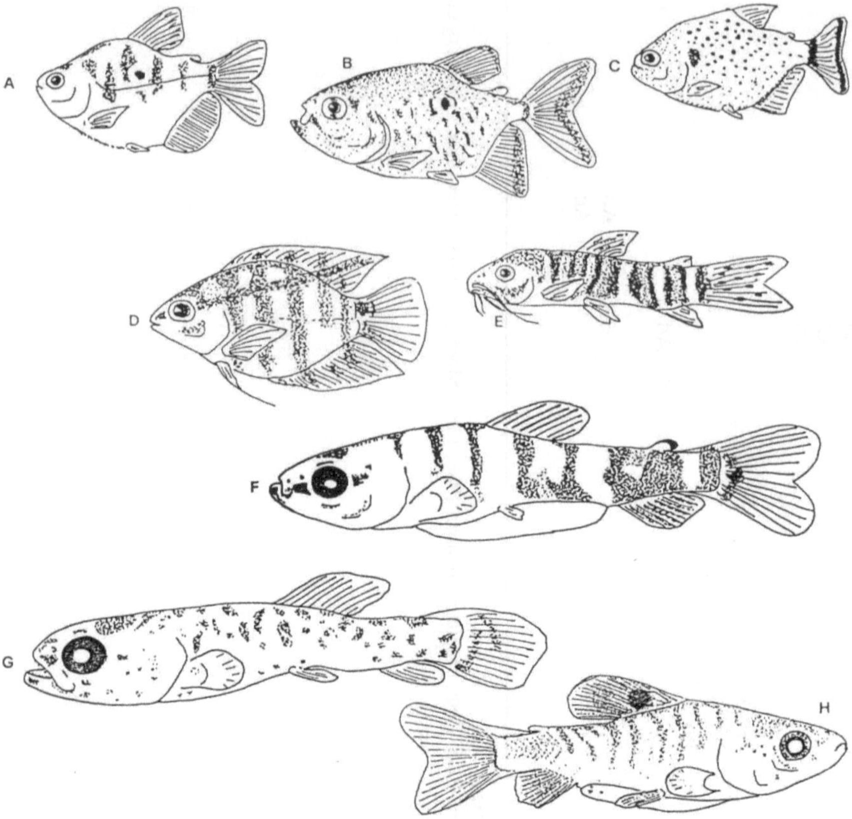

FIGURE 3.4. Color patterns in juveniles of fishes of the Orinoco River: (A) *Mylossoma duriventre*; (B) *Colossoma macropomum*; (C) *Pygocentrus caribe*; (D) *Mesonauta festivum*; (E) *Hoplosternum littorale*; (F) *Leporinus friderici*; (G) *Hoplias malabaricus*; (H) *Prochilodus mariae*. (From Machado-Allison 1987)

camouflage, and mimicry. Studies on tropical freshwater species are too few to establish a generalized explanation, if there is one. So far, we have identified some examples of mimicry, such as the patterns observed in juveniles of *Piaractus brachypommus* (prey) *Pygocentrus caribe* (predator) (Mago-Leccia 1977, Machado-Allison 1987); camouflage, such as the patterns observed in catfishes like *Pseudoplatystoma fasciatum*, *P. tigrinum*, and *Brachyplatystoma filamentosum*; and social behavior, as in *Prochilodus mariae*, *Mylossoma duriventre*, *Colossoma macropomum*, and *Piaractus brachypomus* (Figure 3.4). Changes in feeding habits also occur, such as an increase in the amount of items in the diets, including insect larvae, fins, fish, seeds, and detritus (Table 3.2).

Another alternative grouping proposed recently by Winemiller (1989) presents interesting information on the different life history strategies of the fishes in the Llanos of Venezuela. This author proposes that three different life history strategies are present:

1) "**Equilibrium Strategy**," in species associated with higher juvenile survival as a result of greater parental care of progeny. The principal characteristics are large oocyte size, small clutch size (few offspring), long gestation periods, and late maturation. Freshwater stingrays of the genus *Potamotrygon* provide an extreme example of this strategy. Included in this grouping are almost all the cichlids and several families of the Order Siluriformes, in which relatively large oocytes, nesting, brood protection, and acyclic spawning are evident. As the author points out, the equilibrium-strategist species apparently reproduce with some degree of success, even though predation is intense during the early dry season, when fishes of numerous juvenile classes were found at high densities. To some extent, this association of life history traits agrees with the relative "**K-strategy**" as originally proposed by Pianka (1970).

2) "**Opportunistic Strategy**," which includes small species that lack parental care, have a short generation time, and show minimal investment per offspring. Opportunistic species remain reproductively active despite apparent high juvenile and adult mortality during severe conditions and high predation in the dry season. Examples are the annual fishes and poeciliids (Cyprinodontiformes), and some characids.

3) "**Seasonal Strategy**," which includes species with life history attributes characterized by cyclic (annual) reproduction. These species possess long generation times, concomitant with the reporductive cycle, large clutches, and little parental care. Those fishes exhibit a characteristic burst of reproduction, migrations to and from head waters, entering the "esteros" or "varzeas" during the rains, and/or flooding. They exploit temporal and permanent habitats that secure survival and growth. However, several of these species suffer drastic reductions in population size due largely to predation on the immature during the early dry season. Examples are *Prochilodus mariae*, curimatids (*Curimata argentea*), and a great number of characids.

The last two life history strategies (opportunistic and seasonal) apparently comprise Pianka's (1970) earlier "**r-selected**" attributes. However, the differences among them are great enough to separate them into two groups.

In general, neotropical fishes inhabiting seasonal environments show a variety of life history patterns. The "**equilibrium strategists**" are associated with sedentary local populations, relatively stable adult food resources, prolonged breeding seasons, and parental investment in individual offspring, which results in enhanced juvenile survivorship. The "**opportunistic strategists**" include species that show rapid recolonization of disturbed habitats through rapid maturation and multiple spawning. On the other hand, most fishes of the Llanos (Orinoco Basin) appear to be associated with a "**seasonal strategy**" which is characterized by exploitation of the annual expansion of aquatic and community production. Included in these group of are species that exhibit large adult size, high fecundity, absence of parental care, and long distance spawning migrations to productive, wet season floodplains (Winemiller 1989).

Another important observation is that the Characiforms span the entire spectrum of reproductive patterns. Obviously, the phenotypic variation represents evolutionary divergence from an ancestral protocharaciform that could exhibit a more limited repertoire of life history traits. No hypothesis can be proposed for the South American Characiformes until a phylogenetic analysis of the group is performed.

Trophic Interrelationships and Community Structure

Although food webs in tropical waters are often very complex, when early stages of fish are considered, they may be based on relatively few sources. For example the "aufwuchs" or bentic organic matter, and plancktonic organisms, support numerous species (loricariids, callichthyids, characids, prochilodontids and curimatids) in the flooded plains.

Conditions in tropical waters are quite different from temperate regions with respect to the dominance (biomass) of detritivorous species and the great importance (number and variety) of predatory species. When comparing these two regions in terms of the possible food sources, we conclude that tropical waters are characterized by:

1. Allochthonous vegetable material which is important as direct food for many fish species including their early stages of development. Leaves, fruits and seeds are used by many characoid species which can occasionally act as seed dispersers.

2. Zooplankton (copepods, ostracods and cladocerans) and aquatic insect larvae are important as direct food for many early stages of fish species living in flooded areas of the Orinoco River Basin.

3. Insects are important as fish food, including aquatic stages of Odonata, Ephemeroptera and Diptera and terrestrial insects stranded on the water's surface. More than 40% of food ingested by juvenile and adult fish living in "morichales" and/or small head-water streams, come from terrestrial insects.

4. Detritus and/or mud is important as food for many fishes which strain large quantities for their contained microorganisms. Long intestines, "spiral valves," are found in loricaroids, curimatids, and prochilodontids.

5. Large numbers of individuals and different kinds of piscivorous fishes which occur in these communities such as cynodontids, erythrinids, serrasalmins, and large pimelodids.

6. Several species with special parasitic or semiparasitic behavior (lepidophagous, pterygiophagous and haematophagous) observed in "piranhas" (Serrasalminae) and "catfishes" (Trichomycteridae).

At this stage of our knowledge it would be premature to establish generalized food chains or nets for the fishes of the Orinoco Basin. However, it appears that food chains in the tropics are very complex due to seasonal and ontogenetic changes, although they may be based on few resources. As an example, for a small community in the flooded areas of the Venezuelan Llanos there are at least two identified chains:

(1) detritus —> microorganisms —> benthic crustaceans or insects —> fish —> piscivores (other fish or vertebrate).

(2) Phytoplankton —> Zooplankton —> planktivores fish —> piscivores.

However, nutrient sources are highly variable according to the season. During the rainy season fish feed principally on plankton (phyto- and zoo-), rotifers and insect larvae which are used by early stages. During dry season which corresponds with juvenile or adult stages of fish, they are more specialized (piscivory, detritus, etc.). This in summary shows that most of the species possess considerable plasticity in their diets. Predators change their preferred prey as they grow and change biotopes, to whatever foods are available seasonally. In some cases, predators arrest their feeding activities during severe conditions (Machado-Allison and Royero 1986). These observations indicate to us that our understanding of "growth," as influenced by earlier concepts such as **niche** (Hutchinson 1957), **competitive exclusion** (Zaret and Rand 1971), **food partitioning** (Schoener 1974), **optimum foraging** (Werner and Hall 1974), and others needs to be revaluated when tropical continental fishes are concerned.

REFERENCES

Colonello G, Castroviejo S, Lopez G (1986) Comunidades vegetales asociadas al rio Orinoco en el sur de Monagas y Anzoategui (Venezuela). Memoria, Soc. Cienc. Nat. La Salle, XLVI(125–126):127–166

DeSilva S, Schut J, Kortmulder K. (1985) Reproductive biology of six Barbus species indigenous to Sri Lanka. Envir. Biol. Fish., 12:201–218

Fink S, Fink W (1981) Interrelationships of the ostariophysan fishes (Teleostei). Zool. J. Linn. Soc., 72:297–353

Hamilton S, Lewis W (1987) Causes of seasonality in the chemistry of a lake on the Orinoco Rivers floodplain, Venezuela. Limnol. Oceanogr. 32(6):1277–1290

Hutchinson G (1957) Concluding remarks. Cold Spring Harbor Symp. Quant. Biol. 22:415–427

Joerger MD (1984) Ecology of juvenile Brachyplatystoma flavicans (Siluroidei:Pimelodidae) in the lower Orinoco River of Venezuela. MsC Thesis, California State Polytechnic Univ. Pomona, 90 pp (mimeo).

Kramer DL (1978) Reproductive seasonality in the fishes of a tropical stream. Ecology, 59:976–985

Lopez H, Lundberg J, Marsh E (1984) Design and operation of a small trawling apparatus for use with dougout canoes. Am. J. Fish. Manag. 4:331–334

Lowe-McConnell R (1964). The fishes of the Rupununi savanna district of British Guiana. I. Groupings of fish species and effects of the seasonal cycles on the fish. Journ. Linn. Soc. (Zool.) 45:103–144

Lowe-McConnell R (1975) Fish Communities in Tropical Freshwaters: Their distribution, ecology and evolution. Longman, London and New York

Lowe-McConnell R (1987) Ecological Studies in Tropical Fish Communities. Cambridge Univ. Press, New York

Lundberg J, Lewis W, Saunders J and Mago-Leccia F (1987) A major food web component in the Orinoco River channel: Evidence from planktivorous electric fishes. Science, 237:81–83

Machado-Allison A (1982) Estudios sobre la subfamilia Serrasalminae (Teleostei-Characidae). Parte I. Estudio comparado de los juveniles y adultos de las "*cachamas*" de Venezuela (Géneros *Colossoma* y *Piaractus*). Acta Biol. Venez. 11(4):145–195

Machado-Allison A (1986) Aspectos sobre la historia natural del "curito" Hoplosternum littorale (Hanckock, 1928) (Siluriformes Callichthyidae) en el bajo llano de Venezuela: desarrollo, alimentación y distribución espacial. Acta Cient. Venez. 37(1):72–78

Machado-Allison A (1987) Los Peces de los Llanos de Venezuela: Un ensayo sobre su historia natural. Universidad Central de Venezuela, CDCH, Caracas

Machado-Allison A, Garcia C (1986) Food habits and morphological changes during ontogeny in three serrasalmin fish species of the Venezuelan floodplains. Copeia, 1:123–126

Machado-Allison A, López H (1975) Etapas del desarrollo de Loricariichthys typus (Bleeker, 1864) (Osteichthyes, Siluriformes, Loricariidae). Acta Biol. Venez., 9(1):93–119

Machado-Allison A, Royero R (1986) Biomasa total y habitos alimentarios en peces de un ecosistema riverino restringido en Venezuela. Acta Cient. Venez., 37(1):94–95

Machado-Allison A, Zaret T (1984) Datos sobre la biología reproductiva de Hoplosternum littorale (Siluriformes-Callichthyidae) de Venezuela. Acta Cient. Venez., 35(2):142–146

Mago-Leccia F (1970) Lista de los Peces de Venezuela. Ofic. Nac. Pesca, MAC, Caracas, 283 pp

Mago-Leccia F (1978) Los peces du agua dulu di Venezuela cuadernon Lagoven, Caracas.

Mago-Leccia F (1983) Entomocorus gameroi, una nueva especie de bagre auquenipterido (Teleostei, Siluriformes) de Venezuela, incluyendo la descripcion de su dimorfismo sexual secundario. Acta Biol. Venez., 11(4):215–236

Mago-Leccia F, Zaret T (1978) The taxonomic status of Rhabdolichops troschell (Kaup, 1856), and speculations on gymnotiforms evolution. Env. Biol. Fish., 3(4):379–384

Marrero C (1990) Estudio de los habitos alimentarios de los peces de la Familia Apteronotidae (Pisces Gymnotiformes): un enfoque a traves de su morfologia bucal externa y sus relaciones filogeneticas. Ph.D. Thesis, Fac. Ciencias, Universidad Central Venezuela

Meade RH, Nordin CF, Perez-Hernandez D, Mejia A and Perez JM (1983) Sediments and water discharge in Rio Orinoco, Venezuela and Colombia. In: Proc. 2nd Inter. Symp. River Sedimentation. Water Resources and Electric Power Press. Beijing, China. pp. 1134–1144

Nico L, Taphorn D (1985) Diet of Acestrorhynchus microlepis (Pisces:Characidae) in the low llanos of Venezuela. Copeia, 3:794–796

Nico L, Taphorn D (1988) Food habits of piranhas in the low llanos of Venezuela. Biotropica, 20:311–321

Pianka E (1970) 'On r and k selection.' Amer. Nat., 100:593–597
Prejs A, Prejs K (1987) Feeding of tropical freshwater fishes: seasonality in resource availability and resource use. Oecologia, 71:397–404
Rodriguez M, Lewis W (1989) Diversity and species composition of fish communities of Orinoco floodplain lakes. Nat. Geo. Research, 6(3):319–328
Sanchez L, Vasquez E (1986a) Estudio estacional y longitudinal de la hidrodinamica y fitoplancton en una seccion del Bajo Orinoco (Venezuela). Memoria Soc. Cienc. Nat. La Salle, XLVI(125–126):69–94
Sanchez L, Vasquez E (1986b) Notas sobre las macrofitas acuaticas de la seccion baja del rio Orinoco, Venezuela. Memoria Soc. Cienc. Nat. La Salle, XLVI(125–126):107–126
Schoener T (1974) Resource partitioning in ecological communities. Science 185:27–39
Welcomme R (1969) The biology and ecology of the fishes of a small stream. J. Zool. 158:485–529
Welcomme R (1979) Fisheries Ecology of Floodplain Rivers. Longman, London
Werner E, Hall D (1974) Optimal forraging and size selection of prey by the bluegill sunfish (Lepomis macrochirus). Ecology 55:1042–1052
Winemiller K (1989) Patterns of variation in life history among South American fishes in seasonal environments. Oecologia 81:225–241
Zaret T, Rand A (1971) Competition in tropical stream fishes: Support for the competitive exclusion principle. Ecology, 52(2):336–342

4

Reproductive Characteristics of Characiformes

Naércio A. Menezes and Anna Emilia A. de M. Vazzoler

Introduction

The Characiformes, as defined by Fink and Fink (1981), are a major group of ostariophysean fishes confined to the freshwaters of the Neotropics and Africa. They range in size from miniature, fully mature species under 20 mm standard length (Weitzman and Vari 1988) to species such as the tambaqui (*Colossoma macropomum*) and the dourado (*Salminus maxillosus*) which attain lengths of at least one meter. Considering the diversity and commercial importance of the group, all available information concerning reproduction can be considered scarce.

We follow Greenwood et al. (1966) in recognizing the following South American families: Characidae, Erythrinidae, Ctenoluciidae, Cynodontidae, Lebiasinidae, Parodontidae, Gasteropelecidae, Prochilodontidae, Curimatidae, Anostomidae, Hemiodontidae, and Chilodontidae.

For the families which have representatives in the aquarium trade, there is extensive literature based primarily on artificial conditions. It is not our intention to present or discuss this information. Our main purpose is to focus on the major aspects of reproduction based on our work or on information derived from investigators working under natural field conditions.

A number of commercial species are represented among the families listed above, and recent studies in the Amazon region and in eastern and southeastern Brazil have shown that important food fish, as well as other non-commercial species, undertake seasonal breeding migrations to spawn in particular areas. In the Amazon, the importance of floodplain areas (*várzeas*) as nursery and spawning grounds has been extensively demonstrated. However, only a small fraction of all fish groups have been studied. Virtually nothing is known about the reproductive habits of endemic species found in the headwaters of major Amazonas/Solimões tributaries, and the role that blackwater, lakes, and other lacustrine environments play in life cycles is just beginning to be properly evaluated. Much scientific work remains to be done before the major reproductive patterns of characiform fishes are fully understood. This information is crucial for establishing further policies concerning conservation, management, and farming of natural resources.

Reproductive Habits

The species whose reproductive aspects have been studied belong to the families Anostomidae (16), Characidae (38), Ctenoluciididae (1), Curimatidae (8), Cynodontidae (4), Erythrinidae (4), Hemiodontidae (4), Lebiasinidae (2), Parodontidae (2) and Prochilodontidae (7) (see Vazzoler and Menezes in press, for details).

With the exception of the characid subfamily Glandulocaudinae, all the other species have external fertilization. The Glandulocaudinae are represented by about 17 genera and approximately 50 species belonging to three or four tribes. They are widespread in South America, ranging from Central America to Argentina. The males of all glandulocaudine species (and also the females of members of the Diapomini) have caudal glands and certain scales on the tail modified as different kinds of pumping mechanisms. In the Glandulocaudini, the pumping mechanism involves modification of scales derived from the base of the dorsal caudal fin lobe and modifications of particular caudal fin rays (see Menezes and Weitzman 1990). The caudal glands are assumed to produce pheromones which are dispersed through pumping mechanisms to attract or stimulate females during courtship. For members of the Xenurobryconini, Weitzman and Fink (1985) stated that "the modified structures of the male caudal fin may serve as a pumping mechanism increasing the rate of dissemination of a pheromone or pheromones into the surrounding water from glandular tissue situated in and around the mouth of the pouch air sac." Internal fertilization has been postulated for certain glandulocaudines based primarily on the observations that females can lay viable eggs without the presence of a male for a relatively long time after their last contact. Females of the genus *Corynopoma* kept in aquaria produced eggs and young after being isolated from males for seven months (Weitzman et al. 1985).

Internal fertilization in glandulocaudin fishes has been recently reviewed and discussed by Weitzman et al. (1985). Production of live eggs and subsequent young without the presence of males has been confirmed so far in the tribes Corynopomini, Xenurobryconini and Glandulocaudini.

Nelson (1964) suggested that internal fertilization evolved as an adaptation to the seasonal variation of the environment. This has recently been questioned by Weitzman et al. (1985) who found that many glandulocaudines do not live in habitats with well-marked dry and wet periods. According to Nelson (1964), during the dry season, mating can occur because adults are concentrated in small dry ponds and streams. The females would delay spawning while waiting for the wet season, during which abundance of food under flooded conditions would favor the development of larvae. However, contrary to Nelson's hypothesis, species of *Mimagoniates* and *Pseudocorynopoma* found along the eastern and southeastern Brazilian coast live in streams not subjected to seasonal fluctuations in water level. In the Amazon basin where relatively long-term dry and wet seasons occur, glandulocaudines are represented by very few species. Livebearing is not known for any glandulocaudin and some species may not be internally fertilized.

Among the Characiformes, typically characterized by external fertilization, there are a number of species, including some of the most important South American food fishes, which form large schools during part of the year to migrate to spawning grounds. These spawning migrations are always associated with river level fluctuations. Studies of the reproductive migrations of *Prochilodus scrofa* and *P. platensis* in the Paraná basin and of *P. platensis* in the Paraguay basin are discussed and summarized by Lowe-McConnell (1975) and Welcomme (1985). These two species have similar habits and form large schools before starting migrating upstream at the beginning of the rainy season. The run to spawning grounds, known as "piracema" or "arribação" by the fishermen and local people in Brazil, is characterized by a typical spiraling movement of the fishes locally called "rodada." Godoy (1975) observed the natural spawning of the "curimbatá" (*Prochilodus scrofa*) in a stretch of 1.5 kilometers of the Rio Mogi-Guassu at Cachoeira de Emas, São Paulo. According to him spawning took place in the middle of the river and lasted about 30 hours. During spawning activity the males emitted a characteristic croaking sound, and both males and females moved in wheeling circles near the surface of the river. In the Mogi-Guassu, spawning occurs between November and February and, after spawning, the fishes move downstream to feeding grounds located in the Rio Grande. The semi-pelagic eggs are carried downstream by the current but develop rapidly and the larvae, after spending some time in protected flooded areas, move downstream toward the feeding grounds. Migrations of the same species occur between the Itaipu Reservoir dam and the Rio Paranapanema mouth, in a stretch of 370 kilometers of the upper Rio Paraná (Agostinho et al. in press). In October, adult fish leave the reservoir, migrating upstream to spawn in lotic environments of the river and its tributaries between November and March and subsequently returning to their feeding and gonadal recuperation grounds in the reservoir. The young fishes remain for two years, feeding and growing in lentic environments of flooded areas until attaining the length of 18.9 cm (standard length) at first maturity in May to June when they move downstream to the reservoir occupied only by adult fishes.

Similar reproductive migrations, known as "subienda" in Colombia and "ribasone" in Venezuela, have been described for *Prochilodus magdalenae* in Colombia (Dahl 1971; INDERENA 1973), *P. mariae*, *P. reticulatus*, *P.*

robrotaeniatus, *Semaprochilodus laticeps*, and *S. knerii* in Venezuela (Novoa and Ramos 1978; Machado-Allison 1987; Mago-Leccia 1988).

Seasonal reproductive movements of fishes through the main river channel, of the type mentioned above for prochilodontid species, are also made by other characiform species. Among these species, *Salminus hilarii*, *S. maxillosus*, *Piaractus mesopotamicus*, and *Leporinus elongatus* undertake long spawning migrations in rivers of the Paraná basin. These and *Prochilodus scrofa* have high to very high fecundity and *S. hilarii* and *P. scrofa* are total spawners (Vazzoler and Menezes in press), characteristics which can be generalized for the species that make long spawning migrations. Data from Godoy (1975) indicate that the migratory movement is essential for the development of the gonads and maturation of the eggs. "Curimbatás" and other migratory characiforms kept by him during 10 years under artificial conditions, failed to reproduce in spite of developing gonads.

In the Amazon and other river basins in northern South America, ascending spawning migrations have been reported for *Boulengerella cuvieri*, *Hydrolicus scomberoides*, and *Myleus pacu* in the Rupununi river, Guiana, (Lowe-McConnell 1964), *Prochilodus nigricans*, *Semaprochilodus brama*, *Psectrogaster amazonica* in the Rio Tocantins, Brazil, (Santos et al. 1984; Carvalho and Merona 1986) and *Colossoma macropomum*, *Curimata cesarina*, *Mylossoma duriventre*, and *Piaractus brachypomus* in the llanos of Venezuela (Machado-Allison 1987). Although differing in some aspects, all these migrations are characterized by seasonal movements of large schools to spawning areas upstream. However, in some large tributaries of the Amazon and in central Amazonia along the Rio Solimões-Amazonas, reproductive migrations are more complex and involve components different from those already discussed.

Commercial fisheries in the main Amazonian tributaries and the central Amazon are largely based on characiforms which undertake spawning migrations. In the Rio Madeira, Goulding (1979, 1981) noted that characiforms are the only fishes explored during the reproductive migrations and most of the species involved show remarkably similar migratory patterns. Some characids, prochilodontids, curimatids, anostomids, and hemiodontids are largely concentrated in the right bank tributaries of the Rio Madeira during the low water season and start moving downstream to enter the Madeira by the end of November. Spawning apparently occurs in turbid waters of the Madeira and in floodplain areas lower than the main river at the beginning of the rainy season. Goulding (1981) described a sequence of spawning migrations starting with the "jaraqui escama fina" (*Semaprochilodus taeniurus*) and followed by other characiform species. The last species to descend was *Colossoma bidens*, which started to migrate during the last week of February.

In the case of the Rio Madeira, the longitudinal migratory movement is in the opposite direction of the upstream migrations and not all characiform species behave the same. According to Goulding, the "tambaqui" (*Colossoma brachypomum*) remain in the Rio Madeira until spawning. It seems clear that, during the spawning season, the fishes move down from clearwater tributaries toward the boundary between clear and turbid water to spawn in the latter, either in the main river or in floodplain areas. The exact spawning sites, however, are yet to be determined. Feeding grounds are located in flooded forests. From these areas, the fishes begin schooling when the water level drops and move upstream in the Rio Madeira until they find another affluent. The spawning period is thus critical since commercial fishermen take advantage of the spawning migrations to make catches.

The reproductive migrations in the Rio Negro of the "jaraquis" (*Semaprochilodus insignis* and *S. taeniurus*) described by Ribeiro (1983) and Goulding et al. (1988) essentially follow the pattern described by Goulding (1979, 1981) for the characiforms of the Rio Madeira. From their observations, it is evident that the dynamics of the process are more complex than the previous studies suggested. In this case, there are no "fixed" breeding and feeding sites, as in the case of the prochilodontid species studied in the Paraná-Paraguay basins, but, instead, a continuous migration of fishes involving nutrient-poor and nutrient-rich rivers. At the beginning of the flooding season, adults migrate downstream from the nutrient-poor blackwater river to spawn in the turbid waters of the whitewater river. After spawning, spent adults return to floodplain areas of the blackwater river to feed on allochthonous food. As the water drops, they move out of the feeding areas into another blackwater tributary further upstream through the main channel of the whitewater river. After the eggs hatch, the fry move out into nursery habitats in floodplain lakes. As Ribeiro's (1983) data indicate, recruitment of new schools into nutrient-poor blackwater tributaries occurs through dispersal migration of young during the dry period from the main channel of the nutrient-rich river. Thus, the continuous movement of adults from nutrient-poor tributary to another nutrient-poor tributary located upstream along the main river would be compensated for by the drift of eggs and larvae downstream.

Spawning of *Semaprochilodus taeniurus* takes place in November and December, tapering off into January, while *S. insignis* is known to spawn from January to March in the same area of the Solimões where whitewater mixes with the Rio Negro's blackwater (Ribeiro 1983; Vazzoler et al. 1989a). A study of gonadal development of the two species and of the relatively rare hybrid, the jaraqui açu, demonstrated relatively rapid gonadal growth in *S. taeniurus* and a slower rate in *S. insignis* (Vazzoler et al. 1989b). The hybrids' gonads did not seem to develop beyond an early stage of maturation. In October, prior to the beginning of the rainy season and during low water level in the lower Rio Negro, both species are found together in mixed schools. Beginning in November, segregation of the two-year-old and older fish into monospecific schools takes place. Schools of

S. taeniurus move to the spawning grounds at this time while *S. insignis* schools, also composed of different cohorts, appear at the same spawning grounds from January onwards. Young larvae up to 60 mm length of both species are present in monospecific schools. When they then enter the so-called "dispersal" stage as juveniles, they join with schools of the other species of similar size and form mixed schools during their first year of life. This recent study (Vazzoler and Amadio 1990) examined details of biometrics and age structure of jaraqui schools and proposed a behavioral model, correlating physiological gonadal changes with particular schooling behavior which, when combined, represent an effective reproductive isolating mechanism by separating the times of spawning of the two species.

In the central Amazon basin, many characiform species are known to make reproductive migrations out of floodplain lakes and other waterbodies into the main Solimões-Amazonas during the beginning of the floods (Santos 1978; Schwassmann 1978; Paixão 1980; Araujo-Lima 1984; Zaniboni Filho 1985; Fernandes 1988; Petry 1989). In the Lago do Rei System near Manaus, Fernandes (1988) observed that from December through March *Potamorhina latior*, *Psectrogaster amazonica*, and *Curimata kneri* (Curimatidae), *Leporinus trifasciatus* (Anostomidae), *Prochilodus nigricans* and *Semaprochilodus taeniurus* (Prochilodontidae), *Eigenmannima melanopogon* (Hemiodontidae), *Roeboides myersi, Mylossoma aureum*, *M. duriventre*, *Triportheus albus*, *T. angulatus*, and *T. elongatus* (Characidae) migrate through the main channel ("paraná") which connects the Lago do Rei system with Rio Amazonas. Most of the species examined had ripe eggs, a clear indication that after migrating through the main lake-channel (paraná do Rei) the fishes would be ready to spawn upstream in the Solimões-Amazonas. After spawning the fishes moved to floodplain areas to feed. Spent fishes were found back to the flooded lake system in January.

Field studies of Fernandes (1988) and others carried out in the central Amazon basin indicate that a typical lateral migration with spawning purposes occurs. It is characterized by massive movement of fishes out of floodplain waterbodies upstream into the main river. Where spawning actually occurs is not precisely known, although there is some evidence that it may be restricted to river channels (Araujo-Lima 1984). Recent studies, developed by Petry (1989) in the same Lago do Rei system, revealed that spawning may also occur, at least in part, in the paraná do Rei before mature fishes reach other spawning areas upstream. His data also show that larvae from different age groups drift along the river with the currents and reach the floodplains where food (plankton) is more abundant than in main rivers during the highwater season.

The relationship between spawning and river level fluctuation is quite obvious and, in spite of the great controversy about the trigger mechanism for the spawning process, there is general agreement that rising water is crucial for larval dispersal into nursery floodplain areas. Thus, floodplain areas in the central Amazon basin play an extremely important role in the reproductive cycle of characiforms and other fishes and have been considered by Petry (1989) as essential for the survival of their larvae.

Of the remaining characiforms which are not known to make long reproductive migrations in the rivers, a few developed some kind of parental care (guarders in the classification of Balon 1975, 1981). This is apparently restricted to members of the families Anostomidae, Erythrinidae, Lebiasinidae, and the piranhas of the family Characidae. The piranhas, in contrast to the migratory species, are partial spawners with low fecundity per batch and, as far as is known, also spawn during the high water season (Lowe-McConnell 1975). All the reports on the erythrinid *Hoplias malabaricus* show that it is a typical nest spawner. During the breeding season a shallow nest, about 25 to 30 cm deep, is made on the substratum and the female produces from 6000 to 61,000 eggs (Vazzoler and Menezes in press). After spawning the female leaves the nest and the male takes her place guarding it and protecting the eggs and young. Nest building activity was also reported for *Leporinus affnis* (Mayer 1928) but this kind of behavior has never been observed in nature for any anostomid species.

Breeding activities of some species of piranhas of the genus *Serrasalmus* have been described in aquaria situations (Braker 1963; Myers 1972) and indicated that they make nests and guard the eggs and young. This behavior has never been confirmed in nature. In the Llanos of Venezuela, Machado-Allison (1987) reported that *Pygocentrus notatus* usually uses roots of aquatic plants to lay its eggs during the flood season. Sazima and Zamprogno (1985) found larvae of *Serrasalmus spilopleura* inside the roots of water hyacinth (*Eichornia crassipes*) also during the rainy season. Protection of the eggs and larvae by the parents was not mentioned and spawning near roots of aquatic plants, as Sazima and Zamprogno (1985) suggested, may actually be an indirect means of providing shelter and food to the larvae.

The spawning behavior of lebiasinids (*Copeina guttata*, *Copella arnoldi*, *Pyrrhulina vittata*, and species of *Nannostomus* and *Poecilobrycon*) is extensively described in the aquarium literature (Breder and Rosen 1966). In one species *Copella arnoldi* the eggs are laid out of water on the surface of leaves of aquatic plants or on rocks and are periodically splashed with water by the males. Some of the other species of *Copella*, the species of *Pyrrhulina*, *Copeina* and *Nannostomus* lay their eggs under water. In Venezuela, Machado-Allison (1987) observed the breeding and spawning behavior of *Lebiasina pleurotaenia* in nature. When the fishes were about to spawn, males and females jumped out of the water several times and, on each occasion, the females deposited a small amount of eggs on the leaves of *Eichornia* sp. During the entire spawning process, the female laid about 100 eggs which were later wet regularly by the male.

The available data regarding reproduction in the characiforms, which are non-migratory and without parental care, indicate that their breeding activities usually coincide with the rainy season. Most of the species of the Paraná basin in eastern and southeastern Brazil and Argentina breed during spring and early summer (from October to January; Vazzoler and Menezes in press). They take advantage of favorable conditions for the development of larvae and young due to the increase in temperature and water level. Data from species of the Amazon Basin reveal that their reproductive activities occur during the months of rising and high water (from November to March; Vazzoler and Menezes in press), confirming the general idea that river level fluctuation is the main factor that regulates breeding in the tropics.

The number of ripe eggs produced by females of the non-migratory and non-guarder species is usually very low when compared with the egg production of the migratory species. The fecundity of *Astyanax bimaculatus* (700 to 3200 eggs), a characid from the Paraná basin, for example, is considerably lower than that of *Salminus maxillosus* (2,600,000 eggs), a migratory characid from the same river basin (Vazzoler and Menezes in press).

Conclusions

A little over 1000 characiform species occur in South America. Considering that all the information on reproduction is based on only about 10% of this total, we conclude that reproductive characteristics of the remaining 90% of species are virtually unknown. This lack of knowledge is particularly impressive considering the importance of characiforms to the commercial fisheries in South America. With the rapid increase in the fishing effort due to population growth, government policies concerning protection of stock levels and the definition of alternatives to increase commercial catches can only be effective when based on reliable information about all species. It has been predicted that when the more favored food species become scarcer, the qualitative composition of fish landings in the Central Amazon will change towards the exploitation of species presently not commercially important (Petrere 1978).

Characiforms make up a large percentage of the endemic fish faunas especially in the upper reaches of tributaries of the major river basins in South America. Some of the endemic species in these areas might have already disappeared as a consequence of goldmining activities and damming of some of these tributaries for hydroelectric purposes. Knowledge of the reproductive biology of endemic characiform species is of crucial importance to establish the guidelines necessary to make effective conservation decisions.

Characiformes include some of the most colorful species exploited in the aquarium trade. Although there is abundant information on reproductive habits of many species based on the observation of fishes raised in aquaria, practically nothing is known about the reproduction of ornamental fishes in nature. Such data would be extremely useful and are urgently needed. Fish culture projects have been proposed as possible alternatives to replace overexploited natural stocks and they must rely heavily on reproductive data.

Finally, the data derived from what is known about reproductive characteristics of characiforms have revealed some generalized patterns, particularly with respect to migratory species. It is evident that floodplain areas in the Amazon Basin are critical for reproduction and survival of many species. Within this context, the present rate of deforestation and alteration of the floodplains by human occupation makes the continuation of these and similar studies imperative.

References

Agostinho AA, Vazzoler AEAM, Gomes LC, Okada EK (in press) Estratificación espacial y comportamiento de *Prochilodus scrofa* en distintas fases del ciclo de vida, en la planicie de inundación del alto río Paraná y embalse de Itaipu, Paraná, Brasil. Rev. d'Hydrobiol. Trop.

Araujo-Lima CARM (1984) Distribuição espacial e temporal de larvas de Characiformes em um setor do Rio Solimões-Amazonas, próximo de Manaus, AM. Master's Dissertation, Inst. Nac. Pesq. Amaz., Manaus. 84 pp

Balon EK (1975) Reproductive guilds of fishes: A proposal and a definition. J. Fish. Res. Bd Can. 32(6):821–864

Balon EK (1981) Additions and amendments to the classification of reproductive styles in fishes. Env. Biol. Fishes 6(3–4):377–389

Braker WP (1963) Black piranhas spawned at Shedd Aquarium. Aquarium Philadelphia 32(10):12–14

Breder CM Jr, Rosen DE (1966) Modes of Reproduction in Fishes. Natural History Press, New York. 941 pp

Carvalho JL, Merona B (1986) Estudos sobre dois peixes migratórios do baixo Tocantins antes do fechamento da barragem de Tucuruí. Amazoniana 9(4):595–607

Dahl G (1971) Los peces del norte de Colombia. Ministério de Agricultura, Inst. Des. Rec. Nat. Renov., INDERENA, Bogota, D.E. 390 pp

Fernandes CC (1988) Estudo de migrações laterais de peixes no Sistema Lago do Rei (Ilha do Careiro), AM, Brasil. Master's Dissertation, Inst. Nac. Pesq. Amaz., Manaus. 170 pp

Fink SV, Fink WL (1981) Interrelationships of the ostariophysan fishes (Teleostei). Zool. J. Linn. Soc. 72:297–353

Godoy MP de (1975) Peixes do Brasil: sub-ordem Characoidei; bacia do rio Mogi Guassu. Editora Franciscana, Piracicaba, São Paulo 4:629–847

Goulding M (1979) Ecologia da Pesca do Rio Madeira. Conselho Nacional de Desenvolvimento Científico e Tecnológico, Inst. Nac. Pesq. Amaz., Manaus, Amazonas 172 pp

Goulding M (1981) Man and Fisheries on an Amazon Frontier. Dr W. Junk Publ., The Hague. 137 pp

Goulding M, Carvalho ML, Ferreira EG (1988) Rio Negro, Rich Life in Poor Water. SPB Academic Publishing, The Hague. 200 pp

Greenwood PH, Myers GS, Rosen DE, Weitzman SH (1966) Phyletic studies of teleostean fishes, with a provisional classification of living forms. Bull. Amer. Mus. Nat. Hist. 131:339–456

INDERENA (1973) Operación Subienda 1973. Investigatión pesqueira. Instituto de Desarollo de los Recursos Naturales Renovables, Bogota, Colombia. 113 pp

Lowe-McConnell RH (1964) The fishes of the Rupununi savanna district of British Guiana, South America. Part 1. Ecological groupings of fish species and effects of the seasonal cycle on the fish. J. Linn. Soc. (Zool.) 45:103–144

Lowe-McConnell RH (1975) Fish Communities in Tropical Freshwaters. Longman, New York. 337 pp

Machado-Allison A (1987) Los peces de los llanos de Venezuela. Un ensayo sobre su historia natural. Universidad Central de Venezuela, Consejo de Desarollo Científico y Humanístico, Caracas. 141 pp

Mago-Leccia F (1988) Los peces de agua dulce de Venezuela. Cuadernos Lagoven, Caracas. 35 pp

Mayer F (1928) *Leporinus affinis* (Günther). Aquatic Life, Baltimore, 11(12):212–213

Menezes NA, Weitzman SH (1990). Two new species of *Mimagoniates* (Teleostei:Characidae:Glandulocaudinae), their phylogeny and biogeography and a key to the glandulocaudin fishes of Brazil and Paraguay. Proc. Biol. Soc. Wash. 103(2):380–426

Myers GS (1972) The piranha book. TFH Publications, Inc. Ltd., New Jersey. 128 pp

Nelson K (1964) Behavior and morphology in the glandulocaudine fishes (Ostariophysi, Characidae). Univ. Calif. Publ. Zool. 75(2):59–152

Novoa DF, Ramos F (1978) Las Pesquerias Comerciales del Rio Orinoco, Venezuela. Editora Venegrafica. 161 pp

Paixão IMP (1980) Estudo da alimentação e reprodução de *Mylossoma duriventris* Cuvier, 1818 (Pisces, Characoidei), do lago Janauacá, AM, Brasil. Master's Dissertation, Inst. Nac. Pesq. Amaz., Manaus. 127 pp

Petrere M (1978) Pesca e esforço de pesca no Estado do Amazonas. II. Locais, aparelhos de captura e estatísticas de desembarque. Acta Amaz., Manaus, 8(3):suplemento 2. 54 pp

Petry P (1989) Deriva de ictioplancton no Paraná do Rei, várzea do Careiro, Amazônia Central, Brasil. Master's dissertation, Inst. Nac. Pesq. Amaz., Manaus. 68 pp

Ribeiro MCLB (1983) As migrações dos jaraquis (Pisces:Prochilodontidae) no Rio Negro, Amazonas, Brasil. Master's Dissertation, Inst. Nac. Pesq. Amaz., Manaus. 192 pp

Santos GM dos (1978) Observações limnológicas sobre a asfixia e migração de peixes na Amazônia Central. Ciênc. Cult., São Paulo, 31(9):1034–1040

Santos GM dos, Jegu M, Merona B (1984) Catálogo de peixes comerciais do baixo Rio Tocantins. Projeto Tucurui. Eletronorte/Conselho Nacional de Desenvolvimento Cientifico e Tecnológico, Inst. Nac. Pesq. Amaz. 83 pp

Sazima I, Zamprogno C (1985) Use of water hyacinths as shelter, foraging place and transport by young piranhas, *Serrasalmus spilopleura*. Env. Biol. Fishes 12(3):237–240

Schwassmann HO (1978) Times of annual spawning and reproductive strategies in Amazonian fishes. *In* Thorpe JE (ed.). Rhythmic Activities of Fishes. Academic

Press, London. pp. 187–200
Vazzoler AEA de M, Amadio SA (1990) Aspectos biológicos de peixes amazônicos. XIII. Estrutura e comportamento de cardumes multiespecíficos de *Semaprochilodus* (Characiformes, Prochilodontidae) do baixo Rio Negro, Amazonas, Brasil. Rev. Brasil. Biol. 50(3):537–546
Vazzoler AEA de M, Amadio SA, Caraciolo-Malta MC (1989a) Aspectos biológicos de peixes amazônicos. XI. Reprodução das espécies do gênero *Semaprochilodus* (Characiformes, Prochilodontidae) no baixo Rio Negro, Amazonas, Brasil. Rev. Brasil. Biol. 49(1):165–173
Vazzoler AEA de M, Caraciolo-Malta MC, Amadio SA (1989b) Aspectos biológicos de peixes amazônicos. XII. Indicadores quantitativos do período de desova das espécies do gênero *Semaprochilodus* (Characiformes, Prochilodontidae) do baixo Rio Negro, Amazonas, Brasil. Rev. Brasil. Biol. 49(1):175–181
Vazzoler AEA de M, Menezes NA (in press) Síntese de conhecimentos sobre o comportamento reprodutivo dos Characiformes da América do Sul. Rev. Brasil. Biol.
Weitzman SH, Fink SV (1985) Xenurobryconin phylogeny and putative pheromone pumps in glandulocaudine fishes (Teleostei:Characidae). Smithsonian Contrib. Zool. 421:1–121
Weitzman SH, Fink SV, Menezes NA (1985) A critique of the monophyly of the Glandulocaudinae. *Appendix 3 to*: Xenurobryconin phylogeny and putative pheromone pumps in glandulocaudine fishes (Teleostei:Characidae) by Stanley Weitzman and Sara V. Fink. Smithsonian Contrib. Zool. 421:1–121.
Weitzman SH, Vari RP (1988) Miniaturization in South American freshwater fishes: An overview and discussion. Proc. Biol. Soc. Wash. 101(2):444–465
Welcomme RL (1985) River Fisheries. FAO Fish. Tech. Paper 202:1–330
Zaniboni Filho E (1985) Biologia e reprodução do matrinchã *Brycon cephalus* (Günther, 1869) (Teleostei:Characidae). Master's Dissertation. Inst. Nac. Pesq. Amaz., Manaus. 134 pp

5
Seasonality of Reproduction in Amazonian Fishes

Horst O. Schwassmann

Introduction

In his famous monograph on the fisheries of Amazonia, José Veríssimo (1895) emphasized the need for studying the times of reproduction of different species in relation to annual water level changes. Veríssimo also made a reasonably accurate estimation of about 2000 Amazonian species, one-tenth of which he considered to be well known because of their importance as commercially utilized food fishes.

Interest in spawning habits and seasonal reproduction of Amazonian fishes resurged in 1936 when Carlos Estevão de Oliveira established a fish culture program at the Goeldi Museum in Belém near the mouth of the Amazon River (Menezes 1972). Failure of the curimatã (*Prochilodus* sp.) to spawn in pond culture led to the successful application of the recently developed pituitary injection technique (Cardoso 1934; Von Ihering and Wright 1935), resulting in obtaining up to three broods per year (Fontenele et al. 1946). Azevedo and Canale (1938) introduced the distinction between "total" and "partial" spawners among Amazonian fishes; a list of examples of the two reproductive habits was published by Azevedo (1938).

Reproduction in fishes usually takes place during a particular time of the year. In many species actual spawning is preceded by migratory movements to spawning grounds. Seasonality of reproduction in tropical species became of renewed interest when it was realized that photoperiodic temperature-dependent advance timing of a physiological rhythm of gonadal maturation could not be invoked to account for many examples of strictly seasonal spawning in tropical waters (Schwassmann 1971). Having attempted to treat seasonal reproduction in fishes as an interaction of synchronizing environmental factors with an "endogenous" physiological rhythm of gonadal ripening, I also emphasized the need for more information concerning times of spawning in tropical fish species in nature in relation to annually cyclic changes in climatic-hydrological conditions (Schwassmann 1971). During the following 20 years, extensive field studies have added valuable data; some symposia on fish rhythmic activities

(Thorpe 1978) and fish phenology (Miller 1979a) have been held, including an attempt at conceptualizing the different cybernetic and physiological approaches to the problem of seasonality (Miller 1979b). A recent book by Roberts (1989) contains a chapter on reproductive adaptations which casts doubt on the generally prevailing concept of predominantly "seasonal" reproduction in tropical freshwaters, including the Amazon. Roberts feels that our notion of reproductive seasonality is biased by an emphasis on data from a few large and commercially important species and that "continuous" reproduction, perhaps interrupted by a "non-reproductive" phase, might be a more prevailing pattern in smaller species.

This chapter will review and update an earlier report about spawning patterns in fish populations from central and lower Amazonia (Schwassmann 1978). What appears from inspecting the available information is a correlation between the degree of contrast in seasonally changing habitat conditions and the degree of precision in annually cyclic reproductive behavior in populations of Amazonian fishes. Very much involved in the interdependence of seasonal phase relations is the utilization of fruits and seeds of higher plants during their usual fruiting time of high water level in flooded forests by many species of fish. Although known for a long time (Veríssimo 1895; LeCointe 1945), the ecological implications for seed dispersal and for the cyclic growth process of fishes have been recognized only recently (Gottsberger 1978). Goulding (1980) and Goulding et al. (1988) have investigated details of this animal-plant interaction and provide lists of tree and fish species involved.

The Environment

The Amazon drainage basin covers an area of approximately seven million km^2. The principal headwaters, Apurimac, Ucayali, and all western tributaries, originate in the geologically young Andean mountains and carry large amounts of silt in suspension, giving these waters a milky appearance. These are the "whitewater" rivers—Purús, Napo, Solimões, Marañon, etc. Those tributaries which join the main river along its middle and lower course come from the Precambrian Guiana and Brazilian shields and carry little sediment; they are the "clearwater" rivers—Tapajós, Xingú, Trombetas, and others. Very poor in dissolved nutrients, transparent but brown in color due to dissolved humic substances are the "blackwater" rivers, like the Rio Negro.

Tropical rain forest is the dominant landscape form, covering three-fourths of the basin, most of the higher ground, the "terra firma." At the higher elevations of the Precambrian shields forest merges gradually with a slightly drier savanna vegetation. Along the Solimões/Amazon and other whitewater rivers lies the seasonally inundated floodplain, the "várzea," often as wide as 100 km, consisting of a network of large and small lakes that are interconnected by "furos" and "paranás."

Rainfall is high over the Amazon basin, averaging 2.3 meters per year. It is not evenly distributed over the entire area; highest precipitation is noted in the northwestern portion from the upper Rio Negro to the Andean foothills, with annual values exceeding three meters. A slightly drier belt of less precipitation stretches across the lower Amazon at Santarém in a S–E to N–W orientation.

An uneven distribution of rainfall throughout the year is causing a distinct seasonality of the environment which consists of an alternation of a drier "summer" with a really wet "winter" season, while temperature changes are insignificant. In the northern and southern parts of the basin, these dry and wet periods occur at roughly opposite times of the year. The northern inclination of the sun's arc during June and July brings the continental low-pressure cyclone towards the north, increasing precipitation on the Guiana shield. When the sun's arc is over the southern tropics in December and January, the rains are heaviest south of the equator. Along the middle and lower Amazon, the rainy season lasts from about December to June with a slightly earlier onset of rains near Manaus, beginning in September and October.

The seasonal differences in rainfall cause great changes in volume and level of the main rivers resulting in a cyclic flooding and gradual draining of the immense várzea. The annual level change is about 15 meters near Iquitos on the Peruvian Amazon (Maranõn) and averages 10 meters at Manaus at the Rio Negro/Solimões confluence, while the upper Rio Negro shows less amplitude of water level change due to the more evenly distributed annual rains.

In the estuary, the seasonal alternations in water level are no longer noticeable since they are absorbed by the ocean. However, the great difference in discharged water volume, about 75,000 m^3/s in the low water season and more than 230,000 m^3/s in the rainy part, is causing a horizontal displacement of the mixing zone of river with sea water, estimated to reach 200 km in the Pará River estuary (Egler and Schwassmann 1962; Schwassmann et al. 1989).

Wherever there is some brackish water influence, as along the shore areas of the outer estuary of the Amazon and Pará Rivers and at the shores of the many islands in the bays of Marajó and Guajará, the tidal várzea of the inner estuary is replaced by tidally inundated mangrove areas. The eastern part of Marajó, the largest estuarine island (46,000 km^2) goes through a seasonal change of flooding, usually lasting from February to June, and a dry period with hardly any rain from September to December when most lakes and interior rivers are left completely dry. The seasons on Marajó Island very much resemble those in the várzea of the Solimões, with the difference that the contrast between wet and dry conditions is more severe on Marajó.

A drastically different aquatic habitat is found in the vast number of small- and medium-sized streams of forests and adjacent savannas of the dry ground, the "terra firma."

Many of these small streams show hardly any annual changes in discharge and in physico-chemical characteristics, except for a more frequent daily flooding and water level increase of very temporary nature during the rainy season.

The Fishes

A very diverse freshwater fish fauna is found in the Amazon basin. Roberts (1972) estimated that 1300 species were recorded from the Amazon, and Goulding (1980) thought that about an equal number still needed description. No recent complete listing of all known species exists and an early list by Fowler (1948–1954) included all South American species known at that time. The commercially important species in the Manaus and Itacoatiara regions are contained in accounts by Petrere (1978), Smith (1981) and for the Belém market area by Souza and Imbiriba (1978).

Previous efforts to relate seasonally changing conditions in the Amazon area with the life cycle of fish populations, leading to migration and reproduction (Lowe-McConnell 1969, 1975; Schwassmann 1978), suffered from a lack of detail concerning the ecology of many species. Since then, comprehensive studies of fish communities in the Madeira system (Goulding 1980) and the Rio Negro (Goulding et al. 1988) have been accomplished. Important insights into previously little recognized relationships between fishes and forest trees have come to light. Goulding's (1980) study is of general applicability for most tributary systems in the area because of similarities in composition of the aquatic communities as well as environmental conditions, the Madeira River being largely a whitewater stream while the Rio Machado is of the clearwater type.

With respect to spawning habits, Lowe-McConnell (1969, 1987) distinguished two principal types of fishes. Her "piracema" fishes enter into a mostly up-river directed migration and spawn at the beginning of the rainy season; they are "total" spawners, depositing a very large number of eggs within a brief time. The other type, "small-brood" (partial) spawners, tend to spawn in the dry season before the onset of the annual rains; they would lay a small number of eggs in several successive batches, and many of them might show parental care.

"Piracema" is an Amerindian word for the upstream movement of fish schools. Such spawning migrations are best known for populations of different characin species and are treated in the chapter by Menezes and Vazzoler in this volume.

Migratory movements of several species of catfish, most of them belonging to the pimelodid family, are also described by Goulding (1980). As in the case of migratory characins, all observed movements appear to be directed upstream and, depending on the species, occur during high as well as low water periods. Spawning and feeding migrations seem to be involved.

An upstream directed spawning run, alternating with a down-river feeding migration, has been reported for the planktivorous Mapará (*Hypophthalmus perporosus*) of the hypophthalmid family in the lower Tocantíns River (Carvalho 1978). Similar long-distance migrations are also involved in the life cycle of one of the commercially very important siluroids of the Pará River and Amazon estuary, the Piramutaba (*Brachyplatystoma vaillanti*). It is not known if these migrations extend over thousands of kilometers along the Amazon, or if there exist many separate local populations. The tagging program, which began in 1978 (Godoy 1979), experienced difficulties due to poor tag returns. Information concerning spawning times and life cycles of estuarine species in the Pará estuary was provided by Barthem (1985).

Concerning actual spawning sites of floodplain inhabiting species, Von Spix and Von Martius (1829–31, vol. 3, p. 1360) reported spawning activity to take place in the deep parts of the flooded forest ("Igapówald") from which the young fish returned to the rivers before the end of the flood stage. Junk (1984) reports the absence of an entire age class of Pescada (*Plagioscion* sp.), apparently due to a sudden fall in water level during the rising phase when spawning had been completed. Not all fish reproduce at the time of high waters. Mapará (*Hypophthalmus edentatus*) is reported to spawn from February to April, two to four months before peak flooding, in floodplain lakes near Manaus (Carvalho 1980).

While all migratory characins appear to be total spawners, at least in the rivers and floodplain lakes, floodplain lakes also contain numerous fish populations that are partial spawners and small-brood spawners, such as the many species of cichlids and gymnotoids whose reproductive activity consists of several successive egg-laying periods. Such partial spawning is the rule in fish communities inhabiting small streams and may extend over a period of several months during the dry or the wet part of the year (Schwassmann 1976; Kramer 1978; Lowe-McConnell 1979). Partial spawning could be looked upon as an adaptation to a seasonally less predictable, or to a more constant, environment were it not for the noted massive representation of partial spawners in a seasonally most extremely cyclic environment, the eastern grassland portion of Marajó Island where our most recent study is in progress. During October to December, when severe drought conditions exist, high concentrations of fish are found in a few isolated water holes of the otherwise dried-up Goiapí River. These isolated water pockets serve as estivating places for those fishes which remain on the island throughout the entire year.

Eighteen species of gymnotoids were counted in this river. Some members of the Sternopygidae enter a phase of gonadal recrudescence only after the onset of annual rains in January to February. Others, like *Electrophorus*, show fully developed gonads several months in advance of the rainy season. Some gymnotoids, such as several Apteronotidae, *Rhamphichthys* sp., *Rhabdolichops troscheli*, and one *Eigenmannia* sp., remain near, or even in, the river channel where they were seen estivating. Other species of *Eigenmannia* disappeared and seem to undertake a migration toward yet unknown areas for spawning; they could no longer be found in the actual river bed. *Electrophorus* spawns long before the beginning of the rainy season; their nests with larvae were found from October onwards (Schwassmann and Assunção 1989).

A search for relevant physico-chemical factors of the environment which might be involved in the timing of preparatory physiological phases and gonadal ripening, ultimately resulting in migration and spawning, has been initiated in laboratory-type experiments by Kirschbaum (1975). Kirschbaum succeeded in inducing gonadal recrudescence and spawning in a gymnotoid fish, *Eigenmannia virescens*, by manipulating pH, conductivity, water level, and simulated rain. Lowering of conductivity alone, paralleling the dilution of biotope water by strong rains, appeared to be the most potent factor in initiating spawning (Kirschbaum 1975). After his initial success with *Eigenmannia virescens*, Kirschbaum demonstrated the same responsiveness in control of the reproductive cycle in various other gymnotoid electric fishes as well as in some African mormyrids (Kirschbaum 1979, 1984, 1987).

It is difficult to generalize from these findings in the laboratory because of the great diversity of seasonal changes in physico-chemical parameters in different bodies of water. A significant lowering of the total salt content is noted with the progressing wet season in the estuary where the considerably higher river discharge displaces the mixing zone of sea- and freshwater into the ocean. A correlation of lower conductivity with the rainy season also exists in the tidal rivers of Marajó Island. These rivers have no net discharge of their own during the latter part of the dry season and merely fill and drain with water from the bay. Very little annual change is noted in the conductivity of waters in small streams (igarapés) of the terra firma; their dissolved salt content is always very low. In the floodplain lakes, where many of the gymnotoid species are found, the situation is reversed from that observed in estuarine rivers. Data from a three-year study on Janauacá Lake by Schmidt (1972, 1973) show low conductivity ($20 \mu S$) at lowest water level during the dry season and high values ($50 \mu S$) in the rainy season when the lake is receiving floodwater from the main river. Low dry season conductivity is due to pure water influx from small forest streams.

Another difficulty in specifying environmental factors that might be involved in synchronizing gonadal ripening and spawning is due to the differences in the time of spawning among species that are closely related and also live together in the same aquatic habitat. Of the gymnotoids which we found on Marajó, *Electrophorus* and *Rhamphichthys* spawn from October on, three months before the onset of the rains (Schwassmann and Assunção 1989), while several *Eigenmannia* species, *Rhabdolichops*, and several Apteronotids begin their reproduction only after flooding has occurred.

Most other fish on Marajó Island seem to await the onset of major rains, with the exception of several species of piranhas which begin their reproduction a little earlier. In Panama, the gymnotoid *Hypopomus occidentalis* was also found to reproduce during the dry period of the year (Hagedorn 1988).

After many years of field work with *Gymnorhamphichthys*, another stream-dwelling gymnotoid, it is apparent that their rather prolonged spawning period in small streams shows no consistent seasonal pattern. I have found separate populations living in adjacent streams to be completely out-of-phase with each other (Schwassmann 1978). In addition, while revisiting the same streams one or several years later but during the same months as previously, I was surprised to realize from the size class distribution of the sampled descendent population that apparently no annually cyclic pattern of growth stages and reproduction existed. More than 25 years of intermittent field data on the sandfish, *Gymnorhamphichthys* sp., suggests that this small gymnotoid requires approximately one calendar year to reach maturity. Sexual maturity is easily seen by the spectacular forward progression of the urogenital pore from its juvenile location under the opercular openings to a level underneath the eyes and the simultaneous development of a conspicuous urogenital papilla (Schwassmann 1976). Gonadal ripening has occurred by that time and spawning proceeds in many successive bouts for three to four months in the population, perhaps not quite that long in the individual. Only twice was I able to capture post-spawning individuals in rather moribund condition. It seems likely that the spawners die after one reproductive session of several months duration. The late Willy Schwartz conveyed to me in 1971 a similarly noted absence of post-reproductive specimens of the Cardinal Tetra, *Cheirodon axelrodi*, in its natural habitat which is in stark contrast to an observed total life span of two or more years when kept in aquaria where they did not reproduce.

Intrapopulation synchronization of gonadal maturation and out-of-phase spawning was reported for a mormyrid fish in Lake Victoria by Scott (1974), who noted two different populations of the same species spawning at different locations of the lake during different times of the year.

Discussion

The adaptive value, or ultimate cause, of annually periodic reproduction in fishes is the production of offspring at a time when the aquatic habitat is providing a sufficient food supply for the growing young. A series of specific environmental variables, formerly designated as proximal causes, would induce certain stages of gonadal growth resulting in eventual spawning. When considering seasonal reproduction as biochronological adaptation (Schwassmann 1971, 1980), these environmental periodic factors act as mere synchronizers on an endogenous physiological rhythm. In fishes of higher latitudes, changing daylength and temperature have been documented to regulate the timing of annually periodic spawning. Near the equator, photoperiodic and temperature changes are absent, or greatly attenuated, and some factors related to flooding by seasonal rains are assumed to take over (Schwassmann 1971, 1978; Lowe-McConnell 1975, 1979). Experimental demonstration of the action of four of such factors related to seasonal rains came from recent laboratory experiments by Kirschbaum (1975, 1979, 1984). Reports about species that spawn during the dry season by Kramer (1978) and Hagedorn (1988) and our field studies on Marajó Island (Schwassmann and Assunção 1989), as well as the common observation of many floodplain species with fully developed ovaries, several months prior to the onset of major rains (Honda 1974; Schwassmann 1978), would argue against generalizing Kirschbaum's findings. Either the response to the rain-related factors may be different in certain species or other, as yet unknown, synchronizers may be acting.

While seasonal reproduction appears to be the general pattern in floodplain lakes and large rivers, or generally in areas where conditions of alternating drier and rainier seasons prevail, in the small forest streams of higher ground one finds aseasonality in the now no longer annually cyclic reproduction of many species. It seems that in this very uniform aquatic habitat, fishes, and other organisms as well, have lost an annual periodicity, even though intrapopulation synchronization of gonadal growth and spawning is preserved. The concept of aperiodic reproduction in the tropics used to prevail in the older ecological literature (Hesse 1924; Hesse and Doflein 1943; Bünning 1967) and was discussed recently (Schwassmann 1980; Roberts 1989). Fishes of the Amazon basin show annually cyclic reproductive behavior wherever strong seasonal changes in their aquatic habitat exist. The synchronizing factors that result in gonadal ripening and spawning seem to be related to annually cyclic rains. Loss of phase control of the circannual rhythm exists in the life history patterns of small species living in annually very uniform biotopes.

REFERENCES

Azevedo P de (1938) Da Biologia dos Peixes Nordestinos. In Livro Jubilar Prof. Travassos. Rio de Janeiro III:51–60

Azevedo P de, Canale L (1938) A hipófise e sua ação nas gônadas dos peixes neotrópicos. Arq. Inst. Biol. São Paulo 9:165–186

Barthem RB (1985) Occorência, distribuição e biologia dos peixes da Baia de Marajó, estuário amazonico. Bol. Mus. Para. Emilio Goeldi, Zoologia 2:49–69

Bünning E (1967) The Physiological Clock. (2nd ed.) Springer-Verlag, Berlin

Cardoso DM (1934) Relação gênito-hipofisária e reprodução nos peixes. Arquiv. Inst. Biol. São Paulo 5:132–136

Carvalho FM (1980) Composição química e reprodução do mapará (*Hypophthalmus edentatus* Spix. 1829) do Lago do Castanho, Amazônas (Siluriformes, Hypophthalmidae). Acta Amazônica 10(2):379–398

Carvalho JL de (1978) Contribuição ao conhecimento da biologia do mapará,

Hypophthalmus perporosus Cope, 1878 (Pisces, Hypophthalmidae) no Baixo e Medio Tocantins. Boletim da FCAP Belém 10:37–57

Egler WA, Schwassmann HO (1962) Limnological studies in the Amazon estuary. Bol. Mus. Para. Emilio Goeldi N.S. 1:2–25

Fontenele O, Camacho EC, Menezes RS de (1946) Obtenção de tres desovas anuais de curimatã, *Prochilodus* sp. (Pisces: Characidae, Prochilodinae), pelo método de hipofisação (nota previa). Bol. Mus. Nacional—Zoologia 53:1–9

Fowler HW (1948–1954) Os peixes de água doce do Brasil I–IV. Arq. de Zool. Est. São Paulo 6:1–628; 9:1–400

Godoy MP de (1979) Marcação e migração da piramutaba *Brachyplatystoma vaillantii* (Val. 1840) na bacia amazônica (Pará e Amazonas), Brasil. (Pisces, Nematognathi, Pimelodidae). Boletim da FCAP Belém 11:3–21

Gottsberger G (1978) Seed dispersal by fish in the inundated regions of Humaitá, Amazonia. Biotropica 10:170–183

Goulding M (1980) The Fishes and the Forest. Explorations in Amazonian Natural History. University of California Press, Berkeley, California

Goulding M, Carvalho ML, Ferreira, EG (1988) Rio Negro, Rich Life in Poor Water. Amazonian Diversity and Foodchain Ecology as Seen Through Fish Communities. SPB Academic Publishing, The Hague

Hagedorn M (1988) Ecology and behavior of a pulse-type electric fish, *Hypopomus occidentalis* (Gymnotiformes, Hypopomidae), in a fresh-water stream in Panama. Copeia 1988(2):324–335

Hesse R (1924) Tiergeographie auf ökologischer Grundlage. Verlag Gustav Fisches, Jena, Germany

Hesse R, Doflein F (1943) Tierbau und Tierleben, in ihrem Zusammenhang betrachtet II: Das Tier als Glied des Naturganzen. G Fischer, Jena, Germany

Honda EM (1974) Contribuição ao conhecimento da biologia de peixes do Amazônas. Alimentação de tambaquí, *Colossoma bidens* (Spix). Acta Amazônica 4:47–53

Junk WJ (1984) Ecology, fisheries and fish culture in Amazonia. In Sioli H (ed.). The Amazon: Limnology and Landscape Ecology of a Mighty Tropical River and its Basin. Dr. W. Junk, Publ., Dordrecht, Boston, Lancaster. pp. 443–476

Kirschbaum F (1975) Environmental factors control the periodical reproduction of tropical electric fish. Experientia 31:1159–1160

Kirschbaum F (1979) Reproduction of the weakly electric fish *Eigenmannia virescens* (Rhamphichthyidae Teleostei) in captivity. I. Control of gonadal recrudescence and regression by environmental factors. Behav. Ecol. Sociobiol. 4:331–355

Kirschbaum F (1984) Reproduction of weakly electric teleosts; Just another example of convergent development? Env. Biol. Fish 10:3–14

Kirschbaum F (1987) Reproduction and development of the weakly electric fish, *Pollimyrus isidori* (Mormyridae, Teleostei) in captivity. Env. Biol. Fish 20:11–31

Kramer DL (1978) Reproductive seasonality in the fishes of a tropical stream. Ecology 59:976–985

LeCointe P (1945) O Estado do Pará: A Terra, A Água e O Ar. Brasiliana ser. 5a, vol. 5. Editora Nacional, São Paulo. 307pp

Lowe-McConnell RH (1969) Speciation in tropical freshwater fishes. Biol. J. Linn. Soc. London 1:51–75

Lowe-McConnell RH (1975) Fish Communities in Tropical Freshwaters. Longman, London

Lowe-McConnell RH (1979) Ecological aspects of seasonality in fishes of tropical waters. Symp. zool. Soc. London. 44:219–241

Lowe-McConnell RH (1987) Ecological Studies in Tropical Fish Communities. Cambridge Univ. Press, Cambridge

Menezes RS de (1972) Potencial da pesca e piscicultura na Amazônia. A Amazônia Brasileira en Foco, Bol. 7:34–61

Miller PJ (ed.) (1979a) Fish Phenology: Anabolic Adaptiveness in Teleosts. Symp. zool. Soc. London 44:449

Miller PJ (1979b) A concept of fish phenology. Symp. zool. Soc. London 44:1–28

Petrere M Jr (1978) Pesca e esforço de pesca no estado do Amazônas. II-locais, aparelhos de captura e estatísticas de desembarque. Acta Amazônica 8, (3), Suppl. 2:1–54

Roberts TR (1972) Ecology of fishes in the Amazon and Congo basins. Bull. Mus. Comp. Zool. 143:117–147

Roberts TR (1989) The Freshwater Fishes of Western Borneo (Kalimantan Barat, Indonesia). Cal. Acad. Sciences, San Francisco

Schmidt GW (1972) Seasonal changes in water chemistry of a tropical lake (Lago do Castanho, Amazonia, South America). Verh. Internat. Verein. Limnol. 18:613–621

Schmidt GW (1973) Primary production of phytoplankton in the three types of Amazonian waters. II. The limnology of a tropical flood-plain lake in Central Amazonia (Lago de Castanho). Amazoniana 4:139–203.

Schwassmann HO (1971) Biological rhythms. In Hoar WS, Randall DJ (eds.). Fish Physiology. Academic Press, New York. pp. 371–428

Schwassmann HO (1976) Ecology and taxonomic status of different geographic populations of *Gymnorhamphichthys hypostomus* Ellis (Pisces, Cypriniformes, Gymnotoidei). Biotropica 8:25–40

Schwassmann HO (1978) Times of annual spawning and reproductive strategies in Amazonian fishes. In Thorpe JE (ed.). Rhythmic Activity of Fishes. Academic Press, London. pp. 187–200

Schwassmann HO (1980) Biological rhythms: Their adaptive significance. In Ali MA (ed.). Environmental Physiology of Fishes. Plenum Publ. Co., New York. pp. 613–630

Schwassmann HO, Assunção MIS (1989) Early electric organ activity in newly hatched larvae of the puraqué *Electrophorus electricus* (L.). Bol. Mus. Para. Emilio Goeldi, sér. Zool. 5(2):143–150

Schwassmann HO, Barthem RB Carvalho ML (1989) A note on the seasonally shifting zone of high primary production in the Bay of Marajó, Pará, Brazil, 1983–1984. Acta Bot. Bras. 2(1):165–174.

Scott DBC (1974) The reproductive cycle of *Mormyrus kannume* Forsk. (Osteoglossomorpha, Mormyriformes) in Lake Victoria, Uganda. J. Fish. Biol. 6:447–454

Smith NJH (1981) Man, Fishes, and the Amazon. Columbia University Press, New York

Souza RAL de, Imbiriba EP (1978) Peixes comerciais de Belém e principais zonas de captura da pesca artesanal. Boletim da FCAP Belém 10:1–15

Thorpe JE (ed.) (1978) Rhythmic Activity of Fishes. Academic Press, London

Veríssimo J (1895) A Pesca na Amazônia. Livraria Classica de Alves a Cia., Rio de Janeiro—reprint 1970. Universidade Federal do Pará, Belém. 130 pp

Von Ihering R, Wright S (1935) Fisheries investigations in northeast Brazil. Trans. Am. Fisheries Soc. 65:267–271
Von Spix JB, Von Martius CFP (1823–1831) Reise in Brasilien auf Befehl Sr. Majestät Maximilian Joseph I, Königs von Bayern, vol. 1, 2, 3, q1, Lindauer, München

Amphibia

6

Diversity of Reproductive Modes in Anurans: Facultative Nest Construction in Gladiator Frogs

Janalee P. Caldwell

Introduction

Ten families of anurans comprising at least 1000 species are known from South America. The diversity of reproductive modes among these amphibians is higher than in any other group of vertebrates. Duellman (1985) defined reproductive mode as a composite of factors, including oviposition site, ovum and clutch characteristics, rate and duration of development, stage and size of hatchling, and type of parental care. Of the 29 modes he recognized, 21 are found in the Neotropics, and, of these, eight are unique to that area.

The evolution of reproductive modes of anurans, especially in tropical regions, has tended toward isolating or removing eggs and tadpoles from aquatic environments that have a high diversity of predators on these life history stages. Of the 10 families in South America, many isolate their eggs or tadpoles by depositing clutches on leaves above water, in foam nests, in burrows at the edge of water, or in phytotelmata such as bromeliads or tree holes. Herein I discuss in more detail a group of South and Central American frogs, called gladiator frogs, that isolate their eggs and tadpoles by constructing basins or nests along the edges of streams and small rivers.

Gladiator frogs belong to the *Hyla boans* species group and occur from Costa Rica and Panama into northern South America (Kluge 1979). Currently, seven species are recognized and all exhibit unique morphological and behavioral characteristics. Males possess well-developed prepollical spines and are exceptionally aggressive toward each other during the breeding season. In a study of the life history and social organization of *Hyla rosenbergi*, Kluge (1981) found that individuals survived less than one year, making this species one of the most short-lived anurans known.

The nests in these species are constructed by males, but, at least in *H. rosenbergi*, may be renovated by females prior to oviposition (Kluge 1981). The nests are constructed in mud, sand, or another suitable substrate at the edges of streams. Nests are constructed by the frog pivoting in the substrate and pushing excess material outward with its hands and feet

(*Hyla pardalis*, Lutz 1960a; *Hyla rosenbergi*, Kluge 1981) or by the male diving below the surface and scooping out mud with its snout (Goeldi 1895). Approximately 2,000 eggs are deposited as a surface film in the nest.

Not all species in the group are known to construct nests. Of the seven species currently recognized, nest construction has been reported for *Hyla pardalis* (Lutz 1960a), *Hyla faber* (Lutz 1960b), *Hyla boans* (Duellman 1970; Crump 1974), and *Hyla rosenbergi* (Kluge 1981). Although breeding behavior has been observed, nest construction has not been reported for *Hyla crepitans* (Duellman 1970) or *Hyla circumdata* and is generally believed to be absent in these species. Little is known about the breeding biology of *H. pugnax* (Kluge 1979).

During field work in Brazil and Venezuela, I studied nesting behavior in two species of gladiator frogs, *Hyla boans* and *Hyla crepitans*. In *H. boans*, some individuals constructed nests whereas others did not. In *H. crepitans*, only one nest was found during a three-week period when the frogs were actively breeding; most individuals did not construct nests. Thus, nest construction, at least in these two species, is facultative and therefore provides a system for the study of behavioral plasticity and tradeoffs. The currency most applicable for examination of this trait is survivorship and the growth of eggs and early embryos of individual clutches deposited in nests and as free-floating surface films, i.e., not in basins.

Two hypotheses have been proposed to account for the function of the nest in these frogs. One is that the nest isolates the eggs and developing embryos from potential predators in the stream, including conspecific tadpoles that are cannibalistic (Breder 1946; Crump 1974; Kluge 1981). The other is that by isolating a small amount of water from the main stream, the water temperature is raised, thereby increasing the growth and developmental rate of the embryos (Lamotte and Lescure 1977). I examined these hypotheses by constructing temperature profiles for nests and free clutches in different microhabitats and by examining growth, developmental time, and survivorship of individual clutches.

Materials and Methods

The primary study site in Brazil was a small river surrounded by an area of terra firme tropical rainforest located 2 km E of Santa Cruz da Serra in the state of Rondônia (Fig. 6.1A). Two distinctly different microhabitats were found in the river. The main channel was formed of granitic rock outcrops, among which were many small backwater rock pools. In contrast, several small tributaries of the main channel had a sandy substrate. At the time of the study, *H. boans* (Fig. 6.1B) was breeding in apparently equal numbers in the two microhabitats. In the sandy substrate they constructed nests, whereas in the rocky areas they deposited clutches in small semi-isolated pools among rocks or in areas with dense leaf litter. Nests were studied near the end of the dry season from 19 August to 10 September 1985.

In Venezuela, the study area was the Rio Agua Blanca, a small stream in lowland forest in Guatopo National Park in the state of Miranda. The stream varied in width from 2 to 4 m, and the substrate was rocky with small backwater pools in which sand and silt collected. A large population of *Hyla crepitans* was breeding along this stream. The frogs generally did not construct nests but deposited eggs in small pools along the edge of the stream; however, one nest was located in a sandy spit in the stream. This part of the study was carried out near the end of the dry season from 28 April to 17 May 1987.

In Brazil, data was obtained for 18 clutches, nine of which were placed in basins constructed in sand (Fig. 6.1C), and nine of which were deposited in rock pools or were encircled by floating leaf litter (Fig. 6.1D). For each clutch, a series of temperature measurements was taken, including (1) air temperature, (2) water temperature at the center of the nest or 3 cm below the eggs if there was no nest, and (3) water temperature 5 cm below the surface 30 cm from the clutch or nest in the main stream. Temperatures were recorded hourly from 1030 to 1430 hours, and whether the clutch was in the sun or shade was recorded each hour. Daily collections of approximately 10 to 20 eggs, embryos, or tadpoles were usually made for as long as the clutch could be followed (2 to 10 days) and preserved in 10% formalin. Embryos and tadpoles were staged according to Gosner (1960) for comparisons of growth and development among clutches. Data were obtained on the physical structure of 12 nests, two of which did not have clutches of eggs deposited in them. The data included major and minor orthogonal axes, height of the rampart, depth of the water, and substrate type. For clutches in pools, width, length, depth, and temperature of water of the pool were recorded.

In Venezuela, fourteen pools were studied, eight of which had clutches or tadpoles of *Hyla crepitans*. Width, length, and depth of water of each pool was recorded. Air and water temperature were recorded at least twice for each pool during the study. Collections of embryos and tadpoles were preserved in 10% formalin. Embryos and tadpoles were staged according to Gosner (1960) and were measured to determine growth and development. The same measurements as described above were taken on the single nest of *Hyla crepitans* that was observed.

Results

Of the 12 basins constructed by H. boans, most (nine) were nearly circular (Fig. 6.1C), unless built under or against a log or branch in the stream. The average major orthogonal axis was 36.2 + 4.3 cm (SD) (range, 30–42 cm; n = 12), and the average minor orthogonal axis was 33.1 ± 3.5 cm (range, 28–38 cm; n = 9). Average surface area was 1040.4 ± 242.4 cm^2 (SD) (range, 707–2385 cm^2; n = 12). Depth of water in the center of newly constructed nests ranged from 3.5 to 7.2 cm (mean, 5.2 ± 1.1 cm; n = 12).

FIGURE 6.1. (A) Rio São Domingo in Rondônia, Brazil, breeding site of *Hyla boans*. (B) *Hyla boans*. (C) Nest of *Hyla boans* constructed in sand. (D) Nest of *Hyla boans*, showing eggs deposited as a surface film encircled by fallen leaves.

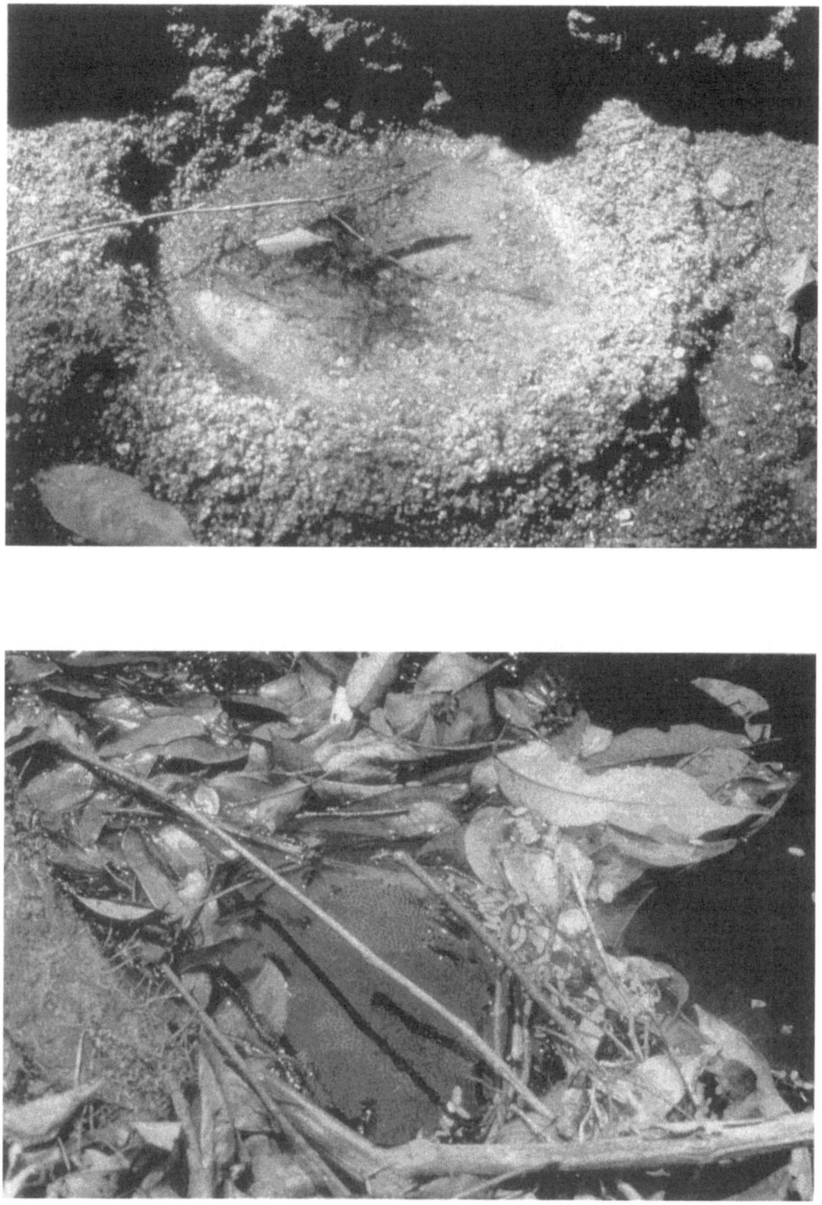

C

D

Figure 6.1 (*Cont.*)

Height of the ramparts was 3.4 ± 1.4 cm (range, 1.5–6.5 cm; n = 12). In this area, nests were constructed in fine to coarse sand, except in two cases where they were constructed in an area with small pebbles.

Of the nine clutches not in basins, six were placed in protected sites (usually encircled by floating leaves; Fig. 6.1D) at the edge of the stream, and three were placed in rocky pools. Depth of the water near these clutches ranged from 5–7 cm (mean, 8.7 ± 4.3 cm; n = 9). The three rocky pools averaged 111 ± 67 cm in length (range, 34–150 cm) and 59 ± 38 cm in width (21–100 cm).

Among the nests, temperature varied considerably, depending primarily on (1) whether water could seep through the substrate into the nest or (2) the amount of sunlight the nest received throughout the day.

Eight basins were paired on the basis of location and having their temperature profiles recorded on the same day (Fig. 6.2). Nests on the left side of this figure were constructed in sand and had no seepage of water. Those on the right side of the figure were generally constructed in areas of coarser sand that allowed the seepage of water. In every case, the nest without seepage of water attained a higher temperature due to warming by the sun.

Most of the eight nests were in areas with large trees along the banks of the stream and were exposed to varying amounts of full sun, filtered sun, or shade depending on their locations. However, Nest 3 was exposed to continuous full sun and had no water seepage; these conditions caused the most extreme difference between water temperature in the nest and that of the stream (Fig. 6.2; Nest 3).

The rate of embryo development increased as temperature increased. Nests 3 and 5 differed in water temperature approximately 8° to 10°C throughout the day. Embryos in Nest 3 reached Stage 23 100 hours after fertilization, whereas those in Nest 5 did not reach Stage 23 until 200 hours after fertilization (Fig. 6.3).

A comparison of Clutches 16 and 17, both with eggs directly exposed to water in the main channel of the stream, and Clutch 19, located in an isolated rock pool, showed that temperature differences also occurred among clutches not deposited in basins (Fig. 6.4). In the isolated pool, where the water temperature was approximately 7°C higher than the main channel, embryos reached Stage 25 by 130 hours after fertilization, whereas embryos in the other two clutches did not reach Stage 25 until 175 hours after fertilization.

Survivorship of early embryos and tadpoles in basins and not in basins was compared. Seven clutches deposited in basins were followed for two to nine days. Only two of these had tadpoles that survived (defined as a nest with eroded ramparts that allowed tadpoles to escape into the adjacent stream). Of the other five clutches followed, all embryos in two were consumed within two and four days by numerous larger conspecific tadpoles, and all embryos in two others disappeared overnight after eight and nine days in the nest, apparently consumed by an unseen predator. These tad-

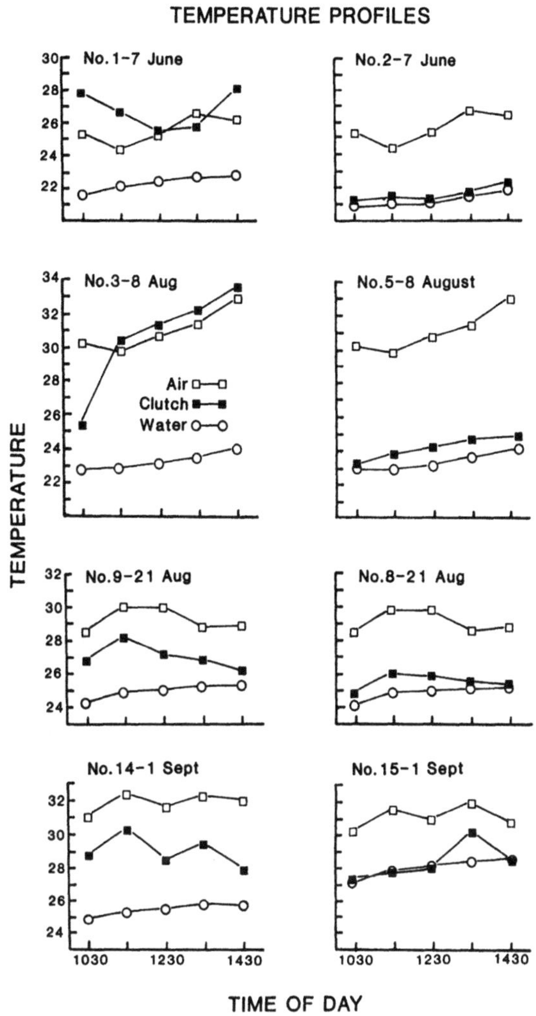

FIGURE 6.2. Temperature profiles of eight nests at the study site in Brazil. Nests on the left side of the figure had no water seepage, in contrast to those on the right side. Nest (clutch) temperature was warmer in all cases in nests shown on the left side.

poles were confined to a very small amount of water as the water level in the stream was gradually dropping. The predator was most likely the aquatic snake *Helicops angulatus*, which was frequently observed in the area (pers. obs.). Tadpoles in another clutch desiccated when the water level in the nest became very low.

Seven clutches deposited in rocky pools or encircled only by leaves (not in basins) were followed for three to ten days. In contrast to the clutches in the nests, the embryos or small tadpoles in all these clutches successfully escaped into the stream.

In Venezuela, *H. crepitans* embryos or tadpoles were found in eight of

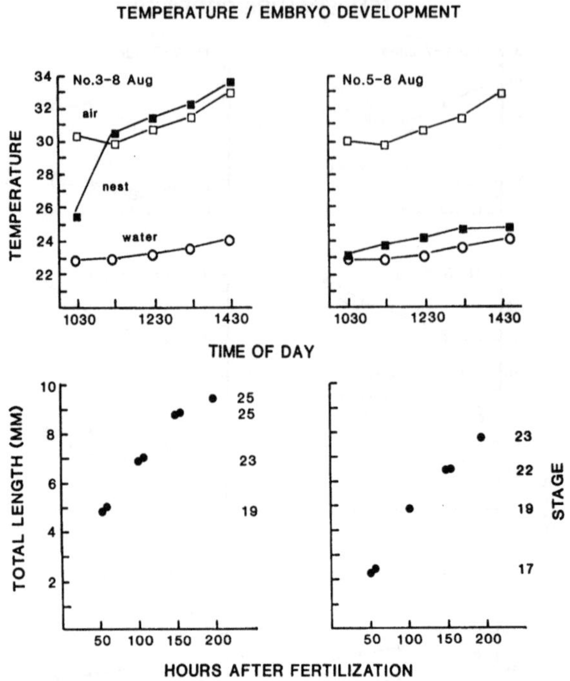

FIGURE 6.3. Embryo development for one of the paired nests in Figure 1. Note that Stage 23 is reached 100 hours later in the nest with the lower temperature.

14 pools examined. The average length of the eight pools was 371.9 ± 283.9 cm (range, 75–850 cm), the average width was 99.4 ± 66.2 cm (range, 43–250 cm), and the average depth was 16.4 ± 9.6 cm (range, 6–34 cm). Daytime temperatures were fairly constant in all pools, ranging from 24.0° to 27.0°C.

The single nest of *H. crepitans* was located at the edge of a mud/sand spit along a portion of the stream where the water was barely flowing. The major and minor orthogonal axes of the nest measured 30 × 27 cm. The rampart was 3.5 cm in width and 2.5 cm in height and was composed of mud, small pebbles (up to 0.75 cm in diameter), and dried leaves, while the bottom substrate of the nest was composed of fine silt and mud. Depth of water in the nest varied from 3.0 to 4.5 cm for the first four days after eggs were deposited in it.

An estimated 1680 eggs were deposited in the nest on the night of 2 May. The tadpoles hatched in four days, and the nest remained intact for five days. As the nest gradually disintegrated, the tadpoles slowly dispersed into the adjacent pool. At the time of dispersal, they were in Stage 23 and were not actively swimming. Many of the tadpoles remained in the discernable basin of the nest six days after fertilization.

Temperature of the nest was taken once from 1100 to 1700 hours on three different days. During this time, air temperature varied from 28.8° to

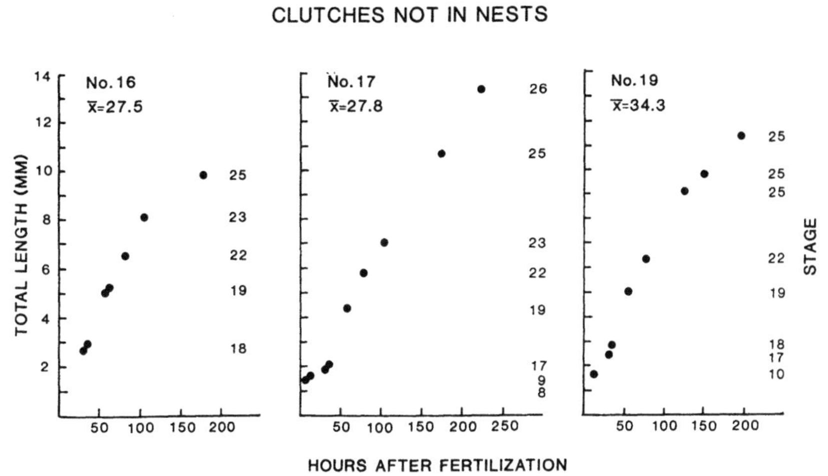

FIGURE 6.4. Clutches not deposited in nests at the study site in Brazil also showed temperature differences. Clutch 19 was in an isolated pool and reached an average of 7° C higher than Clutches 16 and 17, which were exposed to water in the stream. Stage 25 was reached 45 hours earlier in the warmer nest.

32.2°C, nest temperature from 24.8° to 25.7°C, and water temperature in the stream 30 cm from the nest from 24.0° to 25.3°C. Thus, the nest temperature was approximately the same as that of the temperature of the main channel of the stream. The nest was located in an area where it was exposed to shade or filtered sunlight.

In addition to deposition time of eggs in the nest, deposition time of clutches placed in three rock pools was known or could be estimated in order to examine the growth of embryos and small tadpoles. Embryos and tadpoles in the nest were larger at all time periods than those in the other three pools (Fig. 6.5).

Discussion

Both *Hyla boans* in Brazil and *H. crepitans* in Venezuela appear to be facultative nest-builders. In Brazil, whether *H. boans* constructs nests appears to be determined by the substrate. The study area contained habitats that were exclusively rocky or exclusively sandy. In the rocky areas, nests could not be constructed. In the sandy substrate, however, nests were always constructed; no free clutches were found.

In Venezuela, *H. crepitans* also appeared to construct nests depending on whether the substrate was suitable for doing so. Most of the stream in the study area was rocky with backwater pools along the edge, and clutches were deposited in these pools. Only one small area with a sand/mud spit was present, and the single nest was found there.

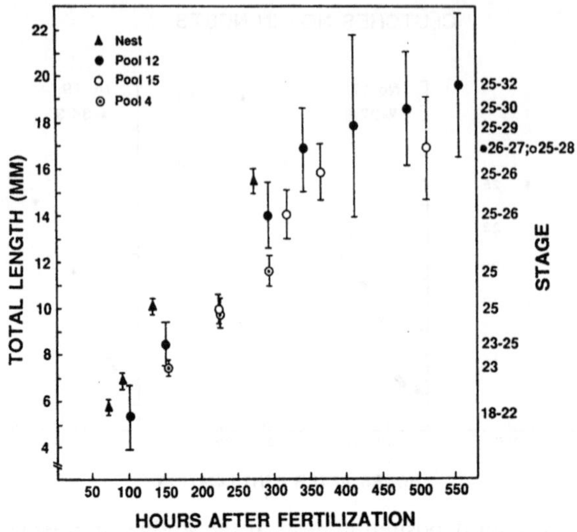

FIGURE 6.5. A comparison of growth and development of four clutches of *H. crepitans* in Venezuela, one of which was deposited in a nest. There was some variation among the clutches deposited in pools; however, the clutch in the nest showed the fastest growth and development.

NEST DESCRIPTION

Most *H. boans* nests were circular or nearly so as was the single *H. crepitans* nest. Both differed from the nests of *H. rosenbergi* described by Kluge (1981) in that they did not contain a calling platform in the nest. Nighttime observations of calling *H. boans* and *H. crepitans* indicated that they call from elevated positions in vegetation above the nest.

GROWTH AND DEVELOPMENT IN RELATIONSHIP TO NEST OR WATER TEMPERATURE

Growth and development of both *H. boans* and *H. crepitans* were similar to those of *H. rosenbergi* described by Kluge (1981). Hatching occurred at approximately Stage 17 in both species.

In Brazil, the growth and development of embryos and tadpoles in nests and not in nests were variable due to environmental conditions. Nests that were in the sun at least part of the day and were constructed in a substrate that did not allow water seepage had the highest water temperatures. Consequently, the embryos and tadpoles under those conditions grew and developed faster than those in cooler water. The same was true of rocky pools in which eggs were deposited. Pools located in the sun experienced a greater degree of warming that those in the shade or through which a current moved.

In Venezuela, there was relatively little variation in temperature among the pools in which the clutches were deposited. This may be partly due to the depth of the water (average, 16.4 cm) and also because the relatively small stream was surrounded by forest and thus shaded most of the day. Embryos in these pools showed relatively little variation in growth and development (Fig. 6.4). The single nest of *H. crepitans* was also in a shaded area and did not experience an increase in temperature above the adjacent water.

Lamotte and Lescure (1977) hypothesized that water temperature in a nest would be higher than that of the adjacent water. Kluge (1981) found for *H. rosenbergi* in Panama that the open water temperature underwent more daily variation than that in the nests, failing to support the hypothesis for *H. rosenbergi* in that location. It was also not true for *H. crepitans* in Venezuela, which bred along a small, shaded stream similar to the *H. rosenbergi* site in Panama. However, in Brazil, the nest and pool temperatures were sometimes much elevated over the adjacent water, thus supporting the hypothesis of Lamotte and Lescure (1977). Kluge noted that only a 2° C difference in temperature could mean a 20% change in developmental time and suggested that strong selection would be expected for choosing a warmer nest. Future work will be needed to learn if nests in warmer areas are used more frequently or if offspring from those nests have a higher survivorship than those that develop more slowly. Previous work has demonstrated that smaller larvae are more susceptible to predation and are at a competitive disadvantage (Formanowicz 1986; Alford 1989).

Embryos in the nest of *H. crepitans* in Venezuela grew and developed faster than those in pools even though there was not a temperature difference (Fig. 6.5). Additional observations on multiple nests will be necessary to determine if this is a general phenomenon or an artifact of the presence of only one nest in the study area. Since all eggs in one nest are from a single sibship, it is possible that these results reflect a sibship effect on size as found in other species of anurans (Travis 1980, 1983).

SURVIVORSHIP

Several authors have hypothesized that the function of the nest is to offset predation on eggs and early embryos (Breder 1946; Crump 1974; Kluge 1981). Kluge (1981) believed that selection for nest-building was most strongly influenced by predation by small fish, tadpoles, and especially conspecific tadpoles.

Predators of *H. boans* embryos and tadpoles observed in Brazil were conspecific tadpoles and possibly the snake *Helicops angulatus*. In addition to several invertebrate and vertebrate predators, Kluge (1981) observed conspecific tadpoles and two other species of tadpoles, *Leptodactylus pentadactylus* and *Physalaemus pustulosus*, preying on the eggs and embryos in nests of *H. rosenbergi*. He noted that the efficiency of cannibalism was impressive, but that it was not clear how predatory tadpoles gained access to nests.

In *H. boans*, at least at the time that this study was done, more mortality occurred in clutches in nests as opposed to those not in nests. In addition to predation, drying of the nest was another source of mortality. The present study was done in Rondônia during the dry season, and water level of the stream was steadily dropping.

Many factors apparently influence the growth, development, and survivorship of eggs and embryos in gladiator frogs. Facultative nest-building may be an adaptation in some of these species for ensuring successful offspring in different habitats and in different seasons. Although *H. rosenbergi* does not breed in the dry season in Panama (Kluge 1981), additional work on the life history of the other species is needed to determine whether they breed in different seasons and if they are behaviorally plastic in terms of their nest-building behavior.

ACKNOWLEDGMENTS

Field work in Rondônia, Brazil, was coordinated by P.E. Vanzolini of the Museum of Zoology at the University of São Paulo. Funding for this work was provided by the Conselho Nacional de Desenvolvimento Científico e Tecnologico, Programa Polonoreste, Subprograma Ecologia Animal. I thank Ross Alford, Steve Richards, Arnold Kluge, and Laurie Vitt for providing comments on the manuscript. Martin L. Cody discovered the nest of *Hyla crepitans* in Venezuela.

REFERENCES

Alford RA (1989) Variation in predator phenology affects predator performance and prey composition. Ecology 70:206–219

Breder CM Jr (1946) Amphibians and reptiles of the Rio Chucunaque drainage, Darien, Panama, with notes on their life histories and habits. Bull. Am. Mus. Nat. Hist. 86:375–436

Crump ML (1974) Reproductive strategies in a tropical anuran community. Univ. Kansas Mus. Nat. Hist. Misc. Publ. 61:1–68

Duellman WE (1970) The hylid frogs of Middle America. Univ. Kans. Mus. Nat. Hist. Monogr. 1:1–427

Duellman WE (1985) Reproductive modes in anuran amphibians: Phylogenetic significance of adaptive stragegies. S. Afr. J. Sci. 81:174–178

Duellman WE, Trueb L (1986) Biology of Amphibia. McGraw-Hill, New York

Formanowicz DR Jr (1986) Anuran tadpole–aquatic insect predator-prey interactions: Tadpoles size and predator capture success. Herpetologica 42:367–373

Goeldi EA (1895) Contribution to the knowledge of the breeding-habits of some treefrogs (Hylidae) of the Serra dos Orgaos, Rio de Janeiro, Brazil. Proc. Zool. Soc. London 1895:89–97

Gosner KL (1960) A simplified table for staging anuran embryos and larvae with notes on identification. Herpetologica 16:183–190

Kluge AG (1979) The gladiator frogs of Middle America and Colombia–A re-evaluation of their systematics (Anura: Hylidae). Occ. Pap., Mus. Zool., Univ. Michigan 688:1–24

Kluge AG (1981). The life history, social organization, and parental behavior of *Hyla rosenbergi* Boulenger, a nest-building gladiator frog. Misc. Publ., Mus. Zool., Univ. Michigan 160:1–170

Lamotte M, Lescure J (1977) Tendances adaptatives a l'affranchissement du milieu aquatique chez les amphibiens anoures. Terre Vie 31:225–311

Lutz B (1960a) The clay nests of *Hyla pardalis* Spix. Copeia 1:61–63

Lutz B (1960b) Nação de territorio em anfibios anuros *Hyla faber* Wied. An. Acad. Bras. Ciênc. 32:143–145

Travis J (1980) Phenotypic variation and the outcome of interspecific competition in hylid tadpoles. Evolution 34:40–50

Travis J (1983) Variation in development patterns of larval anurans in temporary ponds. I. Persistent variation within a *Hyla gratiosa* population. Evolution 37:496–512

7

Reproduction in the Mexican Leaf Frog, *Pachymedusa dacnicolor*

Joseph T. Bagnara and Rakesh K. Rastogi

Introduction

The Mexican leaf frog (*Pachymedusa dacnicolor*) is a neotropical resident of the thorn scrub vegetation found along the coast of western Mexico. It has a remarkable ability to withstand high temperatures and aridity, yet it has evolved a reproductive biology consistent with a more humid ambiance. In fact, its reproductive strategy involves the deposition of eggs on vegetation outside the water, therefore, its success entails the prerequisite of very high humidity. This mode of reproduction fundamentally requires a perception of specific environmental cues, notably temperature and humidity. The Mexican leaf frog is readily available from Delta Biological Supply (Tucson, Arizona) which captures adults in the wild near Los Mochis, Sonora, Mexico. For this reason, and because of the ease in handling in captivity and the ability to alter its reproductive activities in response to the manipulation of temperature and humidity, much has been learned about its reproductive biology (Bagnara et al. 1986; Bagnara 1990). Very likely, this knowledge will be useful in understanding the reproductive biology of other species of phyllomedusine frogs which are significant representatives of the herpetofauna of Central and South America. This chapter provides an account of what is known about the reproductive biology of the Mexican leaf frog, *P. dacnicolor*.

The following brief description outlines the reproduction of *P. dacnicolor* specifically, but it might well hold for other phyllomedusine species. In late spring, when ambient temperatures increase, spermatogenesis begins and males develop a remarkable increase in testicular weight. This is accompanied by darkening of nuptial pads, calling, and reproductive behavior. Because this frog displays a well-defined annual testicular cycle that can be induced in the laboratory, it has been possible to assess morphological, physiological, and behavioral stages throughout the year. A marked annual gonadal cycle is also displayed by females. With cessation of the breeding activity in early fall, a long period of vitellogenic stasis begins.

However, a store of previtellogenic follicles is present in the ovary awaiting spring when vitellogenesis resumes and continues through the summer during a prolonged breeding season. As with the males, elevation in ambient temperature seems to be the cue for the onset of vitellogenesis and subsequent breeding behavior of females. Prolonged periods of amplexus commence before the female is ready to oviposit. When the female perceives appropriate conditions of temperature and humidity exist, she selects an appropriate site for oviposition, usually on leaves hanging just above the water's edge. The total act of oviposition requires four to six hours, during which time the amplectant couple exhibits such ardor that they are not easily disturbed. Accordingly, rather detailed studies and observations are possible, not only during the annual cycle, but during the act of mating itself.

The Annual Cycle

Most of the details about the annual reproductive cycle of *P. dacnicolor* were determined from frogs maintained in greenhouses in Tucson, Arizona. However, the annual cycles of light, temperature, rainfall, and food supply were very much like those in Mexico, where these frogs normally live (Bagnara et al. 1986). They thrive in this artificial environment and can be maintained for generations.

ONSET OF SEXUAL ACTIVITY

There is an obvious sexual dimorphism in this species; the smaller males range from 60 to 80 mm in snout–vent length and vary in weight from 20 to 45 g, while the length of females varies from 70 to 100 mm with weights ranging from 30 to almost 90 g (Fig. 7.1). Aside from differences in size, males and females can be distinguished from one another even during the non-breeding period of the year when secondary sexual characters are less evident. The bodies of the females are more robust and the faces more blunt than those of the male. The more pointed face of the male terminates with a rather square chin that is even more prominent during the breeding season and is especially evident as it rests on the dorsum of the female during amplexus.

As the onset of the breeding season begins, manifestations of sexual dimorphism are more evident. In males, early in the spring following the onset of warm temperatures, triangular nuptial callosities being to develop dorsally at the base of the thumb between it and the first digit. The callosities become black and rough in texture when the males are sexually ready. At the same time, the vocal sacs become more prominent and calling begins. The latter seems to be a manifestation of territoriality, evident only

Figure 7.1. Amplectant pair during water uptake by the female prior to oviposition. (Bagnara et al. 1986)

of loud squawks and wrestling matches followed by the retreat of less aggressive individuals. An additional secondary sexual characteristic of during the breeding season, and is particularly obvious when males encroach upon one another's territory. Such occasions are marked by a series males is the swelling of the cloacal area so that the cloaca almost appears as a kind of suction cup (Fig. 7.2). The swelling in this area of the skin is due to the development of numerous skin glands.

Except for the distention of the abdomen, often not overly striking, there are few physical signs that the females have become sexually ready. Males reach this state earlier than do females because they attempt to amplex females (and other males) rather early in the season. At this time, females usually retreat from males and even when a male successfully mounts a female, her persistent but subtle release calls or vibrations usually cause the male to release her. As the mating season draws near, in late May or early June, females generally accept males more readily and will allow them to remain in amplexus for days. The encumbrance of the amplectant male seems to have little or no effect upon the feeding behavior of the females at this time.

The Mating Process

The time and place of oviposition is apparently determined by the female, cued by persistent conditions of high humidity. Just prior to oviposition,

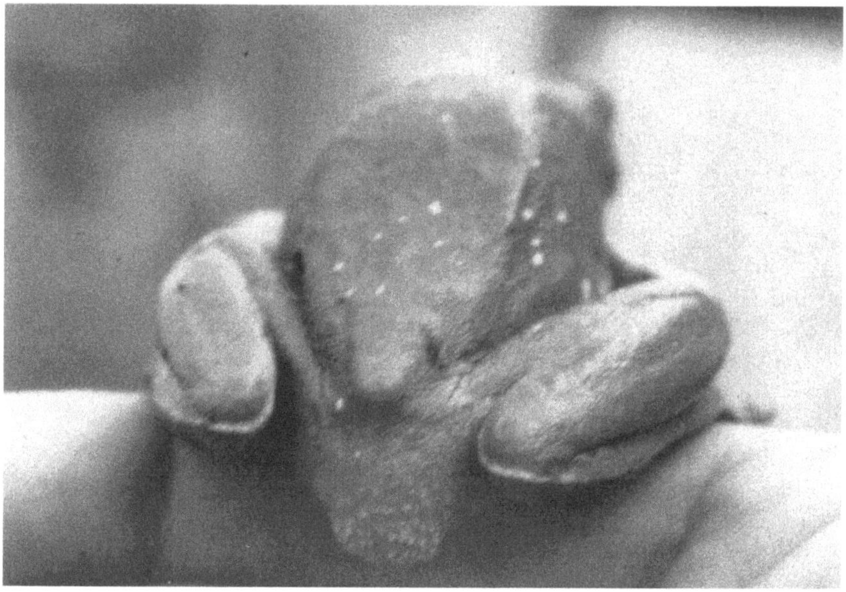

FIGURE 7.2. Cloacal swelling of a sexually active male. (Bagnara et al. 1986)

prepared females exhibit particularly distended abdomens (Fig. 7.3). Apparently, this is the result of the onset of ovulation and the translocation of a number of eggs to the uterus accompanied by filling of the bladder (Bagnara et al. 1986). The exact time for the onset of oviposition is highly variable. In nature, it apparently begins shortly after midnight (Pyburn 1970; Wiewandt 1971); however, in the greenhouse environment, it frequently begins much later. During the months of July and August, females are observed to complete oviposition most often between 7:00 and 9:00 a.m. Since the total period necessary for oviposition may vary from four to six hours, depending upon clutch size, it is concluded that oviposition during this period commences just before dawn. In September, oviposition often begins as late as 8:00 a.m. It is assumed that this late start is due to the lower temperatures that now occur near dawn, causing the females to delay the onset of oviposition until the morning sun has warmed the atmosphere. Such late starting females who are ovipositing large clutches may deposit eggs until early afternoon.

Spawning most often occurs on leaves or branches that are suspended over or near the water's edge, so that the egg mass is located at a position ranging from a few inches to well over a foot above the surface. The female, with an amplectant male aboard, climbs to the chosen site and assumes a secure position by grasping the vegetation in such a way that her

FIGURE 7.3. Attempt by intruding male to dislodge principal male before oviposition. Note distended abdomen of the female. (Bagnara et al. 1986)

cloaca is positioned on the surface of the substrate upon which she intends to oviposit (Fig. 7.4). The male, who holds the female in his grasp by locking his thumbs under her arms beneath her shoulders, is suspended even more precariously as he arches his back so that his swollen cloaca makes contact with her back just above her cloaca. The assumption of this characteristic position for egg laying is often made difficult by the presence of additional males (Figs. 7.3 and 7.5). They follow the path of the female as she climbs to the site of oviposition and attempt to mount her as she secures her position. The involvement of supernumerary males in the act of mating has been observed both in nature (Wiewandt 1971) and in the greenhouse environment, so it seems clear that polyandry is a normal feature of the reproduction of *P. dacnicolor*. To what extent the incidence of polyandry is related to the relative population densities of the two sexes has not been determined. Immediately after the female's position is stable, she immediately extrudes a group of six to ten eggs and, at the same time, an ooze of urine containing spermatozoa is released by the male. This fluid beings to flow down her back toward the extruded eggs. Within a minute or two, an additional small group of eggs is extruded; this continues for about 15 minutes, during which time the male (or males) continue to leak urine on the back of the female. By the end of this period, some 100 to 150 eggs

FIGURE 7.4. Typical act of oviposition. (Bagnara et al. 1986)

are spawned and the female withdraws together with her amplectant male. They then proceed to a source of water into which the female immerses her cloaca and they remain still for about 45 minutes. At the end of this period of water uptake into the bladder of the female, they proceed back to the initial site of egg deposition where the female assumes a position immediately above the previously deposited egg mass. During another 15-minute period, she oviposits another group of eggs containing about the same number as in the previous deposit. Additional fluid flows over the earlier egg deposition which has already begun to swell due to the imbibition of water. The female, with her attached male, again returns to the watering site to refill the bladder over another 45-minute period. This is again followed by another bout of egg laying and the process is continued, so that over a five to six hour period, five to six separate egg depositions occur, forming an egg mass resembling a vertical column. Such egg masses contain an average of about 450 eggs; however, larger masses containing

FIGURE 7.5. An example of polyandry. (Bagnara et al. 1986)

over 1000 eggs are not uncommon. The individual smaller groups of eggs continue to swell as excess moisture flows down over the whole egg mass from the most recently laid egg group.

The importance of water uptake into the bladder during the interim between egg depositions can not be overestimated because this is the source of all water presented to the entire egg mass during oviposition. After this, no additional water is added to the egg mass before hatching. It seems that an entire period of 45 minutes is not needed to fill the bladder. Rather, it appears that this prolonged period is also used for the continued ovulation of oocytes which continues during the early hours of oviposition (Bagnara et al. 1986). At the start of oviposition, only a small percentage of oocytes has been ovulated. The rest are ovulated and transported to the uterus during the prolonged period involved in the oviposition of the early groups of eggs and during the water uptake periods that intervene.

With the completion of oviposition, the amplectant pair separates and each member awaits its next turn at mating. Among males, it maybe as soon as the next day, whereas in females, a longer interim may pass. This longer waiting period in the female is probably related to the time required for vitellogenesis to proceed. Indeed, individual females seem to breed three or more times during the breeding season and this correlates well with the fact that, from spring to early summer, three or more discrete populations of previtellogenic and post-vitellogenic follicles are found in the ovary (Iela et al. 1986). These oocytes are destined to complete vitellogenesis and be ovulated before the end of the seasons except for a few post-vitellogenic eggs that remain in the ovary when the end of the breeding season is terminated by the onset of cold weather.

With the departure of the parents, the deposited eggs are left to the elements. As long as the humidity remains high, the eggs will not dry out and development will proceed until hatching occurs some five to six days later. During this period, the entire egg mass retains its water content, so that the water weight of a clutch of eggs on the first day is like that at hatching (Bagnara and Cohen unpubl.). What does change is the compartments in which that water is located. On the first day, most of the moisture is present in the surrounding jelly but, after a few days, the perivitelline space increases in volume as the water passes from the jelly into the space. Thus, just before hatching, most traces of the jelly are gone, the perivitelline space is large, and individual pre-hatching larvae have developed sufficiently to begin to swim in their own individual "perivitelline sea." The mechanism for this passage of water into the perivitelline space is apparently an osmotic one (Bagnara and Cohen unpubl.).

The reproductive process terminates at hatching, which in *P. dacnicolor* is rather spectacular. With swelling of the egg capsule and vigorous swimming movements of the larvae, it appears that a certain amount of pressure is exerted on the enclosing fertilization membrane. It is also very likely that *P. dacnicolor* embryos, like those of other species, secrete hatching enzymes that weaken the fertilization membrane. After about six days, hatching occurs as the larvae burst forth from their ruptured capsules and, by a flipping action of their tails, literally fly out into space to fall into the water below. The subsequent larval life is like that of most tadpoles and, after about a six-week period, metamorphosis takes place. Young froglets that were derived from eggs laid early in the year have a particularly good start and, with the excellent food supply of mid-summer, they grow so rapidly that they can become sexually mature in time for the next breeding season.

Reproductive Endocrinology

Because of the ease in handling the Mexican leaf frog in captivity, the fact that the actual mating process is so long, and the comportment of the amplectant pair during mating seems unperturbable, it has been possible to obtain unique data about sex homone levels throughout the year. This in-

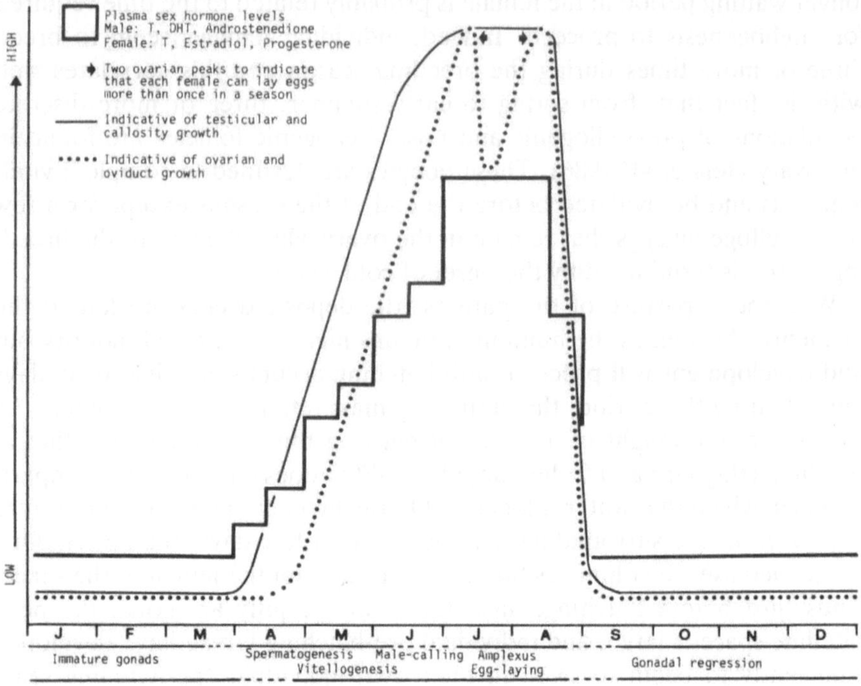

FIGURE 7.6. Schematic representation of annual changes in the plasma sex hormone levels, gonadal condition, and reproductive activity. The curves and the histogram are only indicative of the annual pattern and do not show the absolute values for any of the parameters. T = testosterone; DHT = 5α-dihydrotestosterone. (Bagnara et al. 1986; Iela et al. 1986; Rastogi et al. 1986)

formation is summarized in a general way in Figure 7.6 and more specific data are provided in the individual sections about the male and female, respectively.

THE MALE

A pronounced seasonal cycle in plasma concentrations of the androgens testosterone, 5α-dihydrotestosterone (DHT), and androstenedione is evident (Rastogi et al. 1986). Male *P. dacnicolor* is characterized by relatively high levels of testosterone and DHT and low amounts of androstenedione; estradiol and progesterone appear in low amounts throughout the year. Plasma levels of the three androgens are at a nadir in reproductively inactive males during fall and winter periods. At this time, the testis is composed of collapsed seminiferous tubules containing mainly primary spermatogonia which proliferate as spring approaches. Spermatocytes, spermatids, and spermatozoa are not formed during this proliferative period. The normal spermatogenic cycle shows strong seasonality. Active spermatogenesis begins during the spring and is completed by early summer and is "coupled" with an increase in both plasma androgen levels and testis weight. In the late spring to early summer period, the testis is composed of enlarged seminiferous tubules swollen with sperm bundles. Plasma concentrations of androgens reach their zenith in this period. Such frogs, now furnished with nuptial callosities, start chorusing. The plasma concentration of androgens remains high during the breeding season with androstenedione increasing further in breeding males. After the end of the breeding season, the seminiferous tubules lack sperm bundles and contain only a layer of "quiescent" primary spermatogonia. When breeding activity is completed, plasma androgen levels fall rapidly to the yearly minimum. This is associated with a drastic decrease in testis weight and a reduction of the callosities. Thus, in *P. dacnicolor* males, the seasonal pattern of changes in plasma levels of androgens is closely associated with changes in spermatogenetic activity, testis weight, callosity development, and reproductive behavior.

In terms of relative plasma concentration, testosterone is the predominant androgenic steroid, greatly exceeding that of DHT and androstenedione. On an individual basis, the lowest plasma testosterone level recorded was in a November male and the highest was in a July male in the act of breeding. Likewise, DHT and androstenedione levels vary from a minimum in the autumn to a maximum in the summer. Other sex steroids, like progesterone and estradiol, are present in low concentrations and show little statistical variation during the year. Yearly changes in the plasma concentrations of these two sex steroids can not be correlated with specific phases of the reproductive cycle.

THE FEMALE

Plasma sex hormones manifest profound seasonal changes in the female as well (Iela et al. 1986). Such changes are correlated with those in ovarian composition and weight and oviduct growth. In *P. dacnicolor*, oogonial proliferation occurs throughout the year, but vitellogenesis is restricted to a surprisingly short time during late spring to early summer. A long post-spawning stasis of vitellogenesis occurs, beginning in late summer and lasting until the following spring. From autumn until early spring, the ovary is composed exclusively of previtellogenic follicles. Females are characterized by relatively high plasma levels of testosterone and estradiol and only low concentrations of androstenedione, DHT, and progesterone. Plasma estradiol and testosterone followed a similar seasonal pattern. These two steroids are not only closely correlated with ovarian condition and weight, but also with one another. The yearly minimum is recorded from females with immature, previtellogenic ovaries during the period between post-breeding in late summer and the beginning of spring. During the late spring to early summer, when vitellogenesis takes place, plasma estradiol, testosterone, and the weight of the ovaries and oviducts all increase with remarkable parallelism. Analysis of individual females shows the greater the relative percentage of postvitellogenic follicles within the ovary, the higher the plasma levels of estradiol and testosterone and the ovarian weight. Peak hormone levels, ovarian weight, and oviduct weight occur during actual breeding. Like the plasma levels of estradiol and testosterone, the highest progesterone levels are observed in females about to lay eggs, in amplexus, or actually breeding. The plasma levels of these three sex steroids remain relatively high until about six hours after egg laying and by 24 and 48 hours postovulation drop to low levels. In addition, plasma progesterone levels for all females considered together are positively correlated to ovarian weight. In contrast to plasma estradiol and testosterone, plasma progesterone levels are independent of the relative percentage of postvitellogenic follicles. Plasma levels of androstenedione are lowest in females with previtellogenic gonads and remaining unaltered in females containing less than 20% postvitellogénic follicles. With a much higher percentage of postvitellogenic follicles, the plasma androstenedione levels are significantly higher and remain so during amplexus and egg laying. A significant decline is observed after egg laying. Although, the amplexus/egg laying-related increase in plasma androstenedione level is striking in the female, unlike plasma testosterone, estradiol and progesterone, levels of plasma androstenedione show no relation to ovarian weight or oviduct weight. Plasma DHT is maintained at a relatively low level and does not manifest an ovarian status-related or reproductive activity-related change of note. DHT values represent 5 to 8% of the plasma testosterone levels. As with ovarian weight, there is a positive correlation between oviduct weight and plasma levels of estradiol, testosterone, and progesterone. Plasma androstenedione and DHT did not correlate with oviduct weight. Plasma testosterone levels in the female also reach a zenith during the act of breeding, as in the male. At this time, they tend to become even a little higher than plasma estradiol levels.

The GnRH System

By standard immunohistochemical techniques, evidence of the presence of a characteristic gonadotropin releasing hormone (GnRH) system in the brains of both male and female *P. dacnicolor* has been revealed (Iela et al. 1991). The distribution map of the immunoreactive GnRH elements is extremely simple (Fig. 7.7). With the use of an antiserum against the mammalian GnRH, immunoreactive GnRH perikarya are seen exclusively in the medio-ventral portion of the anterior preoptic area near the anterior limit of the preoptic recess. Immunoreactive GnRH fibers extend anteriorly towards the anterior extension of the diencephalon and posteriorly run below the preoptic recess and the infundibular floor to terminate in the median eminence. A few fibers terminate in the pars nervosa. The pattern

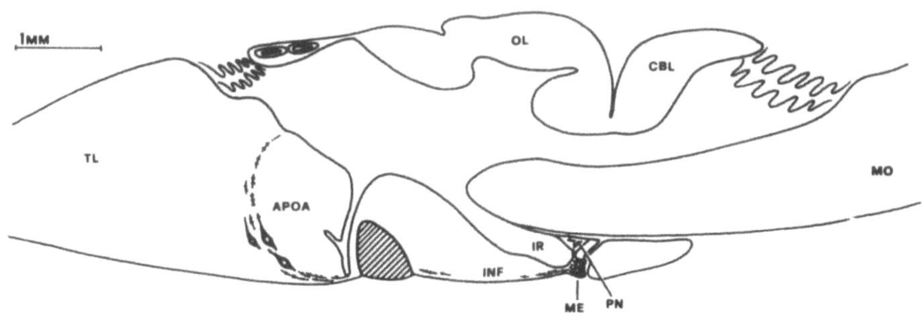

FIGURE 7.7. Schematic illustration of the distribution of immunoreactive GnRH in a mid-sagittal section of the brain. Abbreviations: APOA, anterior preoptic area; CBL, cerebellum; INF, infundibulum; IR, infundibular recess; ME, median eminence; MO, medulla oblongata; OL, optic lobe; PN, pars nervosa; TL, telencephalon. (Iela et al. 1991)

of immunostaining is unchanged with either season or sex. The use of antisera against chicken GnRH-II and salmon GnRH reveals a staining pattern like that of the antimammalian GnRH, indicating the polymorphic nature of the GnRH-like peptides in the brain of this frog.

Additional Consideration

Vitellogenesis and Egg Pigmentation

As is often the case with species that oviposit terrestrially on vegetation, eggs are often green due to the presence of green-pigmented yolk. Presumably, such green coloration affords selective advantage against predation; however, to our knowledge this question has not been studied. The mechanism of green pigmentation and its relationship to vitellogenesis has been studied in *P. dacnicolor* (Marinetti and Bagnara 1983a, 1983b). Incentive for this work was provided by the fact that green pigments are rare among animals, and, indeed, *P. dacnicolor* eggs contain no green pigment but are so colored by the presence of two chemically unrelated pigments, one blue and the other yellow. The blue pigment is an isomer of biliverdin, a hemoglobin derivative, and the yellow is lutein, an acidic xanthophyll of dietary origin.

During vitellogenesis, which occurs rapidly in the spring and continues through the breeding season, biliverdin and lutein are bound to vitellogenin and are then transported to the growing oocytes to impart green coloration (Bagnara 1985). When vitellogenesis is induced by the administration of estradiol, there is an augmentation of biliverdin-bound vitellogenin to such levels that plasma samples are actually blue in color. At the same time, there is a drop in hematocrit, leading to the suggestion that estrogens are involved in hemoglobin metabolism. A similar role for estradiol is suggested in the mobilization of lutein in the liver and its transport to the ovary.

Spermatogenesis

In *P. dacnicolor*, the testis architecture is of the cystic variety, typical of anurans (Rastogi et al. 1988). The primary spermatogonia are distinguished by their large size, 14 μ to 25 μ. They are of two types, distinguished as either light or dark. The former are considered to be stem cells while the latter give rise to the secondary spermatogonia destined to provide spermatocytes for the forthcoming cycle. Spermiogenesis in this species is a fascinating type that deserves further investigation (Rastogi et al. 1988). It generates a spermatozoon composed of an elongated, pointed head-piece and a relatively short motile tail region which consists of a thick axial rod and an axonemal filament connected by a prominent undulating membrane. This is a rather unusual morphology and the question arises as to whether this form is characteristic of the whole sub-family of *Phyllomedusinae*. If so, does this morphology, characterized by the presence of a "corkscrew" like undulating membrane, afford particular advantage in the unique fertilization process of terrestrially deposited eggs?

Conclusions and Perspectives

The Mexican leaf frog, *P. dacnicolor*, is remarkably well-suited for basic research in reproductive endocrinology. It is impossible to find any one species or group that typifies reproductive biology of South American amphibians; however, *P. dacnicolor* is an excellent example of how such an amphibian species copes endocrinologically and behaviorally with the problems it faces in becoming reproductively successful.

P. dacnicolor is an anuran in which seasonal plasma sex hormone levels, gonadal weight changes, and reproductive activity are "coupled." Our sampling adequately indicates that in males, highest levels of testosterone and DHT are maintained through the period when males exhibit reproductive behavior (chorusing, amplexus), whereas androstenedione reaches the highest level only when the male is actually breeding. Likewise, in the female, highest levels of testosterone, estradiol, and progesterone are observed during amplexus and egg laying. Here again, the female shows high levels of testosterone in the plasma, even higher than estradiol during breeding. The seasonal changes in sex hormone levels in both sexes are thus remarkably pronounced, increasing up to 160-fold for testosterone, 50-fold for DHT, and 25-fold for androstenedione in the male and 115-fold for estradiol, 105-fold for testosterone, and 30-fold for progesterone in the female. We lack data on plasma levels of hypothalamic and pituitary hormones (GnRH, FSH, LH) and the pattern of gonadotropin secretion during gonadal recrudescence and during the breeding season. This points out all too well the need for research directed toward understanding reproductive behavior in terms of the integration of nervous and endocrine systems.

Acknowledgments

We wish to express our gratitude to NSF for financial support over many years: We are also grateful to Dr. Forest Morrisett for his devoted work on the developmental stages of *P. dacnicolor*.

References

Bagnara JT (1985) The amphibian egg as a pigment cell. In Bagnara J, Klaus SN, Paul E, Schartl M (eds.). Pigment Cell 1985, Biological, Molecular, and Clinical Aspects of Pigmentation. University of Tokyo Press, Tokyo. pp. 277–281

Bagnara JT (1990) Mexican leaf frog (*Pachymedusa dacnicolor*) as a model in endocrine research. J. Exptl. Zool. 4:145–147

Bagnara JT, lela L, Morrisett F, Rastogi RK (1986) Reproduction in the Mexican leaf frog (*Pachymedusa dacnicolor*) 1. Behavioral and morphological aspects. Occasional Papers, Univ. Kansas Publ. Mus. Nat. Hist. 121:1–31

Iela L, Rastogi RK, Bagnara JT (1991) Reproduction in the Mexican leaf frog, *Pachymedusa dacnicolor*. V. Immunohistochemical localization of gonadotropin-releasing hormone in the brain. Gen. Comp. Endocrinol. 84:129–134

Iela L, Rastogi RK, Delrio G, Bagnara JT (1986) Reproduction in the Mexican leaf frog, *Pachymedusa dacnicolor*. III. The female. Gen. Comp. Endocrinol. 63:381–392

Marinetti GV, Bagnara JT (1983a) Yolk pigments of the Mexican leaf frog. Science 203:410–415

Marinetti GV, Bagnara JT (1983b) Characterization of blue and yellow pigments in eggs of the Mexican leaf frog. Biochemistry 22:5651–5655

Pyburn WF (1970) Breeding behavior of the leaf-frogs *Phyllomedusa callidryas* and *Phyllomedusa dacnicolor* in Mexico. Copeia 2:209–218

Rastogi RK, Bagnara JT, Iela L, Krasovich MA (1988) Reproduction in the Mexican leaf frog, *Pachymedusa dacnicolor*. IV. Spermatogenesis: A light and ultrastructural study. J. Morphol. 197:277–302

Rastogi RK, Iela L, Delrio G, Bagnara JT (1986) Reproduction in the Mexican leaf frog, *Pachymedusa dacnicolor*. II. The male. Gen. Comp. Endocrinol. 62:23–35

Wiewandt TA (1971) Breeding biology of the Mexican leaf frog. Fauna 2:29–34.

8
Reproduction in Caecilians

Marvalee H. Wake

Caecilians (Order Gymnophiona) are limbless, elongate, burrowing or swimming amphibians that inhabit the tropics of most of the world. Their reproductive biology includes several modes. In contrast to other amphibians, probably the majority of caecilians are direct developers (laying eggs on land, with metamorphosis before "hatching," thereby circumventing the larval period) or are viviparous (providing maternal nutrition to the intra-oviductally developing young after the yolk is exhausted and giving birth to fully metamorphosed young). In fact, viviparity is the predominant reproductive mode, with apparently more than half of the species world-wide retaining developing young and nourishing them through a lengthy gestation period (Wake 1977a, 1977b, 1982). However, even such basic information as reproductive mode is not known for a number of the 170 described species and can only be inferred.

Representatives of three of the six families of caecilians occur in South America. The Rhinatrematidae (*Rhinatrema birittata*, *Epicrionops*, eight species) live in northwestern South America from sea level to over 3000 meters in the Andes. They are egg-laying and have free-living larvae that live in streams. In contrast, all members of the family Typhlonectidae, which inhabits South America largely in the Amazon and Orinoco drainages, are viviparous. Most species are secondarily aquatic, though members of some species of the genus *Chthonerpeton* have been collected in moist soil rather far from streams. Typhlonectids include four genera (*Typhlonectes*, three species [Fig. 8.1A]; *Potamotyphlus*, one species, *kaupii*; *Nectocaecilia*, two species; and *Chthonerpeton*, six species; Nussbaum and Wilkinson 1989). Intraoviductal fetuses have been observed in several species and in all genera (see Table 8.1), so I infer that all members of the family are viviparous. Finally, several members of the world-wide family Caeciliaidae occur in South America. Members of the genera *Caecilia*, *Oscaecilia* (Fig. 8.1B), *Microcaecilia*, *Parvicaecilia*, *Siphonops*, *Pseudosiphonops*, *Mimosiphonops*, *Brasilotyphlus*, *Leutkenotyphlus*, and *Dermophis* are terrestrial, burrowing or litter-dwelling animals. Very little is known of the reproductive mode of most of these taxa. Some species of

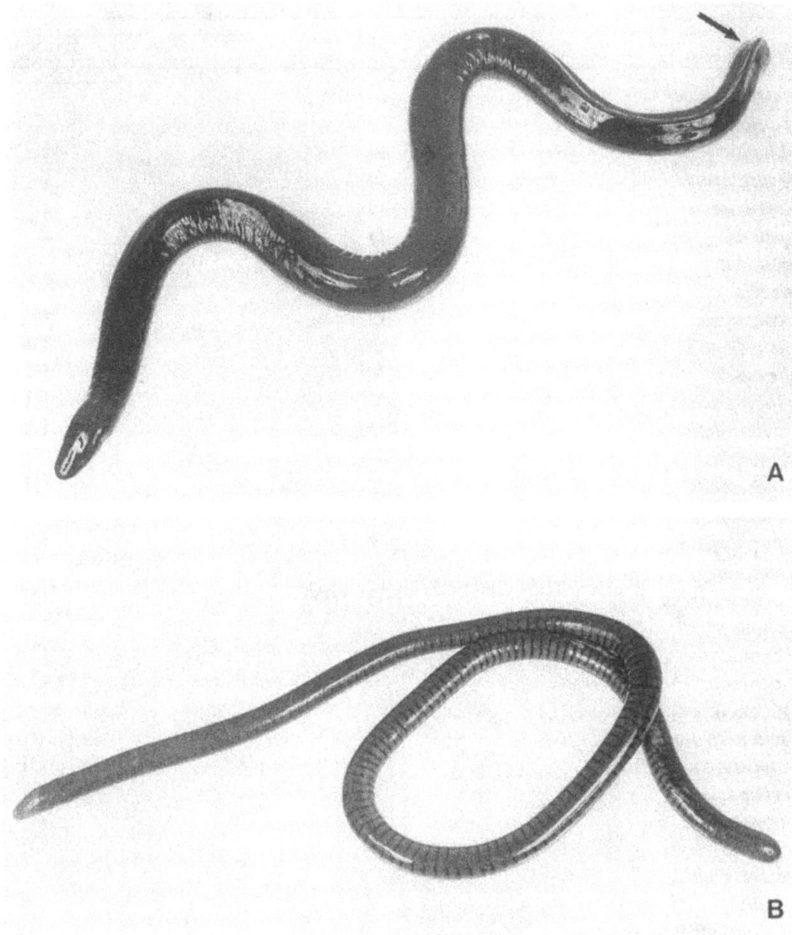

Figure 8.1. A: *Typhlonectes compressicauda* (460 mm total length [TL]). Note the dorsal keel (at the arrow) on this aquatic animal. B: *Oscaecilia ochrocephala* (555 mm TL), a terrestrial, burrowing caecilian.

Siphonops are egg-layers (Goeldi 1899; Gans 1971); *Dermophis* is viviparous (many reports, summarized in Wake 1977a, 1977b, 1980a). Some *Caecilia* may be viviparous and some oviparous, based on the rare collection of apparent newborns that still had a fetal dentition (Taylor 1968), or larvae with gill slits (Tschudi 1839), and on fully yolked ovarian egg diameters (Wake 1977a, unpubl. data). *Oscaecilia* may be a direct developer (Wake unpubl. data).

The limited information available for most aspects of the biology of caecilians is a consequence of several factors: relatively rare collecting of these secretive, usually fossorial, animals until recently; the apparent low

TABLE 8.1. Reproductive modes known for South American caecilians.

Taxon	Mode	Evidence
Rhinatrematidae		
Epicrionops		
bicolor	O	L
marmoratus	O	L
peruvianus	O	L
petersi	O	L
Caeciliaidae		
Caecilia		
corpulenta	O?	OS
occidentalis	O?	OS
pressula	V?	FT
subnigricans	V?	FT
tentaculata	O?	GS
Dermophis		
mexicanus	V	OF
oaxacae	V	OF
parviceps	V	OF
Oscaecilia		
ochrocephala	O? DD?	OS
Siphonops		
annulatus	O	C
paulensis	O	C
Typhlonectidae		
Chthonerpeton		
indistinctum	V	OF
viviparum	V	OF
Nectocaecilia		
petersi	V	OF
Potamotyphlus		
kaupii	V	OF
Typlonectes		
compressicauda	V	OF
natans	V	OF

Key: C = clutch; DD = direct development; FT = fetal teeth; GS = gill slits; L = larvae; O = oviparous; OF = oviductal fetuses; OS = ovum size; V = viviparous

densities in which many populations occur, so that species and even genera are known from only a very few specimens, often only one from a particular location; the difficulty of observing fossorial animals in the field and, to a lesser degree, in the lab; and the difficulty of breeding living animals, though they can be maintained easily in the laboratory. There is not yet a robust hypothesis of the phylogenetic relationships of caecilians (Nussbaum and Wilkinson 1989). It is especially important to recognize that we do not even know, in many cases, the validity of the status of the species that have been described, mostly for the reasons mentioned above. This situation is particularly alarming, for the habitat in tropical forests, grasslands bounded by forests, and streams and stream banks in which caecilians live is rapidly being destroyed. As a result, entire species will be extinct before we have even a limited understanding of their biology. For example, the area near Cochabamba, Bolivia, from which three specimens of a new species (Wake 1984) were collected, is now well known to have been converted to many square kilometers of cocaine poppy fields. It is likely that no biologist will ever see members of this taxon alive. However, certain kinds of habitat disturbance are not detrimental to caecilians and may even provide good conditions for them. For example, in shaded cafetals in which the hulls of coffee beans are discarded in large piles, the mass forms the moist, loose, decaying litter in which caecilians and their prey (earthworms, insect instars, etc.) thrive. Even in such situations, though, efforts should be made to recognize all elements of the flora and fauna and to protect their existence.

The reproductive biology of only two species of caecilians in the world, *Typhlonectes compressicauda* and *Dermophis mexicanus*, is moderately well known to date. The genus *Dermophis*, which is largely Central American, includes one species (*D. parviceps*) that ranges from Costa Rica through northern Colombia. All indications are that the reproductive biology of *D. mexicanus* is generally typical of that of the non-typhlonectid terrestrial viviparous caecilians, so I shall compare the similarities and differences in mode of viviparity based on data for *Typhlonectes* and *Dermophis*. More anecdotal information is available for other taxa, especially the egg laying *Epicrionops*, *Siphonops paulensis* and *S. annulatus*, and the viviparous *Chthonerpeton*.

It is generally accepted that external fertilization, oviparity, and absence of parental care are ancestral conditions of reproductive biology in amphibians (Duellman and Trueb 1986), as well as in other vertebrates. Therefore, all caecilians share two derived states of these three characters (internal fertilization and means of parental care [see below]) and many have a third (direct development or viviparity). Members of the three orders of amphibians have performed many "experiments" in reproductive biology; they differ among the orders. Very few frogs have internal fertilization and only two species are viviparous (Wake 1989); however, they have more than twenty different modes of parental care (Duellman and Trueb 1986). Only primitive salamanders have external fertilization. In most species, internal fertilization is accomplished by the female picking up with the lips of her cloaca a spermatophore deposited by the male during the courtship ritual. Females of a number of species in different families guard their laid clutches until the young hatch. Many species of salamanders (including all of the South American taxa) are direct developers, thus circumventing an aquatic larval period. Only one species of salamander, *Salamandra atra*, is obligately viviparous, though perhaps three more, all salamandrids, are ovoviviparous (developing young maintained in the oviduct, no significant maternal nutrition other than yolk provided, young "born" at diverse stages of development). In contrast, all caecilians practice internal fertilization by means of insertion of the male phallodeum into the vent of the female for sperm transport (nothing is known of the means of mate attraction or recognition). Apparently, all caecilians practice one of two modes of parental care: viviparity, or, in all egg-layers and direct developers observed to date (mostly Old World, but also *S. paulensis* [Gans 1971]), the female remains with the clutch.

Breeding appears to be synchronous and dependent on season in *Dermophis* and *Gymnopis*. In a single population of *Dermophis*, both fetuses and juveniles were of similar sizes at a particular time of the year (Wake 1980a; this occurs in *Gymnopis* from many localities throughout Costa Rica [Wake 1968]). Two to fourteen young are born at the inception of the first rains, apparently when prey abundance increases. The gestation period for the species studied in these genera is approximately 11 months, with fertilization in June to July and birth in May to June. The data indicate that females breed every other year, spending the year after giving birth yolking the next clutch (Wake 1980a). However, some aquatic *Typhlonectes* may breed throughout the year, for pregnant females appear in shipments received at virtually any time of year. Exbrayat (1983, 1986) and Exbrayat et al. (1981, 1982) report that the gestation period is six to seven months, with 2–11 young born, and that females also reproduce every other year. I infer that gestation for *C. indistinctum* may be six months, based on Barrio's (1969) observations. Exbrayat et al. (1983) comment that copulation in *T. compressicauda* occurs from January through April, with birth July through September. There are few data for oviparous species. My sample of larvae of *Epicrionops* collected at the same time contains specimens of very different sizes (66 mm TL to 212 mm TL) and states of development, suggesting either different rates of development or, more likely, reproduction over much of the year.

Data on oviductal egg diameters of South American species (Gana 1961; Wake 1977a, unpubl.) suggest that the egg diameters of non-viviparous species are larger than those of viviparous taxa of all caecilians (4 to 5 mm dia meter vs. 1 to 2 mm dia meter). Clutch sizes are known for only a few taxa, including South American species, and are summarized by Wake (1977a). Some species apparently have no more than six (*S. annulatus*, Goeldi 1899 and *S. paulensis*, Gans 1961) while others may have up to 14, with much variation among females within the same species (Wake 1977a). It is not known how long the developmental period to hatching or the length of the larval stage is for any New World egg-laying species, though the latter may be approximately one year, based on inference from Old World taxa. The free-living larvae of all species observed have exhausted their yolk and actively forage, based on my analysis of gut contents. They have the tooth structure and arrangement of adults, though numbers are fewer.

The viviparous species, with their lengthy gestation periods and provision of nutrient material after the yolk is exhausted, reflect a very different mode of maternal investment. In *Dermophis mexicanus*, the yolk is exhausted approximately three months into the gestation period. At approximately that time, crowns of the fetal teeth and the articular components of the lower jaw mineralize, adductor and depressor musculature differentiates (well in advance of any other head muscles), and the epithelium of the oviduct proliferates, becoming hyperemic and secretory (Wake 1980b; Wake and Hanken 1982). The morphology of the pregnant and non-pregnant oviducts has been described by Welsch et al. (1977) for *Chthonerpeton indistinctum*, by Wake (1970, 1982) for *Gymnopis multiplicata* and *Dermophis mexicanus*, and by Exbrayat et al. (1983) for *Typhlonectes compressicauda*. The secretion (and epithelial cell contents) early in pregnancy are rich in free amino acids and carbohydrates; in mid-pregnancy conjugated amino acids and carbohydrates predominate; some two months before birth, the epithelium, which is now encrypted, apparently due to the activity of the fetuses, produces an extremely lipid-rich secretion (Welsch et al. 1977; Wake unpubl. data). The fetuses clearly ingest this material orally, for it fills the mouth and pharynx of a number of specimens that have been sectioned (Wake unpubl.). Fetuses have a dentition that is characterized by tooth crown shape and distribution that is very different from that of the adult. It is lost at birth and replaced by the adult configuration and is apparently used in obtaining the nutrient material supplied by the mother and perhaps to stimulate secretion of that material by mechanically stimulating cell turnover (Wake 1976, 1977a, 1977b, 1980b).

A major difference in morphology exists between embryos and fetuses of typhlonectids and those of other viviparous (and oviparous) species. The gills of non-typhlonectids, whether of free-living larvae or of intra-oviductal embryos, are paired, triramous, fimbriated structures that are lost well before metamorphosis or birth (Wake 1967, 1977a). The typhlonectids, however, have gills that are paired, large sac-like structures. They are highly vascular, and the supply is from the same three aortic arches as in the triramous gills (Wake unpubl.). Delsol et al. (1981, 1983, 1986) contend that these gills function as pseudoplacentae and that gaseous exchange and possibly nutrient uptake occurs across the epithelial membrane into the circulation. They describe in detail the structure of the gills as an ectotrophoblast.

Virtually nothing is known of the reproductive endocrinology of caecilians. Large corpora lutea are maintained during the entire pregnancy in the viviparous forms studied (Wake 1968; Exbrayat 1983). Circulating progesterone levels at mid- and late pregnancy in *Dermophis mexicanus* are twice those of non-pregnant females (Wake unpubl.). However, Wake (1968) noted that corpora lutea are present in the ovaries of oviparous taxa as well. Nothing is known of hormone levels in those species. Morphology of the ovarian cycle of *G. multiplicata* was described by Wake (1968; that of *Dermophis* is the same [Wake unpubl.]), and of *T. compressicauda* by Exbrayat (1983).

Male caecilians appear to be actively spermatogenic throughout eleven of the twelve months of the year (Wake 1968), whether or not reproduction is synchronous and seasonal. The multi-lobed testes contain lobules that have cell nests that are in diverse spermatogenic stages during most of the cycle, so that there is apparently a supply of mature sperm during much of the year. The situation is similar in *T. compressicauda* (Exbrayat 1986) and *C. indistinctum* (de Sa and Berois 1986). There is no evidence for spermathecae or sperm storage in any females. Males have Müllerian glands, derived from the embryonic tissue that gives rise to the oviducts of females. The posterior several millimeters that join the cloaca are large, tubular, glandular organs that are secretory during active spermatogenesis and regress during testis regression (Wake 1970, 1972, 1981 for many taxa; corroborated by Exbrayat [1985] for *T. compressicauda* and by de Sa and Berois [1986] for *C. indistinctum*). Wake (1981) suggested that this phenomenon is correlated with increased terrestriality and internal fertilization, and that the Müllerian glands are homologues of the prostate (which incorporates in its central portion Müllerian duct components in turtles and mammals).

Parturition has been observed only in the aquatic viviparous taxa. Heinroth (1915) observed birth in *T. natans*; recently, there have been several reports of reproduction in the laboratory (Barrio 1969; Murphy et al. 1977; Exbrayat and Laurent 1983; Exbrayat et al. 1983; Billo et al. 1985), some including observations of copulation. At birth, young emerge tail first; the sac-like gills are often retained, but lost soon after birth.

Much remains to be learned about the reproductive biology of these fascinating amphibians. This account summarizes what is known of their biology; it reveals the many areas that remain to be investigated. It is important that workers get into the field in the tropics to observe these animals in their native habitats, as long as those habitats survive. It is equally important to collect them and work with them in the laboratory. Caecilians clearly are experimenting with terrestriality, internal fertilization, and viviparity. Better understanding of the evolution of their reproductive biology likely will yield major clues to the evolution of viviparity and the import of selection for it in diverse fish, amphibians, and amniote taxa.

Acknowledgements

I thank the National Science Foundation for funding my research on the biology of caecilians. I appreciate the opportunity to present this information in a volume that considers the reproductive biology of South American vertebrates and the need for conservation of these animals.

References

Barrio A (1969) Observaciones sobre *Chthonerpeton indistinctum* (Gymnophiona, Caecilidae) y su reproduction. Physis 28:499–503

Billo RR, Straub JO, Senn DG (1985) Vivipare Apoda (Amphibia: Gymnophiona), *Typhlonectes compressicaudus* (Dumeril and Bibron, 1841): Kopulation, Tragzeit, und Geburt. Amphibia-Reptilia 6:1–9

Delsol M, Flatin J, Exbrayat J-M, Bons J (1981) Développement de *Typhlonectes compressicaudus*, Amphibien Apode vivipare. Hypothèses sur sa nutrition embryonnaire et larvaire par un Ectotrophoblaste. Comptes Rendus Academie des Science, Paris 293:281–285

Delsol M, Flatin J, Exbrayat J-M, Bons J (1983) Développement embryonniare de *Typhlonectes compressicaudus*, (Dumeril et Bibron, 1841). Constitution d'un Ectotrophoblaste. Bulletin Société Zoologique de France 108:680–681

Delsol M, Exbrayat J-M, Flatin J, Gueydan-Baconnier M (1986) Nutrition embryonnaire chez *Typhlonectes compressicaudus* (Duméril et Bibron, 1841), Amphibien Apode vivipare. Memoirs Societe Zoologique de France 43:39–54

De Sa R, Berois N (1986) Spermatogenesis and histology of the testes of the caecilian *Chthonerpeton indistinctum*. Journal of Herpetology 20:510–514

Duellman WE, Trueb L (1986) Biology of Amphibians. McGraw Hill, New York

Exbrayat J-M (1983) Premieres observations sur le cycle annuel de l'ovaire de *Typhlonectes compressicaudus* (Duméril et Bibron, 1841), Batracien Apode vivipare. Comptes Rendus Academie des Science, Paris 296:493–498

Exbrayat J-M (1985) Cycle des canaux de Müller chez le male adulte de *Typhlonectes compressicaudus* (Duméril et Bibron, 1841), Amphibien Apode. Comptes Rendus Academie des Science, Paris 301:507–512

Exbrayat J-M (1986) Le testicule de *Typhlonectes compressicaudus*: Structure, ultrastructure, croissance et cycle de reproduction. Memoir Societe Zoologique de France 43 :121–132

Exbrayat J-M, Delsol M, Flatin J (1981) Premieres remarques sur la gestation chez *Typhlonectes compressicaudus* (Duméril et Bibron, 1841), Amphibien Apode vivipare. Comptes Rendus Academie des Science, Paris 292: 417–420

Exbrayat J-M, Delsol M, Flatin J (1982) Observations concernant la gestation de *Typhlonectes compressicaudus* (Duméril et Bibron, 1841), Amphibien Apode vivipare. Bulletin Société Zoologique de France 107:486

Exbrayat J-M, Delsol M, Lescure J (1983) La viviparité chez *Typhlonectes compressicaudus*, amphibien apode. Bulletin Société Herpetologique de France 26:23–24

Exbrayat J-M, Laurent M-T (1983) Quelques observations concernant le maintien en elevage de deux amphibiens apodes: *Typhlonectes compressicaudus* et *Ichthyophis*. Reproduction de *Typhlonectes compressicaudus*. Bulletin Société Herpetologique de France 26:25–26

Gans C (1961) The first record of egg-laying in *Siphonops paulensis* Boettger. Copeia 1961:490–491

Goeldi EA (1899) Uber die Entwicklung von *Siphonops annulatus*. Zoologisches Jahrbuch Systematik 12:170–173

Heinroth O (1915) Geburt von *Typhlonectes natans* (Blindwuhle) im Aquarium. Blatter fur Aquarien- und Terrarienkunde 26:34–35

Murphy JB, Quinn H, Campbell JA (1977) Observations on the breeding habits of the aquatic caecilian *Typhlonectes compressicaudus*. Copeia 1977:66–69

Nussbaum RA, Wilkinson M (1989) On the classification and phylogeny of caecilians (Amphibia: Gymnophiona): A critical review. Herpetological Monographs 2:1–42

Taylor EH (1968) The Caecilians of the World: A Taxonomic Review. University of Kansas Press, Lawrence

Tschudi JJ (1839) Classification der Batrachier. Memoirs Societe des Science naturelle de Neuchatel 2:1–100

Wake MH (1967) Gill structure in the caecilian genus *Gymnopis*. Bulletin of the Southern California Academy of Sciences 66:109–116

Wake MH (1968) Evolutionary morphology of the caecilian urogenital system. Part I. The gonads and fat bodies. Journal of Morphology 126:291–332

Wake MH (1970) Evolutionary morphology of the caecilian urogenital system. Part II. The kidneys and urogenital ducts. Acta Anatomica 75:321–358

Wake MH (1972) Evolutionary morphology of the caecilian urogenital system. Part IV. The cloaca. Journal of Morphology 136:353–366

Wake MH (1976) The development and replacement of teeth in viviparous caecilians. Journal of Morphology 148:33–64

Wake MH (1977a) The reproductive biology of caecilians: An evolutionary perspective. In Taylor DH, Guttman SI (eds.). The Reproductive Biology of Amphibians. Plenum, New York. pp. 73–101

Wake MH (1977b) Fetal maintenance and its evolutionary significance in the Amphibia: Gymnophiona. Journal of Herpetology 11:379–386

Wake MH (1980a) Reproduction, growth, and population structure of *Dermophis mexicanus* (Amphibia: Gymnophiona). Herpetologica 36:244–256

Wake MH (1980b) Fetal tooth development and adult replacement in *Dermophis mexicanus* (Amphibia: Gymnophiona): Fields versus clones. Journal of Morphology 166:203–216

Wake MH (1981) Structure and function of the male Müllerian gland in caecilians (Amphibia: Gymnophiona), with comments on its evolutionary significance. Journal of Herpetology 15:17–22

Wake MH (1982) Diversity within a framework of constraints: Reproductive modes in the Amphibia. In Mossakowski D, Roth G (eds.). Environmental Adaptation and Evolution, A Theoretical and Empirical Approach. Gustav Fischer Verlag, Stuttgart. pp. 87–106

Wake MH (1984) A new species of *Caecilia* (Amphibia: Gymnophiona) from Bolivia. Amphibia-Reptilia 5:215–220

Wake MH (1989) Phylogenesis of direct development and viviparity. In Wake DB, Roth G (eds.). Complex Organismal Functions: Integration and Evolution in Vertebrates. John Wiley & Sons, Chichester England pp. 235–250

Wake MH, Hanken J (1982) The development of the skull of *Dermophis mexicanus* (Amphibia: Gymnophiona), with comments on skull kinesis and amphibian relationships. Journal of Morphology 173:203–223

Welsch U, Müller M, Schubert C (1977) Elektronenmikroskopische und histochemische Beobachtungen zur Fortpflanzungsbiologie viviparer Gymnophionen (*Chthonerpeton indistinctum*). Zoologische Jahrbuch von Anatomie 97:532–549

Reptilia

9

Follicular Population Dynamics: Its Relation to Clutch and Litter Size in Chilean *Liolaemus* Lizards

Víctor Leyton and José Valencia

Among reptiles, the Iguanidae is one of the most diversified families of lizards, containing a large number of species and a variety of body sizes ranging from small to large. Most are terrestrial and many are scansorial.

Reptilian reproduction has high variability to match environmental conditions. Many iguanids of both temperate and tropical climates have well-defined breeding seasons; others may breed year-round while some reproduce once each season (Fitch 1970; Guillette and Casas-Andres 1980; Vitt and Goldberg 1983).

Beyond the tropics, lizards tend to become reproductively active when the temperature and number of daylight hours increase. The physiological mechanisms by which environmental changes stimulate the gonads and initiate the breeding cycle is not well understood, but the stimulation of the hypothalamus and pituitary axis may be involved.

Resource availability or its correlates, rainfall and temperature, are usually identified as determinants of reproductive tactics variability (Vitt 1982).

Most lizards are oviparous, but viviparity has been attained independently by different genera, seemingly in reponse to cold climates, low moisture, and body size (Porter 1985). Approximately 15% of extant reptiles exhibit viviparity (Tinkle and Gibbons 1977).

There are also large differences of age at maturity, body size of breeding adults, degree of parental care of the eggs, and clutch or litter size. Whereas one or two eggs may be laid by gekkonids, some lizards produce more than twenty eggs per clutch and some produce more than one clutch per season (Tinkle et al. 1970). Nevertheless, both the number of eggs per clutch and the number of clutches per season per female vary from species to species and from population to population.

The interacting anatomical and physiological factors that modulate the number of ovulations from each ovary and clutch or litter size in reptiles include: (1) the number of germinal beds per ovary; (2) the number of growing and atretic follicles during the period of ovarian recrudescence; (3) the number of vitellogenic follicles undergoing synchronous growth and

TABLE 9.1. Composition of reptilian fauna in Chile.*

Family	Genus	Species and subspecies	Endemics
Gekkonidae	*Garthia*	3	3
	Phyllodactylus	3	1
Iguanidae	*Ceiolaemus*	1	1
	Centrura	1	—
	Ctenoblepharis	3	2
	Diplolaemus	3	—
	Liolaemus	59	42
	Phrynosaura	2	2
	Prystidactylus	4	4
	Tropidurus	6	5
	Velosaura	2	2
Teiidae	*Callopistes*	3	3
Colubridae	*Alsophis*	1	1
	Philodryas	4	4
	Tachymenis	2	2
	Total	97	72

*From Veloso and Navarro 1988.

reaching maturity; and (4) the age and body size of adult females (Jones et al. 1982).

The Chilean *Liolaemus* lizards are the most common (Table 9.1) and diverse (Veloso and Navarro 1988) in body and clutch size and reproductive modes. To understand how the clutch and litter size is generated, we have investigated (1) ovarian organization, (2) follicular population at different stages of the ovarian cycle by means of serial sections of the ovaries, (3) the number of growing and atretic follicles in relation to body size classes or age, and (4) the incidence of follicular atresia and body size on clutch and litter size. Comparisons of ovarian cycles of lizard species inhabiting lowlands and mountains were made.

Specimens (*Liolaemus gravenhorsti*, *L. lemniscatus*, and *L. tenuis*) used for this study were obtained near Santiago, Chile, (latitude 33°S) at 700 m elevation and others (*Liolaemus altissimus*, *L. leopardinus* and *L. nigroviridis*) from mountains at 2500 m. Lizards were measured, and ovaries and oviducts removed. For the ovaries, the number and diameter of non-vitellogenic and vitellogenic follicles during the reproductive period and the number of corpora lutea and oviducal eggs were recorded. Ovaries were fixed in Bouin's solution, embedded in paraffin, serially sectioned at 7 μm, and stained with hematoxylin-eosin-alcian blue mixture.

For our data set, four types of follicles were defined according to histological criteria–viz primordial, non-vitellogenic, vitellogenic, and atretic. Females were considered reproductive if the ovaries contained enlarged vitellogenic follicles, oviducal eggs, or embryos, and non-reproductive

Table 9.2. Lizard species used for reproductive analysis.

Species	N	Collection site (altitude)	Reproductive condition
Liolaemus gravenhorsti	28	Santiago, Chile	Non-vitellogenic
	30	(700 m)	Vitellogenic
	30		Egg retention
Liolaemus lemniscatus	14	Santiago, Chile	Non-vitellogenic
	10	(700 m)	Vitellogenic
	4		Egg retention
Liolaemus tenuis	36	Santiago, Chile	Non-vitellogenic
	41	(700 m)	Vitellogenic
	24		Egg retention
Liolaemus altissimus	15	Farellones, Chile	Non-vitellogenic
	10	(3000 m)	Vitellogenic
Liolaemus leopardinus	10	Farellones, Chile	Non-vitellogenic
	4	(3000 m)	Vitellogenic
Liolaemus nigrviridis	7	Farellones, Chile	Non-vitellogenic
	6	(3000 m)	Vitellogenic

when small non-vitellogenic follicles and corpora lutea were found. The number of individuals sampled, locality, and reproductive condition are shown in Table 9.2.

To estimate clutch size, we counted only ovulatory follicles and, to determine litter size, the number of embryos or eggs retained in the oviducts were considered. Minimum body size (snout-vent length, SVL) at sexual maturity is the SVL of the smallest female found containing oviducal eggs or embryos. To evaluate the relationship of clutch and litter size to SVL, females were classified into size categories, with intervals of 5 mm, from sexual maturity onwards. Serially sectioned ovaries for each body size category were used to study the incidence of follicular atresia.

Ovarian Morphology and Follicular Histology

Ovaries are ovoid structures covered by a thin epithelium and having numerous follicles and corpora lutea embedded in a scarce connective stroma. Two areas of embryonic-like tissue or germinal beds are present on both sides of the mesovarium and contain oogonia, epithelial cells, and naked oocytes at different stages of meiosis. When primary oocytes emerge from the germinal beds, they become primordial follicles (Fig. 9.1A and B). As the oocytes grow, they are surrounded by multilayered follicular cells and the theca envelope with blood vessels (Fig. 9.1C and D). When there is yolk deposition in oocytes, the follicles become vitellogenic

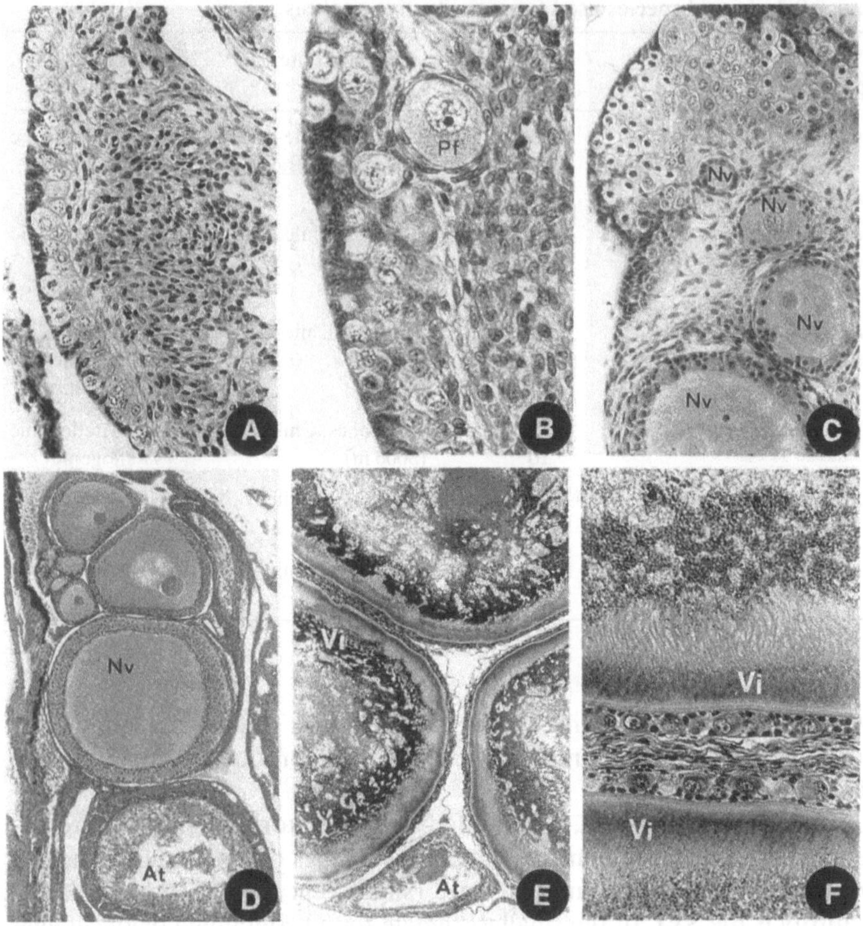

FIGURE 9.1. Ovarian microscopic morphology of *Liolaemus* lizards. A. Germinal bed; B. Primordial follicle, Pf, emerging from germinal bed; C. Non-vitellogenic follicles, Nv; D. Two atretic, At, and three non-vitellogenic follicles; E. Vitellogenic, V, and atretic follicles; F. Follicular envelope during vitellogenesis.

(Figure. 9.1E and F). Follicles at different stages of involution throughout the ovarian cycle were considered atretic (Fig. 1D and E).

The follicular components of the lizard ovary undergo extensive changes of size; the average primary oocyte diameter is 23.6 ($\pm$ 0.76) μm, the primordial follicle 33.7 ($\pm$ 1.9) μm, the non-vitellogenic 272.7 ($\pm$ 37) μm, and the vitellogenic 2 to 7 mm for *L. gravenhorsti*. Naked oocytes of the germinal beds are enclosed by granulosa and theca cells when they incorporate in the ovarian stroma and begin to grow. A zona pellucida is present in multilayered follicles and is clearly thicker during vitellogenesis.

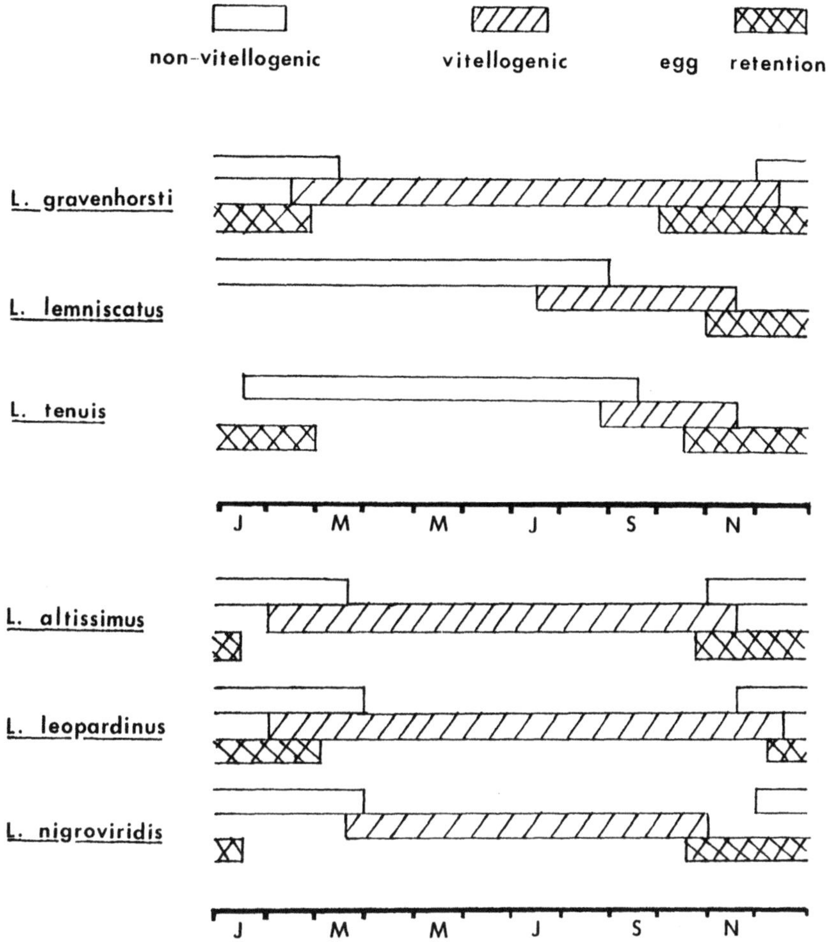

FIGURE 9.2. Population duration of the ovarian cycle of *Lioaemus* lizards.

Ovarian Cycle

Ovaries from post-hatching or post-partum females contain corpora lutea, small non-vitellogenic follicles below 1.5 mm in diameter, and atretic follicles; their abundance is species specific. Meiotic activity of the germinal beds and early follicular organization can be found at any stage of the ovarian cycle. Yolk deposition occurs from late summer to midspring for nearly all species studied, except for *L. lemniscatus* and *L. tenuis* in which vitellogenesis is a spring event (Fig. 9.2.). Nevertheless, each species population females are not synchronous to initiate or culminate vitellogenesis, begin pregnancy, retain eggs, hatching or partum. The duration of the reproductive cycle was estimated by considering the first and last sampled individual containing vitellogenic follicles, eggs inside the oviducts, and the presence of resorbed corpora lutea after egg retention or pregnancy. Events related to the ovarian cycle can be studied macroscopically and microscopically. Thus, during the resting stage, the ovaries contain translucent and small non-vitellogenic follicles and involuted corpora lutea. When vitellogenic egg retention occurs, their total number is discernable under low magnification and also the degree of embryo development after dissecting eggs.

Breeding during the spring and summer seasons is the rule for all the species here studied and gestation is about 60 days before hatching or birth. The average number of eggs produced varies among *Liolaemus*, ranging from 3.2 (±0.4) in *L. leopardinus*, 4.7 (± 0.8) in *L. altissimus*, 5.3 (± 1.5) in *L. tenuis*, 5.7 (± 2.8) in *L. nigroviridis*, 5.9 (± 1.3) in *L. lemniscatus*, and up to 7.3 ± 2.2 in *L. gravenhorsti*.

Body Size at Maturity and Reproductive Mode

Mountain species females have larger SVL than lowland species, but reach maturity at similar body sizes of their life, except in *L. nigroviridis* in which breeding is attained later. SVL at maturity for *L. gravenhorsti* is 46 mm, for *L. lemniscatus* 44 mm, for *L. tenuis* 48 mm, for *L. altissimus* 62 mm, for *L. leopardinus* 68 mm and for *L. nigroviridis* 58 mm. Reproductive mode has no strict correlation with habitat preference.

Follicular Population and Clutch Size

During the ovarian cycle, the average follicular count of the ovaries varies from 28 (±3.5) in *L. tenuis*, 31 (± 2.4) in *L. lemniscatus*, 36 (± 2.3) in *L. altissimus*, 36 (±3.7) in *L. leopardinus*, 37 (± 1.6) in *L. nigroviridis* up to 51 (± 3.6) in *L. gravenhorsti*. There is a tendency to maintain the total counts at three stages of the ovarian cycle for all the animals sampled.

Non-vitellogenic follicles predominate throughout most of the annual cycle and are the main component of the ovary. Primordial follicles are a reserve population. In *L. gravenhorsti* and *L. lemniscatus*, their renewal is not seasonal. For other species, a new generation of follicles may be produced in some seasons such as summer and autumn-winter (see Fig. 9.2). Vitellogenesis approaches 15 to 20 percent of the total follicular population, without differences between lowland and mountain species.

Clutch size corresponds to the total number of ovulated eggs or individuals emerging from hatching or pregnancy. If we compare macroscopically the follicular population and vitellogenic follicles with the clutch size or number of developing embryos in the oviducts, a decrease in the final product is found for *L. lemniscatus* and *L. nigroviridis* from 30 to 3 and from 35 to 3, respectively.

Table 9.3. Percentage of follicular atresia per ovary in three stages of the ovarian cycle.

Species	Non-vitellogenic	Vitellogenic	Egg retention
L. gravenhorsti	21.1	21.6	34.6
L. lemniscatus	0.2	7.8	19.4
L. tenuis	11.7	10.0	15.6
L. altissimus	30.6	7.3	44.8
L. leopardinus	27.3	5.6	41.5
L. nigroviridis	11.8	(*)	19.4

(*) No data available.

Table 9.4. Percentage of follicular atresia related to total follicular population in immature and mature lizards.

Species	Immature	Mature
L. gravenhorsti	1.3	22.4
L. lemniscatus	5.6	6.7
L. tenuis	1.9	12.5
L. altissimus	18.8	40.6
L. leopardinus	(*)	35.3
L. nigroviridis	(*)	13.4

(*) Reduced sample size.

Follicular Atresia

Atresia of follicles can be accurately studied from histological sections for any stage of the reproductive cycle. Follicular atresia of lowland species is lower in non-vitellogenic and vitellogenic females and clearly higher in pregnant females. In mountain species, atresia diminishes during vitellogenesis and then is strongly elevated during pregnancy (Table 9.3).

When follicular atresia is compared between immature and mature animals, it is found that, in young lowland lizards, atresia is lower than for adults at any stage of their cycle; values are higher for mountain species (Table 9.4).

Clutch, Litter, and Body Size

An increase of clutch or litter size is observed at all species studied according to their body size increase from small to large lizards, as shown in Figure 9.3. A higher mean clutch size is found in the lowland group, whereas for the mountain group its clutch is low. Increased snout-vent length in *L. gravenhorsti* is related to litter size, ranging from 5 ± 1 embryos at

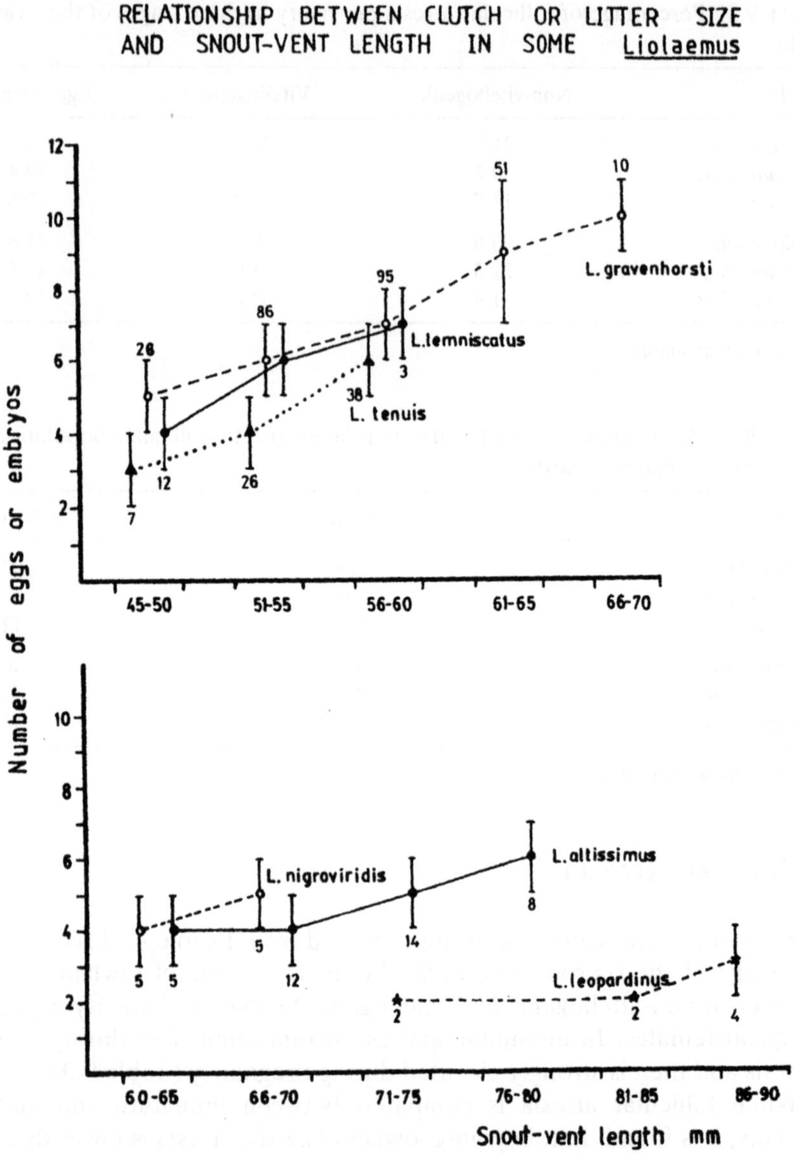

FIGURE 9.3. Clutch and litter in *Liolaemus*. Bars indicate standard deviation and numbers above and below are number of animals sampled. Symbols; open circles, triangles and black circles are for each species. Gaps between numbers are size classes.

maturity to 10 ± 1 in largest females. The increase in the mean number of eggs or embryos doesn't increase in proportion to the body size of *L. altissimus* and *L. leopardinus*. In all cases the variability of egg or embryo number is similar, in spite of conspicuous differences in body sizes.

Conclusion

From our results we can conclude that the reproductive process of six species of the genus *Liolaemus* is markedly seasonal. All these species reproduce during spring and early summer for both valley and mountain populations. Females of these lizard species do not reproduce synchronically. It was found that larger females of each species reproduce early in the season and smaller ones reproduce near the end of season. By the end of summer, all postgravid females were in the resting stage of the ovarian cycle. In Santiago, Chile, both photoperiod and temperature allow lizards a short breeding season, and females will regularly produce only a single clutch or litter per year. Tropical species have an extended breeding season and will produce several clutches per year. Vitt (1982) states than, during the last 20 years, a diversity of reproductive tactics among lizard species have been identified; these are responses to resource availability or its correlates, rainfall and temperature. During the resting stage there is slow follicular growth, that Tinkle and Irwin (1965) attribute adaptative value and is a process of fat accumulation for the next reproductive season. The ovarian recrudescence occurs during the fall and winter, and the non-vitellogenic follicles change to vitellogenic for viviparous species. For oviparous and ovoviviparous lizards, yolk accumulation occurs during late winter or early spring (Guillette 1981).

In lizard ovaries, the first follicle cells originate from somatic cells of germinal beds that surround a primary oocyte. This follicle moves into the ovarian stroma and begins to grow, becoming multilayered. Some non-vitellogenic follicles accumulate yolk into the oocyte of vitellogenic follicles and grow to a critical size as ovulation approaches. Then, during egg retention, a distinct corpus luteum develops (Leyton et al. 1980).

Since viviparity is considered an adaptation to life at high latitudes and elevations, the proportion of viviparous species should be the lowest at low latitudes (Guillette and Gasas-Andreu 1980). Nevertheless, in South American dry temperate areas and mid-latitudes, oviparous and ovoviviparous species coexist. Thus, as was hypothesized by Tinkle Gibbons (1977), egg retention has evolved in relation to the selective advantage of unpredictable environments so that eggs would be laid under optimal environmental conditions for the newborn.

Microscopical analysis of the ovary provided the opportunity to determine the number of germinal beds, the total follicular population, and morphological classification. These data were useful in quantitative studies of the follicular population during female growth and reproductive condition.

Microscopical analysis of lizard ovaries (Jones et al. 1982) have shown that the number of germinal beds varies with age or reproductive condition among individuals of a given lizard species. Some species have one, two, or several small and scattered germinal beds. This number is proportional to the amount of ovulated eggs and growing and atretic follicles (Jones 1978a,b). In *Sceloporus undulatus*, *S. jarrovi*, and *S. occidentalis*, the preovulatory ovaries contain about 34 follicles, and ovaries of *Xantusia vigilis* contain about 53 follicles. All the species sampled in this work have two germinal beds per ovary and total follicular population counts ranging from 28 to 51.

Serial counts of follicles in different stages of the ovarian cycle in the six species of *Liolaemus* lizards studied show that follicular population is stable and that the total number of follicles is species specific. Nevertheless, there is no correlation between the total follicular population and the lizard reproductive mode. The high elevation species include both oviparous and viviparous species with an average total follicular population of 37. This figure is similar to the average found for low elevation species, but this group included a viviparous species with an average higher than all other oviparous and ovoviviparous species.

The analysis of the average distribution for categories of follicles reveals the following: (a) there is a large variability in the percentage of both primordial and non-vitellogenic follicles without correlative reproductive mode or habitat elevation; (b) the percentage of vitellogenic follicles is slightly higher in viviparous than in oviparous species; and (c) there is a higher count of atretic follicles in the viviparous, oviparous, and ovoviviparous species and the highest increase occurs during egg retention.

In our work, the ovaries of all lizards studies showed a high percentage of non-vitellogenic follicles (about 66 to 84%) and a low percentage of vitellogenic or preovulatory follicles (between 11 and 20%). Mean follicular atresia of Chilean species fluctuates from 7 to 27 percent, is clearly elevated during egg retention, and has a lower incidence during the vitellogenic and non-vitellogenic phase. These ranges of atresia are similar to those described in *Sceloporus undulatus* and in *Anolis carolinensis* and *Hemydactilus frenatus* (Jones et al. 1973 and Jones et al. 1976), where the mean value is about 13 to 16 percent. The number of eggs per clutch and embryos varies with the size and age of the female; smaller species generally have smaller clutches, and viviparous species produce fewer embryos than similarly sized and closely related oviparous species. In spite of its body size, *L. leopardinus* produces only one to two embryonated eggs.

The increase in the clutch and litter size in *Liolaemus* lizards can be explained by a progressive reduction of the follicular atresia as the females grow and a higher abdominal capacity to carry eggs and embryos. Comparisons between the number of ovulatory follicles as indicators of clutch size and the number of embryonated eggs as indicators of litter size show differences between 50 to 100 percent, and the error in predicting the litter size using only the number of vitellogenic follicles is mainly due to the high degree of egg losses by some species.

Habitat environmental variables can regulate the reproductive performance because they favor the survival of the young. Seasonal reproduction is affected by changes in food supply, temperature, and photoperiod which, in turn, influence endocrine organs (Licht 1972). The main research questions are related to control of the ovarian cycle, number of follicles reaching maturity during clutch intervals, clutch size, and role of atretic follicles.

Follicular and oocyte growth depend on estrogen-induced yolk protein synthesis by the liver (Jones 1975; Tata 1976). Follicular stimulation hormone controls the transport and delivery of the yolk protein to the ovary. The synthesis of vitellogenin by turtle hepatocytes is inhibited by prolactin and progesterone (Callard and Lance 1977). In this way, cyclical follicular growth may be retarded or suppressed by progesterone secretion, as is the case when fully developed corpora lutea occur in pregnancy or egg retention.

If a refractory period occurs between successive ovulations, the ovarian recrudescence can not be induced due to the absence of gonadotropin receptors. Crews and Licht (1974) hypothesized the secretion of a factor by atretic follicles in *Anolis carolinensis* which can be correlated to the lack of follicular vitellinization and growth stimulation. The response of oviparous and viviparous species can be different depending on elevation and latitude.

Acknowledgements

The authors thank Susana Domínguez and Francisco Cortés for their kind technical assistance. This work was partially supported by project B-3190-9013, University of Chile. Dr. Hamlett made useful suggestions to an earlier version of the manuscript.

References

Callard I, Lance V (1977) The control of reptilian follicular cycles. In Calaby JH, Tyndale-Biscoe CH (eds.). Reproduction and Evolution. Australian Academy of Sciences, Canberra. pp. 199–209

Crews, D, Licht L (1974) Inhibition by corpora atretica of ovarian sensitivity to environmental and hormonal stimulation in the lizard *Anolis carolinensis*. Endocrinology 95:102–106

Fitch HS (1970) Reproductive cycles in lizards and snakes. Misc. Publ. Mus. Natur. Hist., Univ. Kans. 52:1–247

Guillette LJ (1981) On the occurrence of oviparous and viviparous forms of the Mexican lizard *Sceloporus aeneus*. Herpetologica 37:11–15

Guillette LJ Jr Casas-Andreu G (1980) Fall reproductive activity in the high altitude Mexican lizard, *Sceloporus microlepidtus*. J. Herpetol. 14:143–147

Jones RE (1978a) Control of follicular selection. In Jones RE (ed.). The Vertebrate Ovary. Comparative Biology and Evolution. Plenum Publ. Corp., New York pp. 763–787

Jones R, Roth J, Gerrard A, Kiely R (1973) Endocrine control of clutch size in reptiles. I. Effects of FSH on ovarian follicular size-gradation in *Leiopisma laterale* and *Anolis carolinensis*. Gen. Comp. Endocrinol. 20:550–555

Jones RE, Swain T, Guillette L Jr, Fitzgerald RT (1982) The comparative anatomy of lizard ovaries, with emphasis on the number of germinal beds. J. Herpetol. 16(3):240–252

Leyton VA, Miranda E, Bustos-Obregón E (1980) Gestational chronology in the viviparous lizard *Liolaemus gravenhorsti* (Gray) with remarks on ovarian and reproductive activity. Arch. Biol. (Bruxelles) 91:3347–3361

Licht P (1972) Environmental physiology of reptilian breeding cycles: role of temperature. Gen. Comp. Endocrinol. Suppl. 3:477–488

Porter KR (1985) Herpetology. W.B. Saunders Company, Philadelphia-London-Toronto pp. 378–432

Tata JR (1976) The expression of vitellogenin gene. Cell 9:1–14

Tinkle DW, Gibbons JW (1977) The distribution and evolution of viviparity in reptiles. Misc. Publ. Mus. Zool., Univ. Mich. 154:1–55

Tinkle DW, and Irwin LN (1965) Lizard reproduction: Refractory period and response to warmth in *Uta stansburiana* female. Science 148:1613–1614

Tinkle DW, Wilbur HM, Tilley SG (1970) Evolutionary strategies in lizard reproduction. Evolution 24:55–74

Veloso A, Navarro J (1988) Lista sistemática y distribución geográfica de anfibios y reptiles de Chile. Bil. Mus. Reg. Sci. Nat. Torino 6(2):481–539

Vitt LJ (1982) Reproductive tactics of *Ameiva ameiva* (Lacertilia:teiidae) in a seasonally fluctuating tropical habitat. Can. J. Zool. 60:3113–3120

Vitt LJ, Goldberg SR (1983) Reproductive ecology of two tropical iguanid lizards: *Tropidurus torguatus* and *Platinotus semitaeniatus*. Copeia 1983:131–141

10

Diversity of Reproductive Strategies Among Brazilian Lizards and Snakes: The Significance of Lineage and Adaptation

Laurie J. Vitt

Introduction

At one time, it was believed that the local environment was the most important determinant of reproductive strategies in squamate reptiles (Tinkle et al. 1970). Long-term field studies demonstrating geographic differences in reproductive characteristics among populations of individual species occurring in different habitats (e.g., Tinkle and Ballinger 1972; Ballinger 1979) lended support to this notion. If the local environment were the overriding variable contributing to life history variation, it would be expected that species at any given locality would be more similar to each other in reproductive characteristics than to species or populations occurring in very different habitats. More recently, it has been demonstrated that lineage (phylogeny), foraging mode, body size, and body shape, as dictated by habitat selection, account for more of the variation in life history characteristics among squamate reptiles than the local environment (Vitt and Congdon 1978; Vitt 1981; Vitt and Price 1982; Dunham and Miles 1985; Dunham et al. 1988), even though the latter certainly influences reproductive characteristics (see below). In order to understand the evolution of reptilian life histories, it will be necessary to consider all components of variation (Ballinger 1983). The greatest gap in our knowledge of reptilian reproductive patterns stems from a lack of data on tropical species, which comprise a vast majority of the species on earth.

The taxonomically diverse reptilian fauna of Brazil is extremely complex in terms of biogeography, ecology, and life history patterns. In this report, I examine reproductive patterns among Brazilian lizards and snakes based primarily on my own research in the three predominant morphoclimatic domains (see Ab'Saber 1967, 1977): caatinga, cerrado, and the Amazonian portion of hylaea. Descriptions of important features of the squamate faunas of these domains and their biogeography have been recently presented (Vanzolini 1976, 1981, 1988; Rodrigues 1988). The squamate faunas of Brazil are ideal for examining reproductive strategies of tropical reptiles because of their taxonomic and ecological diversity, and because peculiarities of species with limited geographic distribution can be compared with geographic variation of species with extensive geographic distributions.

What I present will be based partially on published data and partially on my own unpublished studies. Data will appear in the form of summaries directed toward three general topics: (1) diversity of reproductive strategies among squamate reptiles at single localities; (2) patterns within species and among closely related species occurring in drastically different habitats; and (3) local adaptation and habitat selection. My treatment of these topics is by no means exhaustive; rather, I have selected examples to illustrate these points.

Diversity of Reproductive Strategies at Single Localities

I provide one example to demonstrate the high diversity of reproductive strategies at single localities. Among squamate reptiles studied at the same locality in semi-arid tropical caatinga during the same time period, every known reproductive strategy (sensu Tinkle et al. 1970) is represented (Table 10.1). Several of the lizard species are illustrated in Figure 10.1. All caatinga lizards except *Mabuya heathi* are oviparous. *Tropidurus*, *Polychrus*, and *Tupinambis* have a large clutch size which varies with female body size. Clutch size is genetically fixed in the four species of geckos (one egg in one species, two eggs in the other three) and the one species of gymnophthalmid (two eggs), and is very low (one to three, usually two) in *Tapinurus*, a species most closely related to *Tropidurus*. *Mabuya*, *Polychrus*, *Tupinambis*, and *Diploglossus* are highly seasonal in reproduction (not necessarily the same season). *Ameiva*, *Cnemidophorus*, *Gymnophthalmus*, and the four gecko species reproduce more or less continuously. The reproductive season is extended in *Tropidurus* and *Tapinurus*, but most reproduction takes place in the dry season.

Among sympatric caatinga snakes, all species in the families Boidae and Viperidae are viviparous whereas the remainder (Colubridae and Elapidae) are oviparous (Vitt and Vanglider 1983). Clutch or litter size varies considerably among species from relatively low in *Micrurus* to relatively high in *Waglerophis* and *Crotalus*. In species exhibiting variable offspring number, the number produced generally increases with the size of females (e.g., *Liophis*, *Waglerophis*; see Vitt 1980, 1983b).

The actual determinants of seasonality in tropical squamates remain unclear. It is clear, however, that no single environmental variable accounts for the incredible diversity in seasonality of snakes and lizards at the caatinga locality. Data I have collected at two Amazonian sites (Howland et al. 1990; Vitt 1991a; Vitt and Blackburn 1991; Vitt unpubl.) and one cerrado site (Vitt 1991b) in Brazil reveal similar diversity in reproductive characteristics. Two recent studies have suggested that certain reproductive characteristics of tropical Australian reptiles reflect characteristics of particular lineages (James and Shine 1985, 1988). My studies in Brazil reveal much the same pattern, i.e., phylogeny supersedes adaptation at the local level. The ultimate cause of this apparent phylogenetic inertia (Ballinger 1983) remains unclear, but at least a portion of it may be attributable to foraging mode (Vitt 1990). Foraging mode has been identified as an important determinant of numerous aspects of the ecology, life history, and behavior of squamates (Vitt and Congdon 1978; Huey and Pianka 1981; Huey et al. 1984; Dunham and Miles 1985; Cooper et al. 1990) and may predate family level effects.

Patterns Within Species and Among Closely Related Species Occurring in Drastically Different Habitats

A number of lizard and snake species and genera are widely distributed in South America and their distributions transect major habitats. I have selected data on the widespread teiid lizard *Ameiva ameiva* and the widespread scincid genus *Mabuya* to demonstrate phylogenetic conservatism in reproductive characteristics with evidence of local adaptation.

An examination of *Ameiva ameiva* (Fig. 10.1A) living in very different habitats (caatinga of northeastern Brazil and lowland tropical forest of Rondonia) reveals that populations of these lizards are more similar to each other in reproductive characteristics than they are to their respective sympatric species. The diversity of reproductive strategies among sympatric caatinga lizards (Table 10.1) demonstrates this, and similar differences among sympatric species exist in Rondônia (see Howland et al. 1990; Vitt 1991a). Nevertheless, there are differences between the two *Ameiva* populations that indicate the importance of local adaptation. These differences are summarized in Table 10.2 and Figure 10.2 (see also Colli 1991). There is a significant relationship between clutch size and SVL (snout-vent length) in both populations with SVL explaining 39.9% of the variation in clutch size of caatinga females (the regression equation for this relationship is: clutch size $= -7.80 + (0.10)(\text{SVL})$, $F_{1,104} = 69.1$, $p < 0.0001$ where F is the critical value from the F distribution (see Sokal and Rohlf, 1981), subscripts are degrees of freedom, and p is the level of significance) and 46.7% of the variation in clutch size of Rondônia females (clutch size $= -3.14 + (0.06)(\text{SVL})$, $F_{1,124} = 108.5$, $p < 0.0001$). Analysis of covariance with SVL as the covariate and locality as the class variable revealed that slopes of the SVL to clutch size regressions differed between populations ($F_{1,228} = 13.7$, $p < 0.001$) with caatinga *Ameiva* increasing clutch size more rapidly with SVL than Rondônia *Ameiva*. Moreover, clutch size, egg mass, and relative clutch mass (RCM, defined here as total clutch mass/total body mass including the clutch) are lower in Rondônia *Ameiva*. The most surprising result is the difference among populations in RCM, which usually is relatively constant among populations of lizard species (see Vitt and Congdon 1978). Lower RCM in Rondônia *Ameiva* may reflect a more streamlined *Ameiva* morphology and concomitant reduced clutch mass as

A

B

C

Figure 10.1. Photographs of six species of lizards that inhabit caatinga of northeastern Brazil: (A) *Ameiva ameiva*; (B) *Gymnophthalmus multiscutatus*; (C) *Diploglossus lessonae*; (D) *Polychrus acutirostris*; (E) *Phyllopezus pollicaris*; (F) *Gymnodactylus geckoides*.

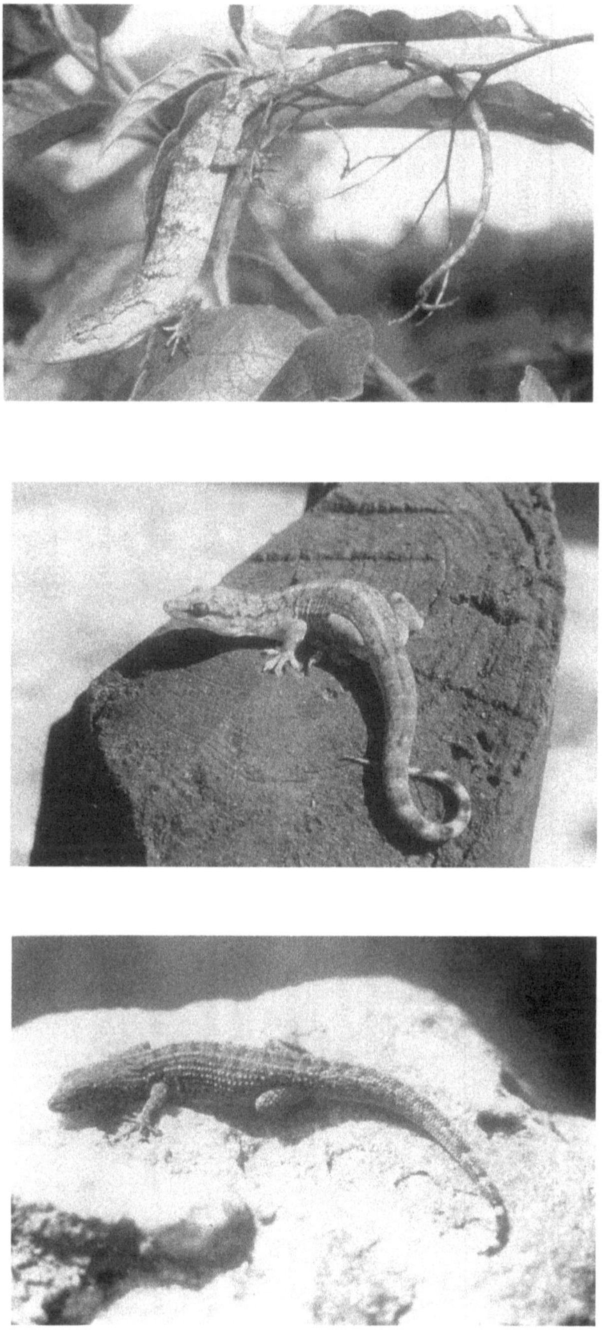

D

E

F

FIGURE 10.1. (*Cont.*)

TABLE 10.1. Summary of data on reproductive characteristics of squamate reptiles from caatinga of Brazil. Abbreviations in the Life History Strategy column are from left to right; age at maturity (E = early maturing, L = late maturing); clutch production (I = seasonally iteroparous, S = seasonally semelparous); and longevity (SL = short-lived, LL = long-lived). SVL is snout-vent length. A majority of the data in this table can be found in the following references: Vanzolini et al. 1980; Vitt 1980, 1981, 1982a, 1982b, 1983a, 1983b, 1985, 1986, 1990; Vitt and Lacher 1981; Vitt and Blackburn 1983; Vitt and Goldberg 1983; Vitt and Vangilder 1983; Blackburn et al. 1984, 1985. Families within the Iguania follow Frost and Etheridge (1989).

Family species	Reproductive Season	Clutch size	Clutches per year	Approx. age of maturity	Maximum SVL (mm)		Life history strategy
					males	females	
Caatinga Lizards							
Gekkonidae							
Gymnodactylus geckoides	continuous	1	> 2	<6 months	48	46	E, I, SL
Lygodactylus klugei	continuous	2	> 2	<6 months	31	34	E, I, SL
Phyllopezus pollicaris	continuous	2	> 2	<9 months	84	85	E, I, SL
Hemidactylus mabouia	continuous	2	> 2	<6 months	68	62	E, I, SL
Tropiduridae							
Tropidurus hispidis	dry season	3–13	> 2	6–10 months	129	107	E, I, SL
Tapinurus semitaeniatus	dry season	1–3	> 4	8–10 months	98	83	E, I, SL
Polychridae							
Polychrus acutirostris	wet season	7–31	1	9 months	124	145	E, S, LL
Anguidae							
Diploglossus lessonae	late dry season	5–7	1	24 months	155	155	L, S, LL
Scincidae							
Mabuya heathi	dry season	2–9	1	12 months*	77	84	E, S, LL
Gymnophthalmidae							
Gymnophthalmus multiscutatus	continuous	2	> 2	<6 months	39	40	E, I, SL
Teiidae							
Ameiva ameiva	continuous	1–9	≥ 2	≤12 months	183	159	E, I, SL
Cnemidophorus ocellifer	continuous	1–5	≥ 4	≤ 6 months	117	84	E, I, SL
Tupinambis teguixin	late dry season	13–29	1	24 months	400	406	L, S, LL

Caatinga snakes**							
Biodae							
Epicrates cenchria	late dry, early wet	8–15	1	>12 months	1334	1405	L, S, LL
Colubridae							
Clelia occipitoluea	dry season	9–14	1	>12 months	1550	1918	L, S, LL
Leptophis ahaetulla	late dry, early wet	4–7	1	12 months	825	973	L, S, LL
Liophis dilepis	dry season	3–10	≥ 2	<12 months	473	576	E, I, SL
Liophis miliaris	nearly continuous	4–7	> 2	<12 months	655	755	E, I, SL
Liophis poecilogyras	nearly continuous	3–17	> 2	<12 months	611	855	E, I, SL
Liophis viridis	nearly continuous	2–6	> 2	<12 months	468	565	E, I, SL
Oxybelis aeneus	wet season	4–6	1	≥12 months	1034	851	L, S, LL
Oxyrhopus trigeminus	wet season	6–7	> 1	≤12 months	432	738	L, S, LL
Philodryas nattereri	late dry season	4–13	1	≥12 months	981	1284	E, S, LL
Philodryas olfersii	late dry season	1–10	1	≥12 months	775	997	L, S, LL
Waglerophis merremii	dry season	4–30	> 1	≤12 months	640	1052	E, S, LL
Elapidae							
Micrurus ibiboboca	late dry season	1	5	≥12 months	916	738	L, S, LL
Viperidae							
Crotalus durissus	late dry season	21–31	1	≥24 months	—	—	L, S, LL

* Approximate age at which first paturition takes place (see Vitt and Blackburn 1983)

** I have only included snake species for which the data are reasonably reliable. I caution that the categorical summaries of life history stategies for snakes are only best estimates and some may prove to be inaccurate. Taxonomy for *Liophis* follows Dixon (1983) and Michaud and Dixon (1987).

Table 10.2. Summary of reproductive characteristics between populations of *Ameiva* from caatinga of northeastern Brazil and lowland tropical forest of western Brazil (Rondônia).

Variable	Caatinga	Lowland Forest
Female SVL (mm)	132.0 + 1.0 (106)	113.5 + 1.1 (126)
SVL at maturity (mm)	95	88
Maximum female SVL (mm)	151	148
Clutch size	5.60 + 0.16	3.24 + 0.09*
Range of clutch size	1–9	1–7
Egg width (mm)	—	22.05 + 0.18 (55)
Egg length (mm)	—	11.58 + 0.12 (55)
Egg mass (g)	1.978 + 0.038	1.546 + 0.030*
Relative clutch mass	0.159 + 0.007 (25)	0.094–0.006 (4)*

* Significantly different at $p < 0.01$ based on Mann-Whitney U test.

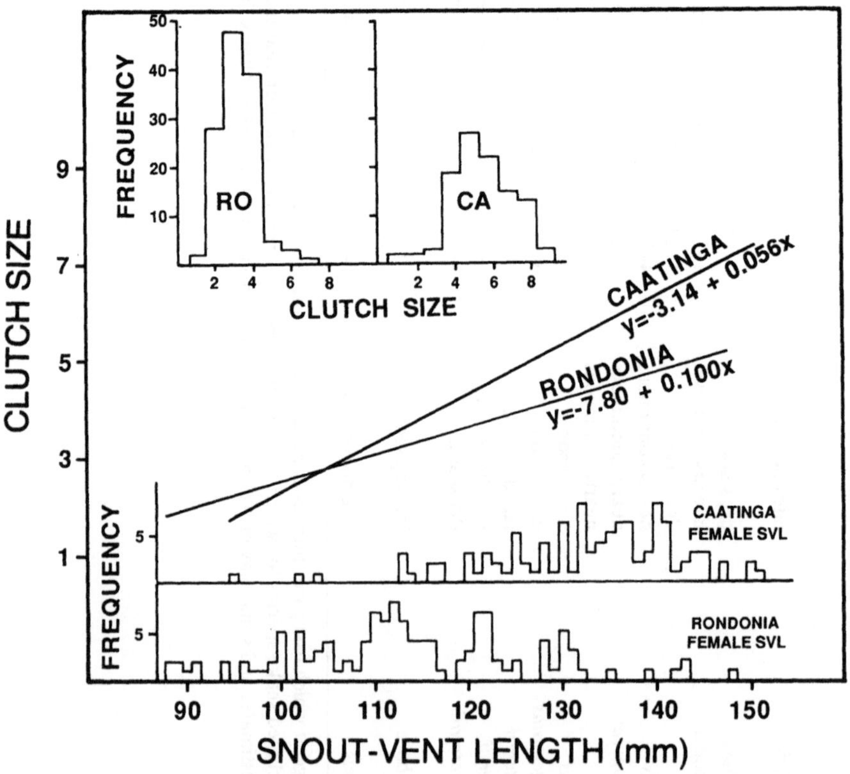

Figure 10.2. Summary of clutch size characteristics for two Brazilian populations of *Ameiva ameiva*. The regressions for clutch size on SVL are shown by solid lines with distributions of SVL and clutch size for females from each locality. See text for statistical results.

TABLE 10.3. Summary of reproductive characteristics for females of Brazilian species of *Mabuya* occurring in different habitats. Data summarized from Vanzolini and Rebouças-Spieker (1976), Vitt and Blackburn (1983, 1991), and Vitt (1991b). All are viviparous.

Domain Species	Maximum SVL (mm)	Litter size	Reproductive season	Climate
Caatinga				
Mabuya heathi	87	2-9	Oct.-Nov.	semi-arid
Cerrado				
Mabuya frenata	80	1-8	Dec.-Jan.	seasonal tropical
Amazonian rainforest				
Mabuya bistriata	109	2-9	Aug.-Dec.	tropical wet
Coastal islands of southern brazil				
Mabuya caissara	90	2-9	Nov.-Dec.	complex—see text
Mabuya macrorhyncha	81	2-4	Dec.-Feb.	complex—see text

an adaptation to escape predators via enhanced locomotor capabilities. Predator diversity is undoubtedly greater in Rondônia and may have influenced the evolution of escape behavior in *Ameiva* (Schall and Pianka 1980). Alternatives to evolutionary adjustments in morphology and reproductive characteristics include modification of escape behavior during pregnancy (Shine 1980; Cooper et al. 1990).

Species in the genus *Mabuya* occurring in the New World tropics exhibit a number of similarities in reproductive characteristics regardless of where they occur (Table 10.3). Similarities include: (1) viviparity with a highly specialized placenta; (2) extremely long gestation period; (3) production of a single brood per year; and (4) attainment of sexual maturity (defined by age at first ovulation) at a "juvenile" body size (Vitt and Blackburn 1983, 1991; Blackburn et al. 1984, 1985; Blackburn and Vitt, this volume). It is also notable that three of the species, *M. heathi*, *M. frenata*, and *M. bistriata*, produce offspring near the end of the wet season in three drastically different habitats. The situation in *M. caissara* and *M. macrorhyncha* is much more complex because they occur on islands at an interface of equatorial tropical air masses and polar air masses. Variation exists within species among the islands as well as between the two species, and ecological correlates are not clear (Vanzolini and Rebouças-Spieker 1976).

Local Adaptation and Habitat Selection

I have already touched on local adaptation in considering geographic variation within species and genera. I now consider variation in reproductive characteristics over evolutionary time as a consequence of habitat selection and expand on my findings on the flat lizard, *Tapinurus semitaeniatus* (Vitt 1981).

Tapinurus semitaeniatus is a dorsoventrally flattened tropidurid that occurs on crystalline granitic extrusions (lageiros) in caatinga of northeastern Brazil. The flattened morphology enhances the utilization of narrow crevices as refuges from predators. Consequences of flattened morphology include: (1) reduced clutch size compared to all other close relatives (genus *Tropidurus*), nearly always two eggs; (2) elongated oviductal eggs and elongated ovarian follicles compared to *Tropidurus*; (3) ovulation of one egg per ovary; and (4) as a consequence of 1 to 3, females containing enlarged vitellogenic follicles or oviductal eggs are as flat as non-reproductive females, i.e., the advantages of being flat are maintained. This set of reproductive characteristics is amazingly convergent with those of an ecologically similar lizard, *Platysaurus*, which is in a different family and lives in South Africa (Vitt 1981).

I have recently discovered several populations of at least two undescribed species of *Tropidurus* occurring on crystalline granitic outcrops in the state of Rondônia in western Brazil. These populations are isolated from all other *Tropidurus* populations in Brazil and from each other by tropical lowland rainforest which acts as a complete barrier to dispersion away from the rock outcrops. No species of *Tropidurus* occur naturally within the lowland rainforest of Rondônia even though as many as 40 lizard species may occur in the forest adjacent to the outcrops (Vanzolini 1986; Vitt unpubl.). Similar to *Tapinurus* of northeast Brazil, these isolated *Tropidurus* are scansorial on the surface of the rock extrusions and use crevices as refuges from predators and to avoid extremes of temperature. The timing of isolation is unknown, but paleoecological data from a site near the outcrops indicates that as recently as 5000 years ago the habitat was

Table 10.4. Comparisons of morphology among Rondônia *Tropidurus*, *Tropidurus hispidis* from caatinga, and *Tapinurus semitaeniatus* from caatinga. For comparison I have presented head width (HW), head height (HH), body width (BW), and body height (BH) as proportions of snout-vent length (SVL), ignoring the potential for allometric effects. The three Rondônia *Tropidurus* populations are from isolated rock outcrops near Santa Barbara, Ariquemes, and Santa Cruz da Serra. All three were completely surrounded by lowland rain forest at the time data were collected. Note that the three Rondônia populations are slightly flattened in morphology, particularly in body height, as compared to *T. hispidis*, but not nearly as flat as *Tapinurus*.

Sample	HW/SVL	HH/SVL	BW/SVL	BH/SVL
Santa Barbara	0.1976	0.1202	0.2855	0.1494
Ariquemes	0.1943	0.1190	0.2862	0.1486
Santa Cruz	0.1948	0.2413	0.1236	0.1542
T. hispidis	0.1979	0.1200	0.2741	0.1606
T. semitaeniatus	0.1858	0.0849	0.3163	0.1053

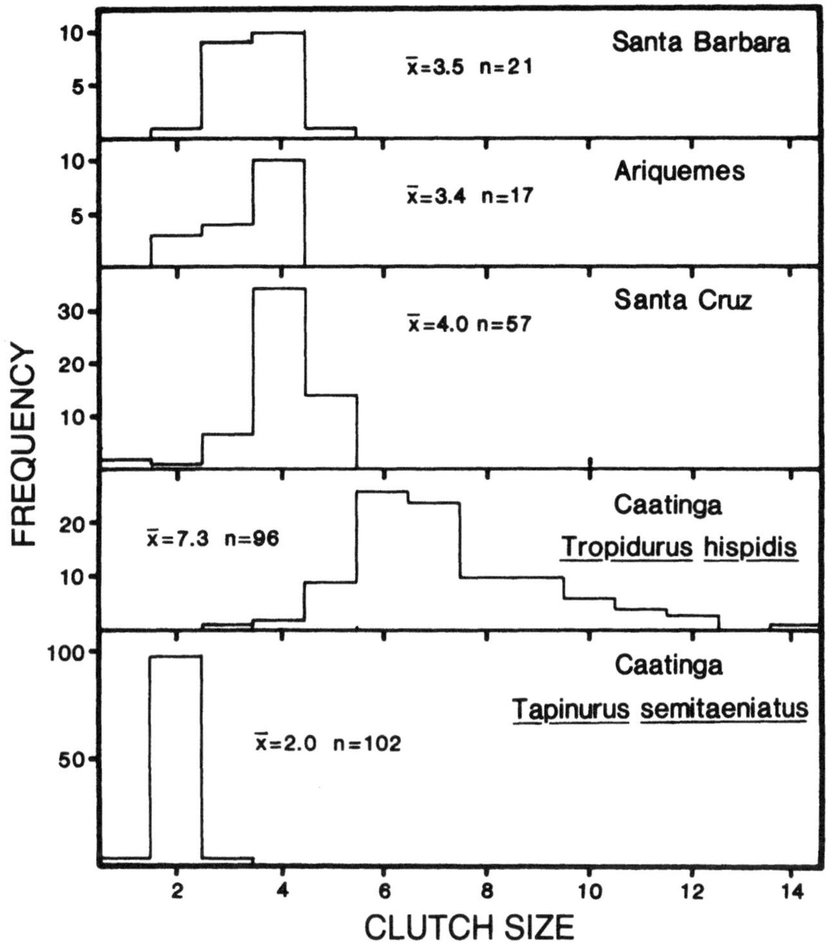

Figure 10.3. Clutch sizes for three Rondônia populations of *Tropidurus*, caatinga *Tropidurus hispidis*, and caatinga *Tapinurus semitaeniatus*. There is a significant correlation between clutch size and snout-vent length in three of the populations; *T. hispidis* from caatinga, *Tropidurus* sp. from Santa Barbara, and *Tropidurus* sp. from Santa Cruz ($p < 0.05$). There is no relationship between clutch size and SVL in caatinga *Tapinurus* and Ariquemes *Tropidurus* sp. ($p > 0.10$). There are significant differences among the three Rondônia *Tropidurus* populations in mean clutch size (Analysis of Variance, $F_{2,97} = 7.7$, $p = 0.0008$) with the Ariquemes and Santa Barbara populations constituting one homogeneous group and the Santa Cruz population another group (based on the Scheffe multiple range test).

cerrado (Absy and van der Hammen 1976) and presumably *Tropidurus* were more widespread.

Most interesting are the morphological and reproductive comparisons between typical open formation *Tropidurus*, such as *T. hispidis*, the isolated *Tropidurus* populations (which represent at least two undescribed species), and the flat lizard, *Tapinurus*. If, as I have suggested (Vitt 1981), flattened morphology in *Tapinurus* represents adaptation to the use of rock crevices for escape and the concomitant reproductive adaptations are associated with the effect of habitat selection on morphology, I would expect the isolated rock outcrop *Tropidurus* in Rondônia to exhibit morphological divergence and divergence in reproductive characteristics in the direction of the flat lizard, *Tapinurus*, even though they may have only been isolated for 5000 years. The three Rondônia *Tropidurus* populations are slightly flattened compared to *Tropidurus hispidis* (Table 10.4), but not nearly as flat as *Tapinurus*. More dramatic is the reduction in clutch size in the isolated populations intermediate between those of *T. hispidis* and *T. semitaeniatus* (Fig. 10.3). Reduced clutch size translates into reduced clutch volume and reduced body height of gravid females, thus allowing entrance into narrow crevices for escape.

There are other reproductive adaptations that appear to be consequences of the exclusive use of rock outcrops. *Tapinurus*, which rarely leave the rocks in northeast Brazil, communally nest under sheets of granite caprock, whereas other *Tropidurus* studied, such as *T. hispidis*, nest individually, usually under rocks or other surface objects in contact with the ground. The isolated *Tropidurus* in Rondônia nest communally under granite caprocks. Additional similarities between the Rondonia *Tropidurus* and *Tapinurus* will be discussed in a later paper.

Conclusions

I have only touched on the many reproductive adaptations of South American squamate reptiles. Life history characteristics of these reptiles are influenced by foraging mode, lineage, body size, and morphology, as well as adaptation at the local level. It is important to evaluate data collected on single populations within the context of historical as well as ecological factors. The diversity of reproductive strategies exhibited by tropical reptiles promises to add significantly to our understanding of the evolution of life history characteristics in reptiles.

Acknowledgments

All field work was coordinated and supported by Dr. Paulo E. Vanzolini, Museu de Zoologia da Universidade de São Paulo through the following grants: (1) Rondonia; Conselho Nacional de Desenvolvimento Científico e Tecnológico, Programa Polonoreste, Subprograma Ecología Animal; (2) caatinga; Academia Brasileiro de Ciências, "Estudos Ecologicos no Nordeste Semi-arido do Brasil"; and (3) Rio Xingu in Pará and the Rio Araguaia in Goias and Mato Grosso; Consorcio Nacional dos Engenheiros Consultados, and Eletronorte.

REFERENCES

Ab'Saber AN (1967) Dominios morfoclimaticos e provincias fitogeograficas do Brasil. Orientação (Dept. Geogr. Univ. S. Paulo) 3:45–48

Ab'Saber AN (1977) Os domiínios morfoclimáticos na America do Sul. Primeira aproximação. Geomorfologia (Inst. Geogr. Univ. São Paulo) 52. 21 pp

Absy ML, van der Hammen T (1976) Some paleoecological data from Rondônia, southern part of the Amazon basin. Acta Amazonica, Manaus 6:293–299

Ballinger RE (1979) Intraspecific variation in demography and life history of the lizard, *Sceloporus jarrovi*, along an altitudinal gradient in southeastern Arizona. Ecology 60:901–909

Ballinger RE (1983) Life history variations. In Huey RB, Pianka ER, Schoener TW (eds.). Lizard Ecology. Harvard University Press, Cambridge, Mass. pp. 241–260

Blackburn DG, Evans HE, Vitt LJ (1985) The evolution of fetal nutritional adaptations. In Duncker HR, Fleischer G (eds.). Vertebrate Morphology. Gustav Fischer Verlag Stuttgart. pp. 437–439

Blackburn DG, Vitt LJ, Beuchat CA (1984) Eutherian-like reproductive specializations in a viviparous reptile. Proc. Natl. Acad. Sci. 81:4860–4863

Colli G (1991) Reproductive ecology of *Ameiva ameiva* (Sauria, Teiidae) in the cerrado of central Brazil. Copeia 1991:1002–1012

Cooper WE Jr, Vitt LJ, Hedges R, Huey RB (1990) Locomotor impairment and defense in gravid lizards (*Eumeces laticeps*): Behavioral shift in activity may offset costs of reproduction in an active forager. Behav. Ecol. Sociobiol. 27:153–157

Dixon JD (1983) Taxonomic status of the South American snakes *Liophis milaris*, *L. amazonicus*, *L. chrysostomus*, *L. mossoroensis* and *L. purpurans* (Colubridae: Serpentes). Copeia 1983:791–802

Dunham AE, Miles DB (1985) Patterns of covariation in life history traits of squamate reptiles: The effects of size and phylogeny reconsidered. Amer. Natur. 126:231–257

Dunham AE, Miles DB, Reznick DN (1988) Life history patterns in squamate reptiles. In Gans C, Huey RB (eds.). Biology of the Reptilia, Volume 16, Ecology B, Defense and Life History. Alan R. Liss, Inc., New York. pp. 441–522

Frost DR, Etheridge R (1989) A phylogenetic analysis and taxonomy of iguanian lizards (Reptilia: Squamata). Misc. Publ. Mus. Nat. Hist., Univ. Kansas 81:1–65

Howland JM, Vitt LJ, Lopez PT (1990) Life on the edge: The ecology and life history of the tropidurine iguanid lizard *Uranoscodon superciliosum*. Can. J. Zool. 68:1366–1373

Huey RB, Bennett AF, John-Adler H, Nagy KA (1984) Locomotor capacity and foraging behavior of Kalahari lacertid lizards. Anim. Behav. 32:41–50

Huey RB, Pianka ER (1981) Ecological consequences of foraging mode. Ecology 62:991–999

James C, Shine R (1985) The seasonal timing of reproduction: A tropical–temperate comparison in Australian lizards. Oecologia (Berlin) 67:464–474

James C, Shine R (1988) Life history strategies of Australian lizards: A comparison between the tropics and the temperate zone. Oecologia (Berlin) 75:307–316

Michaud EJ, Dixon JR (1987) Taxonomic revision of the *Liophis lineatus* complex (Reptilia: Colubridae) of Central and South America. Contr. Biol. Geol. Milw. Public Mus. 71:1–26

Rodrigues MT (1988) Distribution of lizards of the genus *Tropidurus* in Brazil (Sauria, Iguanidae). In Vanzolini PE, Heyer WR (eds.). Proceedings of a Workshop on Neotropical Distribution Patterns. Academia Brasileira de Ciências, Rio de Janeiro, Brasil. pp. 305–315

Schall JJ, Pianka ER (1980) Evolution of escape behavior diversity. Amer. Natur. 115:551–566

Shine R (1980) "Costs" of reproduction in reptiles. Oecologia (Berl.), 46:92–100

Sokal RR, Rohlf FJ (1981) Biometry. W.H. Freeman and Co. San Francisco

Tinkle DW, Ballinger RE (1972) *Sceloporus undulatus*: A study of the intraspecific comparative demography of a lizard. Ecology 53:570–584

Tinkle DW, Wilbur HM, Tilley SG (1970) Evolutionary strategies in lizard reproduction. Evolution 24:55–74

Vanzolini PE (1976) On the lizards of a cerrado-caatinga contact: Evolutionary and zoogeographical implications (Sauria). Papéis Avulsos Zool. (São Paulo) 29:111–119

Vanzolini PE (1981) A quasi-historical approach to the natural history of the differentiation of reptiles in tropical geographic isolates. Papéis Avulsos Zool. (São Paulo) 34:189–204

Vanzolini PE (1986) Levantamento herpetológico da área do estado de Rondônia sob a influência do rodovia BR-364. Brasilia: Conselho Nacional de Desenvolvimento Científico e Tecnológico, Programa Polonoreste, Subprograma Ecologia Animal, Relatório de Pesquisa no. 1. 50 pp

Vanzolini PE, Rebouças-Spieker R (1976) Distribution and differentiation of animals along the coast and on continental islands of the state of São Paulo, Brasil. Reproductive differences between and within *Mabuya caissara* and *M. macrorhyncha* (Sauria, Scincidae). Papeís Avulsos Zool. (São Paulo) 29:95–109

Vanzolini PE (1988) Distributional patterns of South American lizards. In Vanzolini PE, Heyer WR (eds.). Proceedings of a Workshop on Neotropical Distribution Patterns. Academia Brasileira de Ciências, Rio de Janeiro, Brasil. pp. 317–342

Vanzolini PE, Ramos-Costa AMM, Vitt LJ (1980) Répteis das caatingas. Acad. Brasileira Ciências, Rio de Janeiro

Vitt LJ (1980) Ecological observations on sympatric *Philodryas* (Colubridae) in northeastern Brazil. Papéis Avulsos Zool. (S. Paulo) 34:87–98

Vitt LJ (1981) Lizard reproduction: Habitat specificity and constraints on relative clutch mass. Amer. Natur. 117:506–514

Vitt LJ (1982a) Reproductive tactics of *Ameiva ameiva* (Lacertilia: Teiidae) in a seasonally fluctuating tropical habitat. Can. J. Zool. 60:3113–3120

Vitt LJ (1982b) Sexual dimorphism and reproduction in the microteiid lizard, *Gymnophthalmus multiscutatus*. J. Herpetol. 16:325–329

Vitt LJ (1983a) Reproduction and sexual dimorphism in the tropical teiid lizard, *Cnemidophorus ocellifer*. Copeia 1983:359–366

Vitt LJ (1983b) Ecology of an anuran-eating guild of terrestrial tropical snakes. Herpetologica 39:52–66

Vitt LJ (1985) On the biology of the little known anguid lizard, *Diploglossus lessonae*. Papéis Avulsos Zool. (Sao Paulo) 36:69–76

Vitt LJ (1986) Reproductive tactics of sympatric gekkonid lizards with a comment on the evolutionary and ecological consequences of invariant clutch size. Copeia 1986:773–786

Vitt LJ (1990) The influence of foraging mode and phylogeny on seasonality of tropical lizard reproduction. Papéis Avulsos Zool. (São Paulo) 37:107–123

Vitt LJ (1991a) Ecology and life history of the scansorial arboreal lizard *Plica plica* (Iguanidae) in Amazonian Brazil. Can. J. Zool. 69:504–511

Vitt LJ (1991b) An introduction to the ecology of cerrado lizards. J. Herpetol. 25:79–90

Vitt LJ, Blackburn DG (1983) Reproduction in the lizard *Mabuya heathi* (Scincidae): A commentary on viviparity in New World *Mabuya*. Can. J. Zool. 61:2798–2806

Vitt LJ, Blackburn DG (1991) The ecology and life history of *Mabuya bistriata*: A viviparous Amazonian lizard (Scincidae). Copeia. 1991:917–927

Vitt LJ, Congdon JD (1978) Body shape, reproductive effort, and relative clutch mass in lizards: Resolution of a paradox. Amer. Natur. 112:595–608

Vitt LJ, Goldberg SR (1983) Reproductive ecology of two tropical iguanid lizards: *Tropidurus torquatus* and *Platynotus semitaeniatus*. Copeia 1983:131–141

Vitt LJ, Lacher TE Jr (1981) Behavior, habitat, diet, and reproduction of the iguanid lizard *Polychrus acutirostris* in the caatinga of northeastern Brazil. Herpetologica 37:53–63

Vitt LJ, Price HJ (1982) Ecological and evolutionary determinants of relative clutch mass in lizards. Herpetologica 38:237–255

Vitt LJ, Vangilder LD (1983) Ecology of a snake community in northeastern Brazil. Amphibia-Reptilia 4:273–296

11

Reproduction in Viviparous South American Lizards of the Genus *Mabuya*

Daniel G. Blackburn and Laurie J. Vitt

Introduction

Reproduction in most vertebrates requires that the female construct an egg containing all of the nutrients needed to sustain development and that she deposit that egg in an environment where it can develop and hatch. Like the seed of a plant, a newly-laid egg can be viewed as a compact package of nutrients and energy, housed with detailed genetic intructions on their use, and provided with a protective covering. The success of egg-laying reproduction, i.e., "oviparity," in terrestrial environments is revealed by its presence in such amniotes as birds, monotremes, turtles, tuataras, crocodilians, and most squamates (i.e., lizards, amphisbaenians, and snakes). Nevertheless, developing eggs that have been laid on land can be subject to the vicissitudes of the terrestrial environment—temperature extremes, thermal fluctuations, dehydration, flooding, and predation by animals, fungi, and bacteria that may be genetically programmed to exploit egg nutrients for their own growth and reproduction.

Amniotes have two alternatives to the abandonment of their eggs in a terrestrial environment. One is for one or both parents to remain with or carry the developing eggs, protecting them and possibly regulating egg temperature by physiological and behavioral means. Such egg-tending behavior characterizes birds and monotremes but is rare in reptiles, being present in crocodilians and a small percentage of the squamates (Shine 1988). The other reproductive alternative is for the female to retain the developing eggs in her reproductive tract and to give birth to her young. Such a pattern is termed "viviparity" or "live-bearing reproduction." Among extant amniotes, viviparous reproduction characterizes only the therian mammals and approximately 19% of the squamate species (Blackburn 1985a).

In oviparous squamates, as well as in typical live-bearing squamates, the large, yolk-rich ovum provides the bulk of the nutrients for embryonic development, a nutritional pattern known as "lecithotrophy" (Wourms 1981). Lecithotrophy contrasts markedly with the "placentotrophy" of viviparous mammals, in which the ovulated ovum is miniscule and nearly all of the nutrients for development are supplied by placental means. Lecithotrophic viviparous squamates do develop placentae that transfer gases and water between maternal and fetal tissues; however, at most, only small amounts of organic and inorganic nutrients are supplied by placental means in such forms (Blackburn in press).

Only a few amniote lineages are known to have converged on therian mammals with respect to both viviparity and the provision of large quantities of nutrients by placental means (Blackburn in press). One of these lineages is represented by an Old World lizard, *Chalcides chalcides*, and another by South and Central American lizards of the genus *Mabuya*. Of these, the *Mabuya* are of particular interest evolutionarily because of certain similiarities to therian mammals in terms of placental structure and function. The presence within Old World *Mabuya* of oviparous species, as well as viviparous lecithotrophic forms, raises the prospect that we may be able to reconstruct in detail the morphological and physiological changes leading to the evolution of viviparity and placentotrophy among representatives of a single genus.

The Genus Mabuya

As a large, cosmopolitan genus of the family Scincidae, *Mabuya* has a southern distribution in Africa, Asia, and the New World. In the Old World, both oviparous and viviparous species are represented, but only viviparity has been reported in *Mabuya* of South and Central America (see Vitt and Blackburn 1983). Phylogenetic relationships in the genus are not well-understood, but available information suggests that viviparity has originated four or more times within *Mabuya* (Blackburn 1982; cf. Shine 1985). Although not unprecedented, for viviparity to have evolved so often within a single genus is unusual, despite documentation of nearly 100 origins of viviparity among squamates (Blackburn 1982, 1985a, 1985b, 1991; Shine 1985).

Publications on reproduction in *Mabuya* are scarce and widely-scattered. Reproductive studies on the Old World species have dealt mainly with natural history and ecology, and almost nothing is known about their reproductive physiology, development, and placentation (see Robertson et al., 1962; Fitch 1970; Pasteels, 1970; Barbault 1974, 1976; Simbotwe 1980; Vitt and Blackburn 1983; Raynaud and Pieau 1985; Yaron 1985). For example, placental morphology has been described in only one Old World species, *Mabuya multifasciata*, in a very brief account that suggests similarities to typical viviparous squamates with lecithotrophic nutrition (Weekes 1930).

The South American *Mabuya*, comprising 14 to 15 nominal species, are inferred to represent a single radiation derived from an African stock

(Greer 1970, 1977; Horton 1974). Most or all of the species are viviparous (Vanzolini and Reboucas-Spieker 1976; Vanzolini et al. 1980), with definitive evidence being available for *M. cassaira* (Vanzolini and Reboucas-Spieker 1976), *M. bistriata* (Vitt and Blackburn 1991), *M. macrorhyncha* (Reboucas-Spieker and Vanzolini 1978), *M. heathi* (Vitt and Blackburn 1983), *M. unimarginata* (= *M. brachypoda*; Fitch 1970), *M. frenata* (Vitt 1991; Blackburn unpubl.), and the widespread *M. mabouya* (Fitch 1973; Somma and Brooks 1976; Duellman 1978; Blackburn pers. obs.), as well as *M. guaporicola*, *M. croizati*, *M. ficta*, and *M. dorsivittata* (Blackburn unpubl.).

Comparative Reproduction in Two Brazilian Species

Of the New World *Mabuya*, two South American species have been studied most extensively in terms of their reproductive ecology, development, fetal nutrition, and placentation. The following information is based largely on studies of *M. heathi* from Pernambuco, Brazil, (Vitt and Blackburn 1983; Blackburn et al. 1984) and *M. bistriata* of the Brazilian states of Rondonia, Para, Acre, and Amazonas (Vitt and Blackburn 1991).

Mabuya bistriata is distributed in tropical lowland rainforest in the Brazilian Amazon, gallery forest in the Brazilian cerrado, and north into Venezuela and Surinam. *Mabuya heathi* has a more restricted distribution in northeastern Brazil in a region of thorn scrub forest known as the "zona da caatinga." The areas where these lizards are found show little seasonal fluctuation in temperature, but do experience seasonal patterns of rainfall; the dry season extends from June or July through October or November, and the wet season occupies the remainder the year. Diet was analyzed in *M. bistriata*; like other lizards of their size, these animals subsist on a diet primarily consisting of invertebrates, such as insects, arachnids, and snails.

Lizards of both species are medium-sized, with the sleek, elongate body form typical of New World members of the genus (Fig. 11.1). In our field samples of *Mabuya bistriata*, size ranged from 35.8 mm to 109 mm SVL (snout-vent length), and in *M. heathi* from 31 mm to 88 mm SVL. In both species, adult females were significantly larger than adult males in mean SVL and body mass, as well as in maximum SVL attained; however, males had proportionately larger heads, presumably reflecting evolutionary consequences of male-male aggressive interactions. From the seasonal distribution of size classes, we have inferred that the lizards achieve approximately average adult body size during their first year of life, a fact of considerable significance to female reproduction (see below).

Within each species, reproduction is synchronous (Vitt and Blackburn 1983, 1991). Thus, testis mass, an indicator of spermatogenic activity., undergoes cyclic seasonal changes. For example, in male *Mabuya heathi*,

FIGURE 11.1. *Mabuya heathi* (top) and *Mabuya bistriata* (bottom).

testis mass (relative to body size) attains a maximum in September to October and declines shortly thereafter. The male reproductive cycle in this species parallels cycles of lipid storage and utilization; in our samples, the mass of abdominal fat bodies attained a maximum in July through September and decreased as the testes reached their maximum size. In female *M. heathi*, ovulation occurs in October through December and parturition occurs the following fall, after a gestation period of 8 to 12 months. The reproductive season begins earlier in *M. bistriata*, which ovulate in August though November and give birth the following August to September.

One of the most extraordinary aspects of reproduction in the Brazilian *Mabuya* first became apparent to us from observations of pregnant female reproductive tracts throughout gestation. Females ovulate tiny yolks of 1 mm in diameter and 0.4 to 0.5 mg in dry mass. In our samples, near-term *M. heathi* fetuses exhibited a mean SVL of 30 mm and a dry mass of 154 mg, representing a dry mass increase of 38,400% (and a wet mass increase of 53,800%) over that of the ovum. In *M. bistriata*, which are born at a slightly larger size, the gestational dry mass and wet mass increases exceeded 47,400% and 74,700%, respectively. The gestational increase in dry mass of the developing conceptus can only reflect uptake of material from maternal tissues, i.e., placentotrophy. Thus, these data indicate that the mass of the neonates derives almost entirely from non-yolk (placental) sources. In fact, the data on *M. heathi* probably underestimates the extent of placentotrophy, being based on late-stage fetuses slightly smaller in average size (30 mm SVL) than neonates born in the laboratory (32.4 mm SVL).

Following ovulation and fertilization, a thin shell membrane is deposited around the egg. The eggs remain in the uterine portion of the maternal oviduct for the remainder of development. The developing eggs undergo little growth in size or mass during the first four to seven months of gestation. Most of the dry mass increase occurs in the final three to four months of gestation, during the rapid embryonic growth phase (Fig. 11.2). This period represents the time during which the bulk of the nutrients is being provided to the eggs by the female.

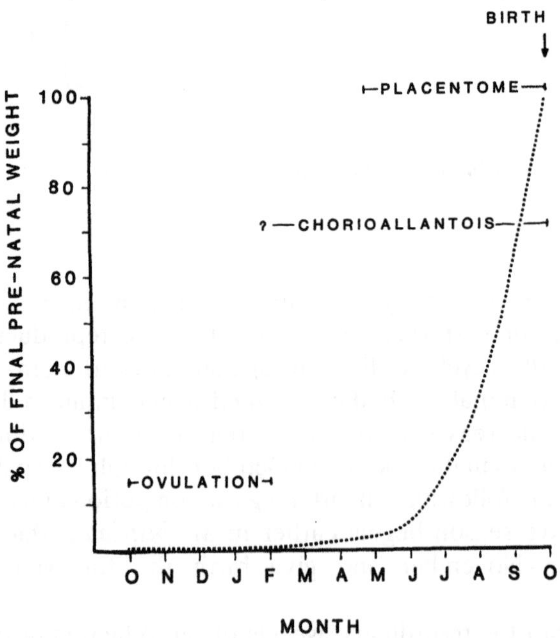

FIGURE 11.2. Dry mass of the conceptus of *M. heathi* during gestation, in relation to reproductive and developmental events. The graph represents a generalized concept based on published data (Vitt and Blackburn 1983; Blackburn et al. 1984; Blackburn 1985b). The increase in dry mass of the conceptus, which is indicative of placentotrophy, mainly occurs following establishment of the placentome.

In both *Mabuya heathi* and *M. bistriata*, ovulation, fertilization, and early development occur at the beginning of the wet season. Because the rapid embryonic growth phase takes place during the dry season, much of the nutrient provision by the female occurs during a period when resource availability may be reduced. However, her ability to store lipid permits the female to buffer the egg from environmental conditions. In female *M. heathi*, the mass of fat bodies begins to increase gradually during the first few months of gestation and decreases rapidly during the rapid embryonic growth phase. Presumably, lipid catabolism reflects the transfer of nutrients to the embryo during this period, coincident with the decline in food availability during the dry season. In *M. bistriata*, lipid cycles are not as marked, due to individual variation, but the mass of the fat bodies in females does decrease slightly during the dry season and rapid embryonic growth phase.

The pattern of extra-vitelline provision of nutrients to the embryos permits females to reproduce during their first year of life, because they can become pregnant within a few months after they are born. In the study population of *Mabuya heathi*, all females were pregnant by January, including those only three to four months of age, and as small as 45 mm SVL. Females of this age are far too small to accommodate a litter of full grown fetuses (which themselves attain 32 mm SVL), but can quite readily hold the tiny 1 mm ova. The first-year females grow rapidly; at six to seven months of age, they average 68 mm SVL. Thus, the pregnant young females become large enough to accommodate their developing fetuses prior to the rapid embryonic growth phase. A similar condition obtains in *M. bistriata*, which ovulate at approximately 2.5 months of age.

Embryonic Nutrition in Squamates

To put embryonic nutrition in *Mabuya* in perspective, comparisons to other squamate species are necessary. As noted above, in oviparous lizards and snakes, embryonic nutrition is lecithotrophic; the nutrients for development derive from the yolk-rich ovum (and sometimes the shell membrane), and physiological exchange between the egg and the external environment probably is limited chiefly to gases and water. Typical viviparous squamates exhibit a similar pattern, although the egg develops within the uterine oviduct, and any physiological exchange therefore occurs with uterine tissues. Again, such exchange is largely limited to gas transfer and water absorption. Small quantities of organic and inorganic nutrients may be taken up from maternal tissues by the egg, but the ovulated yolk supplies the bulk of nutrients for development (e.g., see Panigel 1956; Thompson 1981, 1982; Stewart and Castillo 1984; Stewart 1989; Stewart et al. 1990).

The relative importance of the ovulated yolk in embryonic nutrition in various viviparous squamates is reflected by size of the ovum, and more precisely, by ovum mass relative to mass of the neonate. The tiny 1 mm ovum of *Mabuya heathi*, at a volume of 0.52 mm^3, contrasts strikingly with the large yolk-rich ova of relatively lecithotrophic viviparous species, such as the gekkonid lizard *Hoplodactylus maculatus* (231 mm^3), and the garter snake *Thamnophis sirtalis* (567 mm^3) (Blackburn et al. 1984). In terms of mass of the neonate compared to that of the ovum, lecithotrophic viviparous species (like oviparous species) show a wet mass increase but a dry mass decrease during development. For example, in the viviparous *T. ordinoides*, neonates increase in average wet mass by 41% and decrease in dry mass by 25% as compared to average mass of the ovum (Stewart et al. 1990). The wet mass increase primarily reflects the uptake of water, and the dry mass decrease presumably reflects catabolism and synthesis during development.

The values for lecithotrophic squamates contrast markedly with the enormous developmental increases in dry mass (47,400%) and wet mass (74,700%) exhibited by developing conceptuses of the placentotrophic *Mabuya bistriata*. Aside from *Mabuya*, the only other placentotrophic squamate for which quantitative data exist is the distantly related skink *Chalcides chalcides*, which ovulates ova of 2.5 to 3mm in diameter and 12 mm^3 in volume and exhibits a gestational wet mass increase of approximately 680% (calculated from Ghiara et al. 1987). Thus, no other viviparous squamates are known to approach New World *Mabuya* in terms of the degree of placentotrophy (Blackburn in press). In fact, one can infer from data on neonate mass in mammalian species (Eisenberg 1981) that in terms of quantity of material supplied by placental means, the Brazilian *Mabuya* lie within the range exhibited by eutherians.

Extraembryonic Membranes and Placentation

An understanding of the remarkable reproductive specializations of the Brazilian *Mabuya* also requires some consideration of the structures that function in physiological exchange in both oviparous and viviparous eggs. Like other amniotes, squamates develop extraembryonic membranes that surround the egg and line the inner surface of the shell membrane (see Stewart and Blackburn 1988). Once formed, the extraembryonic membranes are responsible for all of the physiological exchange between the developing egg and its surrounding environment. Among the extraembryonic membranes of squamates is the chorioallantois (Fig. 11.3), a vascularized structure that is homologous to that of other amniotes. As the only highly-vascularized extraembryonic membrane that invariably persists throughout development, the chorioallantois is presumed to be responsible for gas exchange, whether such exchange is with the external environment or with the maternal reproductive tract.

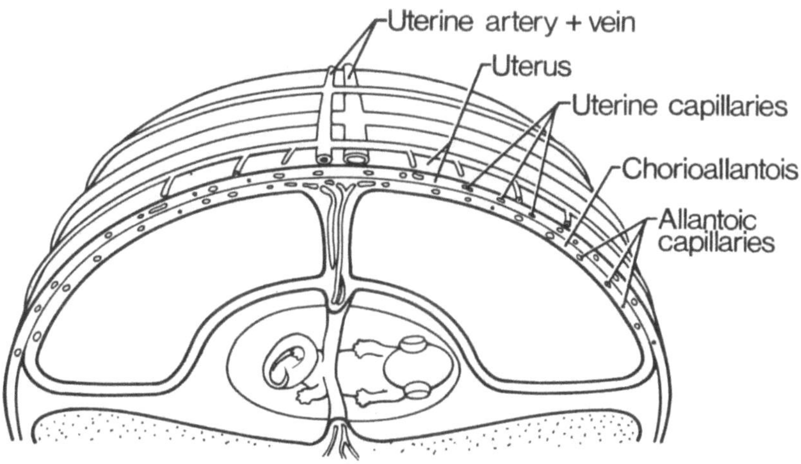

FIGURE 11.3. Transverse section of a pregnant uterus in a generalized lecithotrophic squamate, from Blackburn (1985b), redrawn from Weekes (1930). The embryo lies on its left site on the yolk mass; the uterus extends into the plane of the paper. Simple apposition of the vascularized uterine tissues to the chorioallantois forms the chorioallantoic placenta.

In viviparous squamates, the extraembryonic membranes of the developing egg lie apposed to vascularized uterine tissues, forming the various types of placentae now recognized (Stewart and Blackburn 1988). A squamate "placenta," to paraphrase Mossman's (1937) more widely applicable definition, is an organ that is formed through apposition of extraembryonic and uterine tissues and that accomplishes physiological exchange. As an example, the chorioallantois of the viviparous egg lies in close apposition to the vascularized uterine tissue, comprising the chorioallantoic placenta (Fig. 11.3). Although a thin shell membrane may persist between uterine and chorioallantoic tissues, this membrane typically is very thin as compared to its homologue in oviparous species, such that uterine and chorionic tissues are intimately apposed (e.g., Hoffman 1970; Stewart 1985). Chorioallantoic placentae probably are ubiquitous among viviparous squamates (Blackburn in press), in which they are believed to function in gas exchange.

The morphology of the chorioallantoic placenta has been described in a variety of squamate species that are presumed (from relative sizes of the ovum and neonate) to be lecithotrophic (for reviews see Weekes 1935; Yaron 1985). In such species, the placenta typically consists of unspecialized chorioallantoic tissues apposed to similarly unspecialized uterine tissues. Such a situation has been observed, for example, in the garter snake *Thamnophis sirtalis*, using histological and ultrastructural techniques (Hoffman 1970; Blackburn pers. obs.). On the fetal side of the placenta in this species, a thin layer of chorionic epithelium overlies the allantoic capillaries. Likewise, an attenuate uterine epithelium overlies the uterine capillaries; therefore, maternal blood does not bathe the chorion, unlike the situation in many eutherian mammals. The acellular shell membrane separating the maternal and fetal tissues is reduced to a microscopically thin remnant. Thus, in this species, maternal and fetal capillaries lie in close proximity, being separated only by thin uterine and chorionic epithelia and the vestige of the shell membrane. No glands are evident during gestation, nor are any particular specializations for maternal-fetal transport, such as enlarged cells with specializations for absorption or secretion that are common among therian mammals.

Chorioallantoic placentae with similar morphologies have been described in various other squamates (Yaron 1985), including those shown to be lecithotrophic through detailed quantitative analysis (Stewart and Castillo 1984; Stewart 1985, 1989, 1990); in fact, such placentae probably evolved concomitantly with viviparity (Guillette and Jones 1985; Blackburn in press). The close apposition of maternal and fetal blood vessels in such placentae most likely enhances the diffusion of gases between maternal and fetal circulatory systems.

Placentation in Brazilian *Mabuya*

Because placental morphology in Brazilian *Mabuya* is complex, we shall only summarize features of particular functional and evolutionary significance. In comparison to that of lecithotrophic squamates, the chorioallantoic placenta of *M. bistriata* and *M. heathi* is highly specialized in several respects (Fig. 11.4) (Vitt and Blackburn 1983, 1991; Blackburn and Vitt 1984; Blackburn et al. 1984; Blackburn 1986, in press). The uterus contains numerous multicellular glands that remain active during the rapid embryonic growth phase. Opposite these glands, on the embryonic side of the placenta, the chorion contains populations of tall columnar cells that bear microvilli. These cells are organized in pits on the surface of the chorion, structures known as chorionic areolae, that lie apposed to the mouths of the uterine glands. Histology, ultrastructure, and histochemistry of the specialized chorionic cells of the areolae, as well as their position opposite to the openings of the uterine glands, suggest that they absorb the uterine secretions.

Dorsal to the fetus, there forms a single specialized structure known as the placentome (Fig. 11.4), a region of hypertrophied uterine and chorionic tissues. The uterine endometrium is thrown up into thick folds of tissue that project into a deep invagination into the surface of the chorion. Chorionic cells of the placentome include tall columnar cells with microvilli, much like those of the areolae, a morphology that again strongly suggests an absorptive function. From their morphology, we infer that the placentome and areolae function in maternal-fetal nutrient transfer.

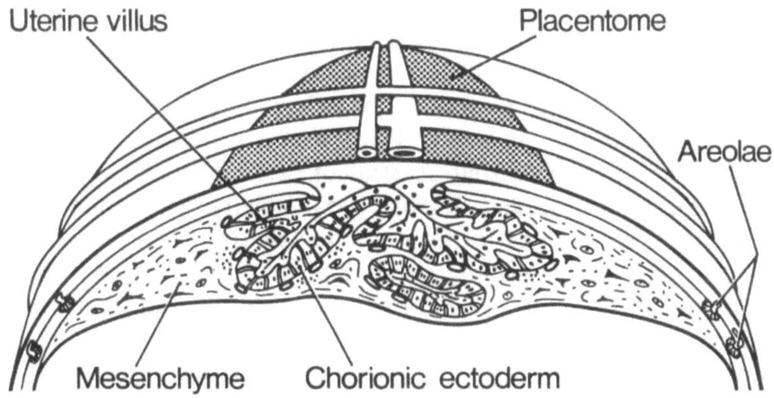

FIGURE 11.4. Dorsal portion of a transverse section through the chorioallantoic placenta in *Mabuya heathi*, from Blackburn (1985b). The placentome consists of interdigitating uterine and chorionic tissues. The chorionic ectodermal cells of the placentome are enlarged and binucleate. Areolae consist of chorionic invaginations that are lined by columnar epithelium and that lie apposed to the openings of uterine glands. Compare the specialized structure of this placenta to the generalized placenta of Figure 11.3.

Like placental morphology, the timing of placental development implicates the chorioallantois in the uptake of nutrients from maternal tissues (Vitt and Blackburn 1983; Blackburn et al. 1984). As outlined above, most of the maternal-fetal nutrient transfer in *Mabuya heathi* and *M. bistriata* (as indicated by the increase in dry mass of the conceptus) occurs during the rapid embryonic growth phase. Developmental studies of an extensive series of embryos reveal that this period does not occur until after the establishment of the chorioallantoic placenta and, in particular, subsequent to differentiation of the placentome and chorionic areolae (Fig. 11.2).

Placental specializations of the two Brazilian *Mabuya* are not confined to those species. Microscopic studies of *M. mabuya* have revealed similar features, as have gross observations of museum specimens of several other New World species (Blackburn pers. obs.). Available evidence indicates that placentotrophy is found in most, if not all, species of *Mabuya* in the Western Hemisphere.

Evolutionary Significance of Reproductive Specialization in *Mabuya*

Placentation of the Brazilian *Mabuya* is highly unusual among reptiles. Species of no other squamate genera are known to supply such a large percentage of the nutrients for development by placental means, nor do any form chorioallantoic placentae with chorionic areolae or other characteristic features of these *Mabuya*. Although a placentome develops in the lizard *Chalcides chalcides* (ten Cate-Hoedemaker 1933; Ghiara et al. 1987; Blackburn pers. obs.), this placentome is only superficially similar to that of the Brazilian *Mabuya* (Blackburn 1986), as are Weekes' (1935) descriptions of the placentome in certain Australian *Leiolopisma*.

The placenta of the Brazilian *Mabuya* is particularly significant evolutionarily because of similarities to the placentae of certain eutherian mammals. A chorioallantoic placenta is ubiquitous in eutherians, in which it is responsible for much, most, or all of the maternal-fetal nutrient transfer, depending on the species. In some eutherian species, the chorioallantoic placenta develops chorionic areolae and placentomes (Mossman 1937, 1987), some of which bear similarities to those of the *Mabuya*.

Selective pressures that have led to the reproductive specializations of New World *Mabuya* are not known. However, these specializations have several functional consequences in *M. bistriata* and *M. heathi* (Vitt and Blackburn 1983, 1991) and, most likely, other species of *Mabuya* as well. First, reliance on placentotrophy, and the associated ovulation of tiny ova, allows females to reproduce during their first year of life. In the absence of this developmental nutritional pattern, newly born females would be far too small to reproduce synchronously with other females, being unable to accommodate eggs the size that would be necessary to produce a neonate by lecithotrophic means. Second, placentotrophy minimizes the physical and physiological burden imposed by the litter until the final months of gestation. Late in pregnancy, litter mass averaged 48.3% and 58.1% of (non-pregnant) female body mass in *M. heathi* and *M. bistriata*, respectively, potentially representing a significant handicap to mobility. However, because the mass of the litter is miniscule for the first several months of gestation, the "cost" of viviparity (Shine 1980) imposed by mass of the litter is minimized.

A third possible consequence of placentotrophy is that it may provide females with the flexibility to alter reproductive investment according to the availability of environmental resources. Whereas lecithotrophic viviparous females invest most nutrients into reproduction at the time of ovulation, placentotrophic female *Mabuya* postpone most of their reproductive investment until late in gestation. If the degree of post-ovulatory nutrient investment is facultative, as has been indicated for some relatively lecithotrophic squamates (Stewart 1989; Stewart et al. 1990), a female may have the capacity to tailor investment to her own nutrient reserves for a considerable portion of gestation. In addition, if female *Mabuya* have the ability to resorb or abort the developing litter when nutrients are scarce, a possibility that has not yet been demonstrated, reproduction could be terminated early in gestation without a significant energetic loss to the female. The extent to which these and other potential benefits of placentotrophy have played a role in the origin and subsequent elaboration of the reproductive specializations of *Mabuya* remain to be determined.

Conclusion

Reproductively, South American *Mabuya* are among the most extraordinary of squamate reptiles. Not only have these scincid lizards evolved a degree of placentotrophy that exceeds that of all other known squamates, but they have converged strongly on a pattern usually assumed to be confined to eutherian mammals. Characteristics of this reproductive pattern include viviparity, a lengthy gestation period, ovulation of small ova, formation of chorioallantoic placentae with chorionic areolae and a placentome, and placental provision of virtually all of the nutrients for development. The independent evolution of such features among mammals and lizards represents one of the most striking cases of evolutionary convergence in reproductive specializations to be documented among terrestrial vertebrates (Blackburn in press).

Aside from their similarities to eutherians, however, South American lizards of the genus *Mabuya* are highly worthy of study in their own right. These lizards represent an extreme in the spectrum of reptilian reproductive features, and their specializations appear to be of functional significance in the environments that they inhabit (Vitt and Blackburn 1983, 1991; Vitt in press). The extent of intra-generic diversity in reproductive mode, placentation, and fetal nutritional mode suggests *Mabuya* to be a model taxon for inquiries into the evolution of viviparity and placentotrophy, as well as the evolution of structural-functional complexes, such as is represented by the placenta.

Unfortunately, survival of at least some of the South American species of *Mabuya* is now threatened over wide portions of their ranges. For example, during the past ten years of habitat destruction in Pernambuco, *M. heathi* has become increasingly scarce, and the range of *M. bistriata* undoubtedly is being diminished by devastation of the Amazonian rainforest. Extinction of these remarkable species would be an incalculable loss, and protection and maintenance of the genus *Mabuya* throughout South America is of the utmost urgency.

Acknowledgments

We thank Dr. Jack McCoy of the Carnegie Museum for permitting the dissection (by Blackburn) of specimens under his curatorship. Lora Miller, Craig Schneider, and Katherine Hewi.. kindly reviewed the manuscript. Michael Simmons (Dept. of Anatomy, New York State College of Veterinary Medicine) rendered the diagrams.

References

Barbault R (1974) Structure et dynamique des populations naturelles du lezard *Mabuya buettneri* dans la savane de lamto (Cote d'Ivoire). Bull. Ecol. 5:105–121

Barbault R (1976) Population dynamics and reproductive patterns of three African skinks. Copeia 1976:483–490

Blackburn DG (1982) Evolutionary origins of viviparity in the Reptilia. I. Sauria. Amphib.-Reptilia 3:185–205

Blackburn DG (1985a) Evolutionary origins of viviparity in the Reptilia. II. Serpentes, Amphisbaenia, and Ichthyosauria. Amphib.-Reptilia 5:259–291

Blackburn DG (1985b) The evolution of viviparity and matrotrophy in vertebrates, with special reference to reptiles. Ph.D. Thesis, Cornell University, Ithaca, New York

Blackburn DG (1986) A review and revision of allantoplacental morphotypes in reptiles. Amer. Zool. 26:115A

Blackburn DG (in press) Convergent evolution of viviparity, matrotrophy, and specializations for fetal nutrition in reptiles and other vertebrates. Amer. Zool.

Blackburn DG, Vitt LJ (1984) Placentation and adaptations for fetal nutrition in a viviparous scincid lizard. Amer. Zool. 24:64A

Blackburn DG, Vitt J, Beuchat CA (1984) Eutherian-like reproductive specializations in a viviparous reptile. Proc. Natl. Acad. Sci. 81:4860–4863

Duellman WE (1978) The biology of an equatorial herpetofauna in Amazonian Ecuador. Univ. Kans. Mus. Nat. Hist. Misc. Publ. 65:1–352

Eisenberg, JE (1981) The Mammalian Radiations. Univ. Chicago Press, Chicago

Fitch HS (1970) Reproductive cycles of lizards and snakes. Univ. Kans. Mus. Nat. Hist. Misc. Publ. 52:1–247

Fitch HS (1973) A field study of Costa Rican lizards. Univ. Kans. Sci. Bull. 50:39–126

Ghiara G, Angelini F, Zerani M, Gobbetti A, Cafiero G, Caputo V (1987) Evolution of viviparity in Scincidae (Reptilia, Lacertilia). Acta Embryol. Morphol. Exper. n.s. 8:187–201

Greer AE (1970) A subfamilial classification of scincid lizards. Bull. Mus. Comp. Zool. 139:151–184

Greer AE (1977) The systematics and evolutionary relationships of the scincid lizard genus *Lygosoma*. J. Nat. Hist. 11:515–540

Guillette LJ Jr, Jones RE (1985) Ovarian, oviductal, and placental morphology of the reproductively bimodal lizard, *Sceloporus aeneus*. J. Morphol. 184:85–98

Hoffman LH (1970) Placentation in the garter snake, *Thamnophis sirtalis*. J. Morphol. 131:57–88

Horton DR (1974) Evolution in the genus *Mabuya* (Lacertilia: Scincidae). Ph.D. Thesis, University of New England, Armidale, Australia

Mossman HW (1937) Comparative morphogenesis of the fetal membranes and accessory uterine structures. Carnegie Inst. Contrib. Embryol. 26:129–246

Mossman HW (1987) Vertebrate Fetal Membranes. Rutgers University Press, New Brunswick, New Jersey

Panigel M (1956) Contribution a l'etude de l'ovoviviparite chez le reptiles: Gestation et parturition chez le lezard vivipare *Zootoca vivipara*. Ann. Sci. Nat. Zool. Biol. Anim. (XI) 18:569–668

Pasteels JJ (1970) Developpement embryonnaire. In Grasse PP (ed.). Traite de Zoologie, Reptiles, Vol 14. Masson et Cie, Paris. pp. 893–971

Raynaud A, Pieau C (1985) Embryonic development of the genital system. In Gans C, Billet F (eds.). Biology of the Reptilia, Vol 15, Development B. John Wiley & Sons, New York. pp. 149–300

Reboucas-Spieker R, Vanzolini PE (1978) Parturition in *Mabuya macrorhyncha* Hoge, 1946 (Sauria, Scincidae), with a note on the distribution of maternal behavior in lizards. Papeis Avulsos Zool. (Sao Paulo) 32:95–99

Robertson IAD, Chapman BM, Chapman RF (1962) Notes on some reptiles collected in the Rukwa Valley, S.W. Tanganyika. Ann. Mag. Nat. Hist. 5:421–432

Shine R (1980) "Costs" of reproduction in reptiles. Oecologia 46:92–100

Shine R (1985) The evolution of viviparity in reptiles: An ecological analysis. In Gans C, Billet F (eds.). Biology of the Reptilia, Vol 15, Development B. John Wiley & Sons, New York. pp. 605–694

Shine R (1988) Parental care in reptiles. In Gans C, Huey, RB (eds.). Biology of the Reptilia, Vol 16, Ecology B. John Wiley & Sons, New York. pp. 275–329

Simbotwe MP (1980) Reproductive biology of the skinks *Mabuya striata* and *Mabuya quinquetaeniata* in Zambia. Herpetologica 36:99–104

Somma CA, Brooks GR (1976) Reproduction in *Anolis oculatus*, *Ameiva fuscata*, and *Mabuya mabouya* from Dominica. Copeia 1976:249–256

Stewart JR (1985) Placentation in the lizard *Gerrhonotus coeruleus* with a comparison to the extraembryonic membranes of the oviparous *Gerrhonotus multicarinatus* (Sauria, Anguidae). J. Morphol. 185:101–114

Stewart JR (1989) Facultative placentotrophy and the evolution of squamate placentation: Quality of eggs and neonates in *Virginia striatula*. Am. Nat. 133:111–137

Stewart JR, Blackburn DG (1988) Reptile placentation: Structural diversity and terminology. Copeia 1988:839–852

Stewart JR, Blackburn DG, Baxter DC, Hoffman, LH (1990) Nutritional provision to the embryos in *Thamnophis ordinoides* (Squamata: Colubridae), a predominantly lecithotrophic placental reptile. Physiol. Zool. 63:722–734

Stewart JR, Castillo RE (1984) Nutritional provision of the yolk in two species of viviparous reptiles. Physiol. Zool. 57:377–383

ten Cate-Hoedemaker NJ (1933) Beitrage zur Kenntnis der Plazentation bei Haien und Reptilien. Zeitsch. Zellforsch. mikrosk. Anat. 18:299–345

Thompson J (1981) A study of the sources of nutrients for embryonic development in a viviparous lizard, *Sphenomorphus quoyii*. Comp. Biochem. Physiol. A, Comp. Physiol. 70:509–518

Thompson J (1982) Uptake of inorganic ions from the maternal circulation during development of the embryo of a viviparous lizard, *Sphenomorphus quoyii*. Comp. Biochem. Physiol. A, Comp. Physiol. 71:107–112

Vanzolini PE, Ramos-Costa AMM, Vitt LJ (1980) Repteis das Caatingas. Academia Brasiliera te Ciencas, Rio de Janeiro, Brasil

Vanzolini PE, Reboucas-Spieker R (1976) Distribution and differentiation of animals along the coast and on continental islands of the state of Sao Paulo, Brazil. 3. Reproductive differences between and within *Mabuya cassaira* and *M. macrorhyncha* (Sauria, Scincidae). Papeis Avulsos Zool. (Sao Paulo) 29:95–109

Vitt LJ (1991) An introduction to the ecology of cerrado lizards. J. Herpetol. 25:79–90

Vitt LJ, Blackburn DG (1983) Reproduction in the lizard *Mabuya heathi* (Scincidae): A commentary on viviparity in New World *Mabuya*. Canadian J. Zool. 61:2798–2806

Vitt LJ, Blackburn DG (1991) The ecology and life history of *Mabuya bistriata* (Scincidae), a viviparous Amazonian lizard. Copeia 1991:916–927

Weekes CH (1930) On placentation in reptiles. II. Proc. Linn. Soc. N.S.W.

55:550–576
Weekes CH (1935) A review of placentation among reptiles, with particular regard to the function and evolution of the placenta. Proc. Zool. Soc. Lond. 2:625–645
Wourms JP (1981) Viviparity: The maternal-fetal relationship in fishes. Amer. Zool. 21:473–515
Yaron Z (1985) Reptile placentation and gestation: structure, function, and endocrine control. In Gans C, Billet F (eds.). Biology of the Reptilia, Vol 15, Development B. John Wiley & Sons, New York. pp. 527–603

12

The Reproductive Biology of the Neotropical Boid Genus *Epicrates* (Serpentes:Boidae)

Peter J. Tolson

Introduction

The boid genus *Epicrates* has a broad geographic distribution, including the islands of the Greater Antilles, the Bahamas, and Central and South America from Costa Rica southward through the Amazon Basin into northern Argentina. Nearly every major island or bank in the Bahamas and western Caribbean contains an endemic species of *Epicrates*, yet in Central and South America only one widely distributed species, the rainbow boa *Epicrates cenchria*, is present. In recent years the reproductive biology of this genus has been examined (Tolson 1980; 1989; Tolson et al. 1984; Teubner 1986; Tolson and Teubner 1987; Teubner et al. in press) and summarized by Tolson and Henderson (in press). Interest in the genus developed in part because three members of the genus are on the Endangered Species List, but also because reproductive activity in *Epicrates* can serve as a model for reproduction in other South American boid snakes. This is critical information, as the reproductive biology of most South American snakes is poorly understood, and the little data that exist generally consist of anecdotal accounts (Mole 1924; Beebe 1946) or limited observations on courtship, litter size, and reproductive timing recorded from captive specimens in zoos or private collections (see Huff 1980; Blody and Mehaffy 1989; Ross and Marzec 1990). The following chapter is a brief review of *Epicrates* reproductive biology, based primarily on the observations and research of the author and his associates over the past ten years and borrowed heavily from Tolson and Henderson (in press). This chapter is divided into three main topics: reproductive timing, reproductive strategies, and reproductive behavior.

Reproductive Timing

Available evidence from both field observations and data collected in captivity indicates that the courtship and mating of *Epicrates* is seasonal and occurs throughout the spring (January through May) depending on the species in question (Murphy and Guese 1977; Huff 1978; Tolson 1980; Bloxam 1982). At the Toledo Zoological Gardens, where all ten species of *Epicrates* have been maintained, the peak month for reproductive activity is February. The earliest breeders, *E. cenchria* and *E. fordi*, normally mate in late December or January, while *E. angulifer*, the species breeding latest in the spring, mates in May or June. There are few records of copulation in the wild, but Mole and Urich (1894) described courtship in *E. cenchria* in January, Grant (1933) reported a pair of Puerto Rican boas, *E. inornatus*, copulating shortly after capture on 19 May 1932, and a native of North Caicos brought the senior author a pair of *E. chysogaster* discovered *in copula* on 8 April 1974. Nearly all gravid female *Epicrates* captured are taken in the summer months (see Sheplan and Schwartz 1974), and when these animals are kept in captivity until parturition, invariably (except for *E. angulifer*, *E. cenchria*, and *E. fordi*) give birth between August and October. There seems to be an inherent degree of plasticity in reproductive timing in *Epicrates*, as both wild and captive populations show variability in the onset of courtship. A series of eight female *E. fordi* collected by the author at Thomazeau, Haiti, on 8 May 1977 all gave birth between 21 and 28 May 1977, whereas an additional gravid female collected at Petionville, Haiti, on 9 May 1977 gave birth on 24 August 1977—three months later than the other conspecifics. If the assumption is made that gestation periods and thermoregulatory behavior were similar for the two populations, then copulation dates could have differed between the two populations by as much as three months. Despite this plasticity, in nearly every case, either direct observations of courtship or courtship dates inferred from parturition dates indicate a clear seasonal pattern of spring courtship in this genus.

Epicrates displays an associated type of reproductive pattern, wherein there is a close temporal synchrony in the timing of mating with an increase in gonadal activity (Crews 1980; 1984). This includes a pattern of prenuptual vitellogenesis; follicle size increases very rapidly over the course of the reproductive season. Much speculative information has been published in regards to the proximate environmental factors which may control gonadal recrudescence in boids; however, most of this information has no supporting data and is not worthy of mention. Teubner (1986) and Teubner et al. (in press) monitored plasma testosterone levels (PTLs) in a sample (N = 35) of male *E. striatus* exposed to different environmental regimes and found that a winter temperature drop (ca. 5 to 10 degrees C) was essential in stimulating testicular recrudescence (Fig. 12.1). Walsh and Davis (1983) reported that mating in Brazilian rainbow boas (*E. c. cenchria*) occurred during increases in temperature fluctuations or a decrease in nighttime low temperatures.

Another environmental cue, increased springtime humidity, had a synergistic effect in elevating PTLs in male *E. striatus* when combined with exposure to lower winter temperatures. This is consistent with the observations

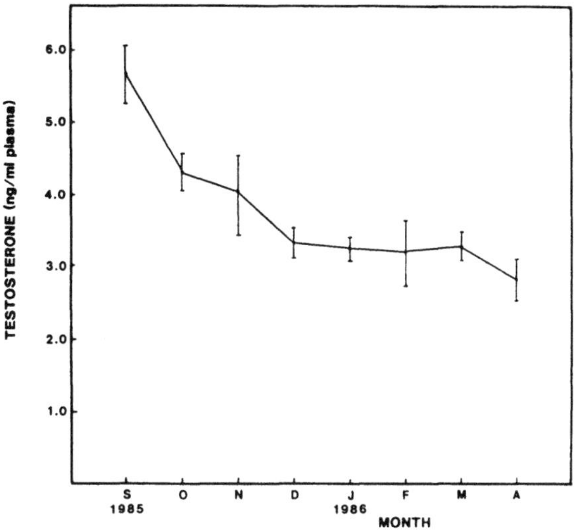

Figure 12.1. Mean plasma testosterone levels (ng/ml plasma) (SE) of male *E. striatus* maintained at a constant ambient temperature of 28°C throughout the reproductive cycle. Figure from Teubner 1986.

of Huff (1978), who reported that increased humidity (particularly heavy misting of the snakes) enhanced reproductive activity of *E. inornatus*. Climatological data (Anon. 1963) indicate that most islands of the Greater Antilles have an annual rainfall peak during the spring, and another in late summer or early autumn, so the onset of spring rains could serve as an important cue for reproduction. As might be expected with crepuscular or nocturnal snakes, photoperiod apparently has little or no effect on reproductive timing in boids (see Ross and Marzec 1990), and alteration of daylength did not affect PTLs in the Haitian boa (Teubner 1986).

Most investigators posit that the evolutionary basis for defined breeding seasons dependent on specific environmental cues is the need for a mechanism to ensure parturition during a period in which survival of the neonates is optimal. In the Antilles, neonate *Epicrates* are usually born during August and September when hatchling *Anolis* populations are substantial and rainfall is abundant. Shine (1977) suggested that spring ovulation is adaptive in snakes of temperate localities, ensuring that the females will be gravid during a season when high body temperatures are easiest to obtain. Although temperatures are warmer during the summer in the West Indies (Anon. 1963), year-round ambient temperatures are probably high enough to ensure adequate rates of embryonic development in Antillean boas, even in winter.

Two lines of research suggest that social interactions also play a crucial role in reproductive timing. Teubner (1986) discovered that isolated male

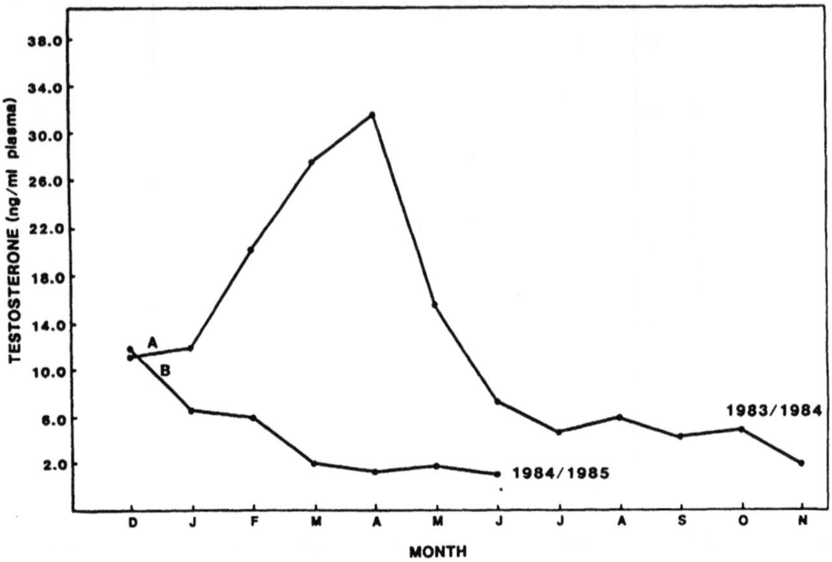

Figure 12.2. Comparison of mean plasma testosterone levels (ng/ml plasma) of male *E. striatus* given a winter temperature drop and (A) exposed to female conspecifics and (B) isolated from female conspecifics. Figure from Teubner 1986.

E. striatus never underwent testicular recrudescence in the absence of conspecific females (Fig. 12.2), even though the appropriate environmental cues were provided. In an additional study, Tolson et al. (1985) using ultrasonography, provided evidence that female *E. angulifer* kept isolated from males during the reproductive period exhibited only partial ovarian recrudescence. Ovarian follicles developed only to a limited degree (e.g., ca. 30 mm diameter) before becoming atretic. The ova were resorbed and never ovulated. Only those females exposed to courtship developed mature ovarian follicles (>100 mm diameter; Fig. 12.3) which were subsequently ovulated.

Reproductive Strategies

There appear to be three basic patterns of clutch and offspring size within the genus *Epicrates*. Smaller species (<1 m snout-vent length [SVL]), such as *E. fordi*, *E. monensis*, and *E. gracilis*, give birth to litters of two to ten small neonates with body masses of 2.0 to 6.4 g and SVLs of 200 to 350 mm. Relative clutch masses (RCMs) for females of these species range from 0.192 to 0.493. Medium sized species (1 to 2 m SVL), such as *E. chrysogaster*, *E. inornatus*, *E. striatus*, and *E. subflavus*, give birth to larger litters of ca. 10 to 40 neonates with masses of 7 to 22 g, with SVLs ranging

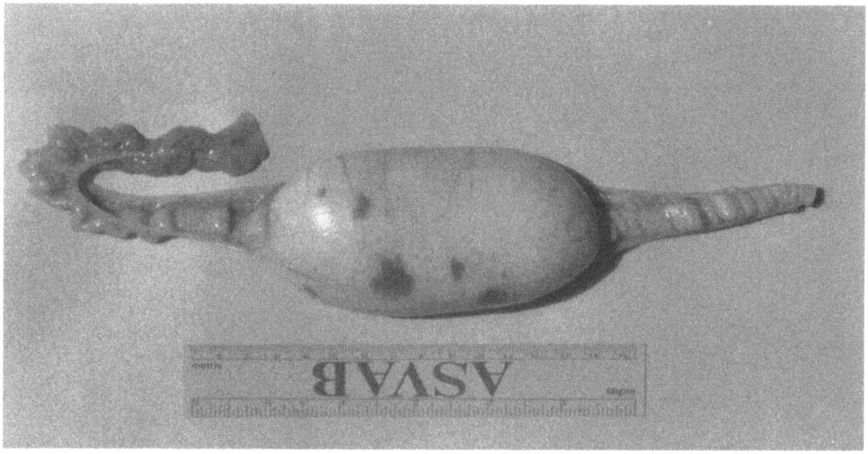

FIGURE 12.3. A recently-ovulated ovum collected post-mortem from a female *E. angulifer.*

in the vicinity of 320 to 470 mm and RCMs of 0.147 to 0.363. *Epicrates angulifer* is the only insular *Epicrates* with truly large neonates with body masses repeatedly exceeding 180 g and SVLs exceeding 600 mm. Tolson and Henderson (in press) reported RCMs of 0.145 to 0.415 for this species, but no data has been reported on RCMs of very large *E. angulifer* (e.g., SVL > 3000 mm; Fig. 12.4). The continental *E. cenchria* produces neonates with body masses of ca. 50 g—larger than any of the insular *Epicrates* except *E. angulifer.*

Although Seigel et al. (1986) hypothesized that small viviparous species may be able to bear a greater RCM than larger species because of differences in longevity, effects of gravity, or habitat utilization, RCMs of larger *Epicrates* in captivity have sometimes approached the maximum RCMs exhibited by the smaller species. No larger *Epicrates*, however, has exceeded the highest RCMs exhibited by the smaller species. Although captive snakes invariably have higher lipid reserves than wild individuals, Ford and Seigel (1989) demonstrated that RCM was fixed relative to diet in a viviparous colubrid.

There appears to be a trend within the genus for larger females of a given species to give birth to a greater number of larger offspring than their smaller conspecifics. However, large females (not necessarily reproductively senescent) may, on occasion, give birth to litters well below their reproductive capacity in terms of clutch size and RCM. In general, *Epicrates* reproduction follows a pattern which might be predicted by evolutionary theory. Smaller species, which are vulnerable to predation from a wide

Figure 12.4. An adult female *E. angulifer* captured on the U.S. Naval Base, Guantanamo Bay, Cuba. The snake was released unharmed.

variety of organisms, would be expected to make a heavier investment in each litter, as the probability of reproduction in subsequent years is lower than that for a species not subject to the same predation pressures.

It might also be predicted that smaller species of *Epicrates* would became sexually mature at an earlier age than larger species, which presumably live longer and would thus have longer reproductive lives. Unfortunately, field data on these parameters are non-existent in the literature. In captive collections, where food intake may be maximized to promote growth, small species of *Epicrates* can reach reproductive size (ca. 100 g mass; 600 mm SVL) within two years (Tolson unpubl.). Males can reproduce at shorter SVLs (and presumably younger ages) than females. One female specimen of *E. monensis* at the Toledo Zoological Gardens reached an SVL of 830 mm (185 mm caudal length [CL]) by two years of age, which is larger than the size record (1010 mm total length) for this species reported by Meerwarth (1901). A male *E. angulifer* reached breeding size (1613 mm SVL; 2.96 kg mass) at three years of age. While this growth rate may seem extreme, certain individual *E. monensis* in a wild population under study have shown equivalent growth (Tolson, unpubl.). Huff (1978) reported sexual maturity for female *E. inornatus* at 1600 mm SVL and ca. 1.0 kg body mass; female *E. subflavus* reach sexual maturity in six to seven years

TABLE 12.1. Selected reproductive data for insular *Epicrates*.

Species	Litter size	Neonte body mass	Relative clutch mass	Sources
E. angulifer	2–28	88.0–203.6 g	0.145–0.415	Huff 1976; Nowinski 1977; Tolson 1983
E. chrysogaster	9–23	7.3–8.4 g	0.241	Tolson 1980; Buden 1975
E. exsul	2–11	4.5–7.3 g	0.277	Tolson unpubl.
E. fordi	2–7	3.6–5.3 g	0.237–0.599	Murphy and Geuse 1977; Tolson unpubl.
E. gracilis	2–5	2.0–4.1 g	0.192–0.300	Murphy et al. 1978; Tolson unpubl.
E. inornatus	12–32	12.0–21.7 g	0.189–0.216	Grant, 1932; Tolson unpubl.
E. monensis	2–10	3.6–6.4 g	0.275–0.343	Tolson 1989; unpubl.
E. striatus	6–51	12.9–19.1 g	0.160–0.319	Hanlon 1964; Tolson unpubl.
E. subflavus	11–33	11.3–9.0 g	0.147	Huff 1979b; Bloxam 1982; Tolson unpubl.

after birth (Huff 1979b). At the Toledo Zoological Gardens, a female *E. striatus* reproduced successfully at four and one-half years of age.

The reproductive value of certain species of *Epicrates* may be considerable. Tolson (1991) reported that individual female *E. fordi*, *E. inornatus*, *E. monensis*, and *E. subflavus* successfully reproduced at ages approaching or exceeding 17 years, and thus the potential lifetime reproductive output of a long-lived female is very great. Table 12.1 reports a summary of selected reproductive data for Antillean *Epicrates*.

Another important aspect of reproductive strategy is the frequency of reproduction. It has been widely reported that *E. cenchria* reproduces annually (Brunner 1979; Vitt and Vangilder 1983; Walsh and Davis 1983), and at least two species of insular *Epicrates* show evidence of annual reproduction. A series of eight female *E. fordi* collected at Thomazeau, Haiti, in May 1977 were all gravid when collected and later gave birth in captivity. Bloxam (1982) recorded three successive years of reproduction for a group of Jamaican boas, *E. subflavus*, housed in captivity at the Jersey Wildlife Preservation Trust.

Gravid females, because of their choice of sites for thermoregulation during the day (such as termitaria and debris piles), are often more easily discovered by collectors, and the preponderance of gravid females in the *E. fordi* sample may be due to sampling error. Other species may reproduce biennially. Huff reported that captive female *Epicrates* (including *E. subflavus*) usually reproduce biennially, and Tolson (1988) observed that ca. 50% of adult (>700 mm SVL) female *Epicrates monenis* were gravid each year on Cayo Diablo, Puerto Rico. This provides evidence that a biennial reproductive cycle exists in the wild for at least one insular species of *Epicrates*. It is quite possible that there are species and/or population differences in reproductive frequency. Many factors can alter female reproductive frequency in snakes, including age, size, amounts of lipid reserves, food availability (Seigel and Ford 1987), and perhaps even environmental cues such as rainfall. More long-term data are needed on the reproductive frequency of wild populations before conclusions can be drawn.

Reproductive Behavior

Male Combat

In certain species of *Epicrates*, males exhibit intense competition for females during the reproductive period. This competition may take one of two forms: a highly ritualized combat similar to that of elapids and viperids, wherein male combatants maneuver to pin the body of the opponent to the substrate or, alternatively, intense aggression, wherein combatants bite and constrict each other with the clear purpose of injury or expulsion from the breeding area. In captivity, only certain elements of these behaviors may be manifested. Tolson (1983) described presumably complete sequences of ritualized male combat in *E. angulifer*. Males crawled together slowly with the posterior portions of the body, including the tails, entwined. This behavior escalated into attempts to pin the opponent to the substrate. Ritualized combat behavior has also been observed in males of *E. exsul* by the author on 24 February, 1 March, 6 March, and 7 March 1984. Extreme agonistic encounters have been observed by the author for *E. inornatus* and *E. subflavus*. Teubner (1986) described intense fighting between male *E. striatus* that deterred smaller males from courtship. When PTLs of these snakes were examined, the dominant alpha males were found to have PTLs much times higher than those of the beta males (70 to 75 ng/ml testosterone vs. 10 to 12 ng/ml testosterone; Fig. 12.5). Plasma testosterone levels ol the beta males increased (with concomitant courtship activity and copulation) after alpha males completed courtship and copulation and aggression ceased. Thus it appears that large dominant males may actually delay gonadal recrudescence in smaller, less aggressive males. Teubner (1986) demonstrated that a sample of 14 Peak I dominant males had mean body masses of 961.6 g, in contrast to the mean body mass of 534.8 g for a sample of 11 beta males. In instances of extremely aggressive combat, males often may have to be separated to avoid injury to one or both snakes. Huff (1978) suggested that placing males together prior to introducing the female enhanced breeding success in *Epicrates*. Interference competition has also been reported for *E. angulifer* by Tolson (1983). It is open to question as to whether these agonistic encounters would be as intense in the wild, as the opportunities for escape of the beta male would doubtless be much greater.

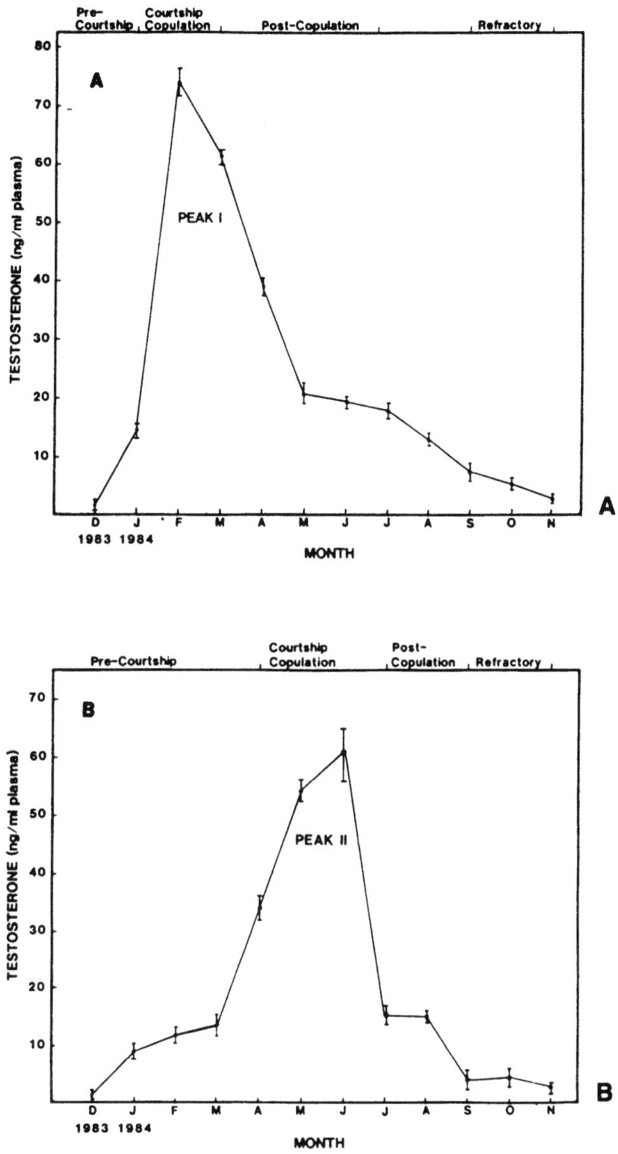

FIGURE 12.5. (A) Mean plasma testosterone levels (ng/ml plasma) ± SE for individual male *E. striatus* that had mean body masses of 961 g 156.8 and were maintained under Haitian climatic conditions. (B) Mean plasma testosterone levels (ng/ml plasma) ± SE for individual male *E. striatus* that had mean body masses of 534.8 g ± 99.1 and were maintained under Haitian climatic conditions.

In captivity males usually signal the onset of reproductive condition by increased restlessness and the shedding of the hemipenial epithelium. The epithelium is easily mistaken for sperm without microscopic examination. The commencement of male courtship activity closely parallels the shedding of this epithelium.

COURTSHIP BEHAVIOR

Boid and tropidophiid snakes are characterized by a complex series of extended courtship behaviors which may last for several weeks. These behaviors were first described for *Epicrates cenchria* by Mole and Urich (1894). Tolson (1980) described the courtship behaviors for several species of *Epicrates*. Following the nomenclature of Gillingham et al. (1977), Tolson (1980) divided *Epicrates* courtship into four distinct phases: tactile chase, tactile alignment, tail search, and copulation.

The tactile chase phase, the initial phase of courtship, seems to focus on the identification of the prospective mate, e.g., species and sexual recognition (Gillingham and Dickinson 1980; Ford 1986). There is usually no intertwinement during this phase; the male slowly crawls alongside the female, making numerous tongue flicks, and the female responds by tongue flicking. Both snakes may be crawling during this phase of courtship. During tactile chase, males may exhibit undulation, dorsal body looping, chin rubbing, and dorsal advance behaviors (see Gillingham 1987 for precise definitions of these acts) but usually in *Epicrates* the tactile chase phase is primarily confined to chase or chase-mount behaviors on the part of the male.

Once the female is more or less stationary, the tactile alignment phase commences. The male drapes several loose coils over the dorsum of the female while keeping his chin in constant contact with the female's dorsal surface. This behavior may serve as an autoerotic function in the male (see Noble 1937). Undulation, dorsal body looping, chin rubbing, and dorsal advance continue with increased intensity. Tolson (1980) hypothesized that, during tactile alignment, the efforts of the male were directed towards keeping the female stationary so that tail search copulatory attempt (TSCA) the behavioral phase preceding copulation, can occur.

Tail search in *Epicrates* differs considerably from other species of snakes, including boids, in both the duration and intensity of the activity; for this reason Tolson (1980) viewed TSCA as a phase of courtship distinct from tactile alignment. During TSCA, all males of the long-tailed insular species of *Epicrates* wrap four to five coils about the female and move them in concertina-like fashion up and down the body of the female in the area of the cloaca. The lone exception is *E. angulifer*. In this species the male merely drapes his tail over the female or encloses her with a single coil. It is during TSCA that the spurs of the male come most prominently into use. During the initial stages of tail search, the spurs are used to stroke the female in a stimulatory fashion, but, during the latter stages, just prior to copulation, the male may use the spurs to dig sharply into the interstitial skin of the female to align her properly for copulation (Murphy et al. 1978; Tolson 1983). After sufficient courtship, the female will usually open the cloaca and allow intromission. In *Epicrates*, copulation cannot be accomplished by the male until the female performs this behavior.

A reluctant female can eventually be stimulated to court by a persistent male, and female receptability plays a large part in the degree of courtship required by the male. In some cases the entire courtship/copulatory sequence may last less than 30 minutes; in others, courtship may continue virtually uninterrupted for over two weeks. Actual copulation may last for as little as 30 minutes or extend over several hours. No copulatory plugs (Devine 1975) have been described for any species of *Epicrates*. The basic elements of *Epicrates* courtship and reproductive timing seem well suited for a solitary snake such as *Epicrates*. A comparison of mating behavior in aggregated vs. dispersed populations of *Epicrates* might be instructive in elucidating the evolution of *Epicrates* reproductive behavior.

Gestation

As in other poikilotherms, the duration of the gestation period is dependent on the thermal activities of the gravid female. Tolson (1988), in a long-term field study of *E. monensis*, found that gravid females would regularly elevate their body temperatures above 30° C by resting in termitaria or under debris exposed to the sun. There is considerable variance in gestation period even within the same species. For example, in *E. angulifer*, gestation varied from a low of 152 days (Huff 1976) to a maximum of 252 days (Nowinski 1977). In the latter case, it was probable that sperm was stored by the female prior to actual fertilization (see Huff 1980). Table 12.2 presents some gestational data for selected species of insular *Epicrates*.

Neotropical boid snakes will undoubtedly be subject to greater degrees of endangerment as Neotropical habitats continue to be relentlessly destroyed. Some species, especially those endemics with limited distributions,

Table 12.2. Gestation periods for insular *Epicrates*.

Species	Gestation	Sources
E. angulifer	152–252 days	Huff 1976; Nowinski 1977; Tolson 1983
E. chrysogaster	238 days	Tolson unpubl.
E. exsul	147–159 days	Tolson unpubl.
E. fordi	139–141 days	Murphy and Geuse 1977
E. inornatus	153–176 days	Huff 1978
E. monensis	132 days	Tolson 1989
E. striatus	216–224 days	Tolson unpubl.
E. subflavus	160–220 days	Huff 1979b

face an uncertain future in the wild. It is therefore incumbent upon managers of boid snakes to expand the base of knowledge dealing with boid reproduction and develop the reproductive technologies necessary to reproduce these species in captivity. Failure to do so may result in the immeasurable loss of some of the most ancient and unique elements of the Neotropical biota.

Acknowledgments

The author is indebted to the many agencies and individuals who collectively have made fieldwork in the Antilles during the past fifteen years a productive experience. Special thanks are due the Departamento de Recursos Naturales of Puerto Rico; the U.S. Fish and Wildlife Service Caribbean Field Office; the Division of Fish and Wildlife, U.S. Virgin Islands; the U.S. Marine Corps, Marine Barracks, Guantanamo Bay, Cuba; the Division of Herpetology, Museum of Zoology, University of Michigan; and the Department of Herpetology, Toledo Zoological Society. Financial support for the author was provided by the Division of Wildlife, U.S. Virgin Islands, and the Institute of Museum Services (Conservation Grants IC-70095-87 and IC-90453-89).

References

Anonymous (1963) Tables of temperature, relative humidity and precipitation for the world. Part II. Central and South America, the West Indies, and Bermuda Her Majesty's Stationery Office, London. viii + 53 pp

Beebe W (1946) Field notes on the snakes of Kartabo, British Guiana, and Caripito, Venezuela. Zoologica 31:11–52

Blody DA, Mehaffey DT (1989) The reproductive biology of the annulated boa in captivity. Int. Zoo Yrbk. 28:167–172

Bloxam QMC (1982) Über Haltung und Zucht der Jamaika-Schlankboa (*Epicrates subflavus*) Stejneger (1901) im Jersey Zoo. Zeitschr. Kölner Zoo 25(4):133–137

Brunner JC (1979) Captive breeding of Colombian rainbow boas, *Epicrates cenchria crassus*. Proc. Third Ann. Reptile Symp., Zoological Consortium, Inc., Thurmont, Maryland. pp. 39–47

Buden DW (1975) Notes on *Epicrates chrysogaster* (Serpentes:Boidae) of the southern Bahamas with description of a new subspecies. Herpetologica 31(2): 166–177

Crews D (1980) Studies in squamate sexuality. Bio. Sci. 30(12):835–838

Crews D (1984) Gamete production, sex hormone secretion and mating behavior uncoupled. Hormones and Behav. 18:22–28

Devine MC (1975) Copulatory plugs in snakes: enforced chastity. Science 187:884–885

Ford N (1986) The role of pheromone trails in the sociobiology of snakes. In Duvall D, Muller-Schwarzc D, Silverstein R (eds.). Chemical Signals in Vertebrates. Plenum Press, New York. 378 pp

Ford N, Seigel RA (1989) Phenotypic plasticity in reproductive traits: Evidence

from a viviparous snake. Ecology 70(6):1768–1774
Gillingham JC (1987) Social behavior. In Siegel RA, Collins JT, Novak SS (eds.). Snakes: Ecology and Evolutionary Biology. MacMillan, New York. pp. 184–209
Gillingham JC, Carpenter CC, Brecke BJ, Murphy JB (1977) Courtship and copulatory behavior of the Mexican milksnake, *Lampropeltis triangulum sinaloae* (Colubridae). Southwest Nat. 22:187–194
Gillingham JC, Dickinson JA (1980) Postural orientation during courtship in the eastern garter snake, *Thamnophis sirtalis*. Behav. Neural Biol. 28:211–217
Grant C (1932) Notes on the boas of Puerto Rico and Mona. J. Dept. Agri. Puerto Rico 16:327–329
Grant C (1933) Notes on *Epicrates inornatus* (Reinhardt). Copeia 1933(4):224–225
Hanlon RW (1964) Reproductive activity of the Bahama boa (*Epicrates striatus*). Herpetologica 20(4):143–144
Huff TA (1976) Breeding the Cuban boa *Epicrates angulifer* at the Reptile Breeding Foundation. Int. Zoo Yrbk. 16:81–82
Huff TA (1978) Breeding the Puerto Rican boa at the Reptile Breeding Foundation. Int. Zoo Yrbk. 18:95–97
Huff TA (1979a) Captive propagation and husbandry of *Epicrates* at the Reptile Breeding Foundation. Proc. 2nd Reptile Symp. on Captive Propagation and Husbandry, Zoological Consortium, Inc., Thurmont, Maryland. pp. 103–112
Huff TA (1979b) Breeding the Jamaican boa (*Epicrates subflavus*) in captivity. Amer. Assoc. Zool. Parks and Aquariums. Reg. Workshop. Proc., Wheeling, West Virginia. pp. 339–345
Huff TA (1980) Captive propagation of the subfamily Boinae with emphasis on the genus *Epicrates*. In Murphy JB, Collins JT (eds.). Reproductive Biology and Diseases of Captive Reptiles. Soc. Study Amphibians and Reptiles Contrib. Herpetology, No. 1. pp. 125–134
Meerwarth H (1901) Die Westindischen Reptilien und Batracher des Naturhistorischen Museums im Hamburg. Mitth. Naturh. Mus. Hamburg 18:1–41
Mole RR (1924) The Trinidad snakes. Proc. Zool. Soc. London 1924:235–278
Mole RR, Urich FW (1894) Biological notes upon some of the ophidia of Trinidad, B.W.I., with a preliminary of the species recorded from the island. Proc. Zool. Soc. London 1894:499–519
Murphy JB, Guese RK (1977) Reproduction in the Hispaniolan boa *Epicrates fordii fordii* at Dallas Zoo. Int. Zoo Yrbk. 17:132-133
Murphy JB, Barker DG, Tryon BW (1978) Miscellaneous notes on the reproductive biology of reptiles. 2. Eleven species of the family Boidae, genera *Candoia*, *Corallus*, *Epicrates*, and *Python*. J. Herpetology 12(3):385–390
Noble GK (1937) The sense organs involved in the courtship of *Storeria*, *Tamnophis*, and other snakes. Bull. Amer. Mus. Nat. Hist. 73:673–725
Nowinski B (1977) Voortplanting van de Cubaanse boa, *Epicrates angulifer*, in het terrarium. Lacerta 35(10):144–147
Ross RA, Marzec G (1990) The reproductive husbandry of pythons and boas. Institute of Herpetological Research, Stanford, California. 270 pp
Seigel RA, Fitch HS, Ford NB (1986) Variation in relative clutch mass in snakes among and within species. Herpetologica 42(2):179–185
Seigel RA, Ford NB (1987) Reproductive ecology. In Siegel RA, Collins JT, Novak SS (eds.). Snakes: Ecology and Evolutionary Biology. MacMillan, New York. pp. 184–209

Sheplan BR, Schwartz A (1974) Hispaniolan boas of the genus *Epicrates* and their Antillean relationships, Ann. Carnegie Mus. 45(5):57–143
Shine R (1977) Reproduction in Australian elapid snakes. II. Female reproductive cycles. Aust. J. Zool. 25:655–666
Teubner VA (1986) Endocrine and behavioral aspects of the reproductive cycle of the Haitian boa (*Epicrates striatus*). Ph.D. Dissertation, University of Toledo, Toledo, Ohio. 86 pp
Teubner VA, Tolson PJ (1984) Behavioral and endocrine aspects of the reproductive cycle in the Haitian boa *Epicrates striatus.* Amer. Zoologist 24:114A
Teubner VA, Tolson PJ, Black CP (in press) The reproductive cycle of the male Haitian boa, *Epicrates striatus.* Gen. Comp. Endocrinology.
Tolson PJ (1980) Comparative reproductive behavior of four species of snakes in the boid genus *Epicrates.* Proc. 4th Reptile Symp. on Captive Propagation and Husbandry, Zoological Consortium, Inc., Thurmont, Maryland. pp. 87–97
Tolson PJ (1983) Captive propagation and husbandry of the Cuban boa, *Epicrates angulifer*. Proc. 6th Reptile Symp. on Captive Propagation and Husbandry, Zoological Consortium, Inc., Thurmont, Maryland. pp. 285–292
Tolson PJ (1988) Critical habitat, predator pressures, and management of *Epicrates monensis* on the Puerto Rico Bank: A multivariate analysis. U.S.D.A. Forest Service General Technical Report RM-166:228–238
Tolson PJ (1989) Breeding the Virgin Islands boa, *Epicrates monensis granti*, at the Toledo Zoological Gardens. Int. Zoo Yrbk. 28:163–167
Tolson PJ (1991) *Epicrates* (West Indian Boa). Reproductive Longevity. Herpetological Rev. 22(3):100
Tolson PJ, Henderson RW (in press) The taxonomy and natural history of West Indian boas. R&A Publishing Ltd., Portishead, England.
Tolson PJ, Reichard TA, Black CP, Slovis T (1984) Determination of litter size and embryo viability in the Cuban boa, *Epicrates angulifer*, using imaging ultrasonography. In Proc. 7th Reptile Symp. on Captive Propagation and Husbandry, Zoological Consortium, Inc., Thurmont, Maryland. pp. 118–123
Tolson PJ, Reichard TA, Terry JM, Roberts CS, Clifford DH (1985) Effects of courtship on follicular growth and development in the Cuban boa, *Epicrates angulifer* (Serpentes, Boidae) Proc. SSAR/HL Ann. Meeting, (abs.), Tampa, Florida. p. 80
Tolson PJ, Teubner VA (1987) The role of social manipulation and environmental cycling in propagation of the boid genus *Epicrates*: Lessons from the field and laboratory. Amer. Assoc. of Zoological Parks and Aquariums Regional Conf. Proc. 1987:606–613
Vitt L, Vangilder LD (1983) Ecology of a snake community in northeastern Brazil. Amphibia-Reptilia 4:273–296
Walsh T, Davis R (1983) Husbandry and breeding of the Brazilian rainbow boa *Epicrates cenchria cenchria* at the National Zoological Park. Proc. 7th Ann. Symp. on Captive Propagation and Husbandry, Zoological Consortium, Inc., Thurmont, Maryland. pp. 108–112

Aves

13

Reproductive Energetics and Physiology of the Desert-Nesting Gray Gull, *Larus modestus*

Lloyd C. Fitzpatrick, Carlos G. Guerra, and Marcos A. Cikutovic

Background

Seagulls (*Larus*, Laridae, Charadriiformes) are one of the most cosmopolitan of avian genera. Their 44 species occupy a wide diversity of habitats between 80° north and 80° south latitude (Harrison 1983). Found from sea level along the coasts of each continent and shores of most islands, to the highest Andean lakes, larids inhabit and reproduce under extremely varied ambient conditions, several in some of the most interesting, challenging, and remote areas on earth. Among those, the gray gull, *Larus modestus*, is truly unique. Dispersed from 0° to 40° south latitude along the Pacific coast of South America during the austral winter, gray gulls congregate during their August to February breeding-nesting season on the coast of northern Chile between Arica (18° S) and Tatal (26° S), where they court, mate, and feed but do not nest. Instead, unlike most other congeners, they nest far from food and water on the barren pampas of the Atacama (Goodall et al. 1945; Howell et al. 1974; Guerra et al. 1988a), a desert so desolate that Moffett (1969) characterized it as "biologically bankrupt." Nesting in the waterless pampas, as far as 100 km from the coast (Guerra et al. 1988a) between 22° and 24° south latitude (Fig. 1), exposes the gray gulls to intense solar radiation and extreme diurnally-variable ambient conditions (e.g., ground temperature = 2° to 61°C, air temperature = 2.5° to 38°C, wind = 0 to 55 km/h; Howell et al. 1974; Fitzpatrick and Guerra 1988; Guerra et al. 1988a) and requires them to commute long distances for food and water.

As early as July or August, large flocks catch the rising thermals and spiral offshore just before sunset to 1000 to 1500 m, heights sufficient to clear the coastal mountains, and then depart between 2100 and 2200 hours, when the desert air is calm, for nesting sites in the Atacama to establish territories (Howell et al. 1974; our observations). They depart the desert at ca. 0400 hours, catching a tailwind from the Andes and arriving just before sunrise. Prior to laying, which normally begins in November and lasts into February, breeding pairs return to the coast before sunrise to forage. Dur-

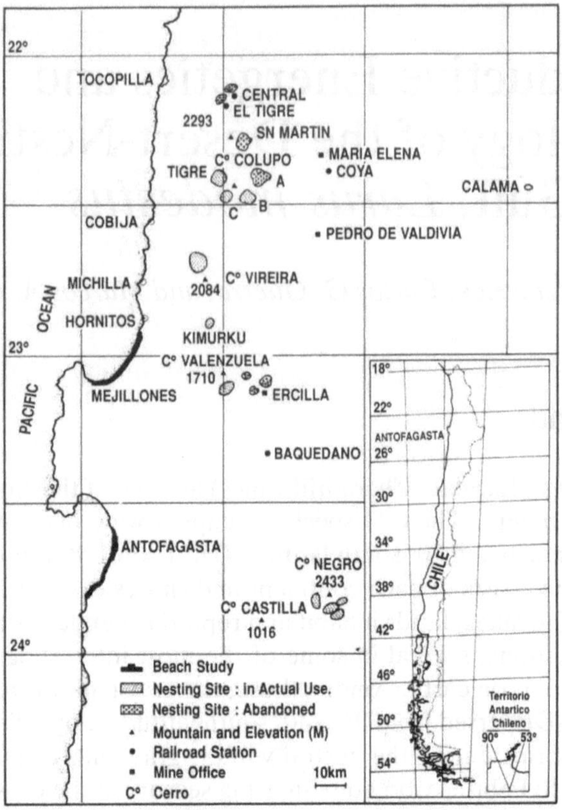

FIGURE 13.1. Map of study areas on the coast and in the Atacama Desert of northern Chile, locating active and inactive gray gull, *Larus modestus*, nesting sites.

ing incubation-brooding (35 to 37 days), members of each pair apparently alternate daily foraging flights. Afterwards, both forage to provision their offspring for about two months. Nesting under severe physiologically challenging desert conditions, at distances from the coast sufficient to limit foraging to once per day, should produce significant constraints on the time, energy, and water budgets in gray gulls and affect their overall life history tactics for resource allocation between somatic maintenance and reproduction. Thus, we expect to find in *L. modestus* some interesting adaptations and evolutionary divergence from other larids, including those inhabiting desert environments.

Regardless of the intriguing nature of their unique breeding biology, gray gulls have received relatively little attention from biologists during the 55 years since Goodall et al. (1945) reported the first Atacama nesting site near the Cerros de Colupo in the Pampa del Miraje (Fig. 13.1). Until Guerra discovered a colony at Cerro Negro (Fig. 13.1) in the early 1980s

(Guerra and Cikutovic 1983a), it was the only known active nesting site for gray gulls in the Atacama. Howell et al. (1974; summarized in Howell 1982) studied the Colupo colony in the early 1970s, providing the only reproductive data on nesting gray gulls before our studies began in 1982. Shortly after their study, the nesting site was abandoned (Devillers and Terschuren 1976). Other published information on gray gulls has been anecdotal, distributional, or about their behavior on the coast (e.g., Murphy 1936; Yañez 1949a, 1949b; Goodall et al. 1951; Cawkell and Hamilton 1961; Philippi 1964; Johnson 1967; Tovar 1968; Moffett 1969; Chapman 1973; Devillers and Terschuren 1976; Braham 1981).

In 1982 we began searching the Atacama for additional nesting colonies, including the large (ca. 10,000 pairs) colony that had abandoned the Colupo site but was still foraging on the beaches between Mejillones and Tocopilla (Fig. 13.1; Guerra et al. 1988a). Our objective has been to locate active colonies so that we could study the life history, behavioral and physiological adaptations, and demographic correlates in the gray gulls that are associated with their desert nesting and provide insight for protection of this unique South American species which is officially considered susceptible to becoming endangered by human activities (e.g., anchovy fishing, mining, egg exploitation). Our purpose here is to summarize what we have learned about the reproductive ecology of the gray gulls, locally called garumas, found on the coast and in the Atacama of northern Chile between 22° and 24° south latitude. We will present principal results of our studies according to a set of hypotheses that has guided our research on *L. modestus* over the past eight years. Since detailed materials, methods, and results are beyond the scope of this review, we refer readers to our previous publications (Guerra and Cikutovic 1983a, 1983b; Guerra and Fitzpatrick 1987; Cikutovic et al. 1988; Fitzpatrick and Guerra 1988; Fitzpatrick et al. 1988a, 1988b; Guerra et al. 1988a, 1988b, 1988c, 1988d; Fitzpatrick et al. 1989; Guerra et al. 1989).

Environmental Setting

Our studies have focused on gray gull demes that forage on the sandy beaches around Mejillones and nest about 35 km inland in the Kimurku-Velenzuela area and those nesting near Cerro Negro ca. 100 km from the rocky coast at Antofagasta where they forage (Fig. 13.1). The hypotheses which have guided our research have derived from the nature of the general environmental setting and location of specific nesting sites and their meteorological profiles.

The typical west-east geographical profile of the study area consists of a 1 to 10 km wide littoral platform, a coastal mountain chain of 600 to 1500 m elevation, an inner desert or nitrate pampas which is the driest area, the Domeyko mountains (up to 3000 m), and, finally, the Andes, rising to ca.

6000 m. The coastal climate is desertic, influenced by latitude, cold Pacific upwelling, and the Humboldt current. Air and coastal water temperatures range normally from 29° to 10°C and 24° to 15°C, respectively, during the summer and winter. Climate of the Atacama, one of the driest deserts on earth, is influenced by latitude, cold upwelling, and rain shadow of the Andes to the east. During nesting (November to February), daily surface and air temperatures range from 2° to 61°C and 2.5° to 38°C, respectively. Diurnal winds from the coast ($\bar{x} = 46$ km/h) occur between 1000 and 1800 hours, and cold dry winds from the Andes ($\bar{x} = 8$ km/h) last from 0300 to 0600 hours. Relative humidity varies between 6 and 99%.

Coastal waters are highly productive because of the nutrient rich upwelling and cold Humboldt current. There are high densities of anchovies, *Engraulis ringens*, and sardines, *Sardinops sagax*, the former representing ca. 80% of the gray gulls' diet. The second most important food item, especially for gray gulls foraging on sandy beaches around Mejillones, is the intertidal sand crab, *Emerita analoga.* Although marine productivity is high and relatively stable, it can be depressed by El Niño-Southern Oscillations (ENSOs). During ENSOs, such as the one during 1982–83, effects on the mortality and reproductive success of numerous seabirds intensify (e.g., Schreiber and Schreiber 1983, 1984; Tovar and Cabrera 1985; Arntz 1986).

Nest Site Location and Characterization

Although general topographic and meteorologic conditions are consistent throughout the pampas, nest sites (Fig. 13.1) are not dispersed randomly; they occur only in areas where the uniform sandy surface is covered with small rocks. Except for a large colony only recently located by Guerra ca. 30 km from the original one nesting at Colupo, active colonies consist of small, widely-dispersed groups or subcolonies as much as 10 km apart, with individual nests frequently separated by several hundred meters. Hyperdispersed nesting within and among subcolonies and asynchronous laying appear to be adaptive responses to anthropogenic exploitation of eggs and increasing predation pressure by turkey vultures, *Cathartes aura*, on eggs and chicks and by peregrine falcons, *Falco peregrinus*, on chicks and adults (see Guerra et al. 1988a).

Nests are of two types: (1) simple shallow scrapes in the sand away from rocks; and (2) shallow depressions surrounded by guano deposits, almost always (>90%) immediately adjacent to the southwest or leeward side (relative to the cold dry night winds from the Andes) of the small rocks (Fig. 13.2). The former are used by adults for incubating eggs (ca. 30 days) and tending chicks for the first week posthatching. During incubation and brooding, nesting pairs alternate daily foraging flights to the coast. Afterwards, both parents forage, leaving the chicks unattended from 0400 to 2400 hours for ca. two months, at which time they are old enough to leave

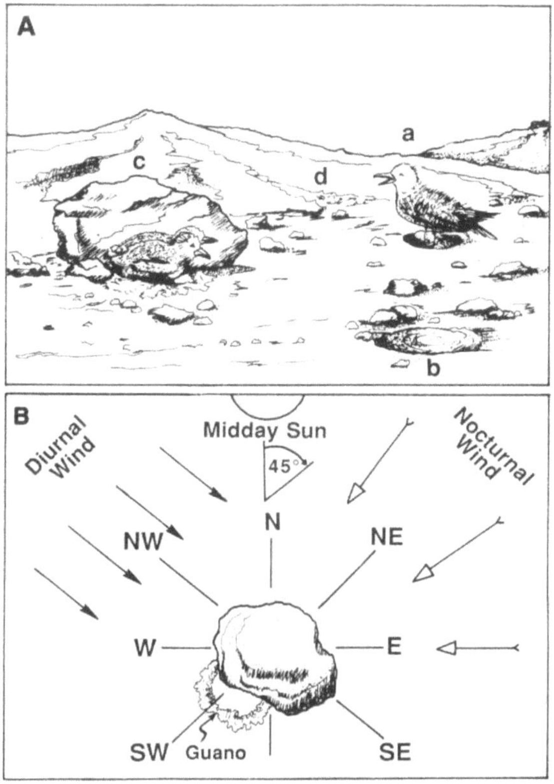

FIGURE 13.2. (A) Schematic view of a typical Cerro Negro desert nesting site, showing (a) an incubating adult thermoregulating and shading eggs, (b) an empty incubation scrape, (c) a chick refuge next to a rock, and (d) a dry channel used by chicks for refuge. (B) Relation of a chick refuge, defined by a ring of guano, to daily winds and sun at midday.

for the coast. Sites next to rocks are used by older chicks as refuges from the cold dry night winds and high daily ground temperatures (T_G). Lowest T_Gs during the hottest period (1100–1400 hours) occur at the southwest side of the rocks, coincident with the chick refuges. Mean T_G between 1000 and 1700 hours, within the chick refuges is 34°C, significantly lower than that (51°C) of the nest scrapes in the open. Between 2300 and 0900 hours, T_Gs are nearly equal within the two nest types. Additionally, during the highest daily temperatures, the refuges receive the greatest exposure to coastal winds, facilitating convective cooling of the chicks. The opposite occurs between 0300 and 0600 hours when the refuges are sheltered from the cold Andean wind. We believe that the gray gulls select rock-strewn areas for their nest sites because of the microclimatic advantages afforded unattended chicks.

Since rocks are usually too small to afford similar protection for the adults, they incubate and tend newly-hatched chicks in the open where they can take advantage of the wind for convective cooling, erecting the feathers of their dark plumage and standing with spread wings during the hottest hours. Erected feathers, especially dark ones, reduce the heat load transmitted to the skin (Walsberg et al. 1978). Unlike Heermann's gulls, *L. heermanni*, which are also gray and reproduce on desert islands off the coast of Baja California, incubating gray gulls must stand for convective cooling over their eggs. Heermann's gulls, which nest close to the water, sit tightly on their eggs and cool evaporatively by panting (Bartholomew and Dawson 1979), an effective thermoregulatory tactic too costly for the water budget of gray gulls. Consequently, in contrast to those of Heermann's gulls, gray gull eggs are exposed during the hottest and driest periods of the day in the Atacama. Additionally, incubating adults leave their nests when turkey vultures approach, apparently for distraction, but thereby exposing the eggs to dry air and direct sunlight.

Adaptations for Desert Nesting

Because of wide daily variation in ambient temperatures, adult nest-tending behavior and long-distance foraging, exposure of the eggs, and long periods during which the offspring are unattended on the desert, we hypothesized that the gray gulls should have one or more of the following when compared to other gulls, including desert nesters: (1) wider thermal neutral zone (TNZ = ambient temperature range within which standard metabolic rates are temperature independent; Schmidt-Nielsen 1985); (2) greater capacity for hypothermia and hyperthermia; (3) lower standard metabolic rates (SMR = rate for inactive post-absorptive gulls within their TNZ; Dawson and Bennett 1973; King 1974), which should decrease evaporative water loss, energy expenditure, and food requirements in nesting inactive gulls; (4) lower eggshell water vapor conductance, which should reduce water loss in exposed eggs; and (5) thermoregulatory behavior which is more efficient in water and energy conservation.

We measured SMR, TNZ and temperature tolerance of gray gulls (see Guerra et al. 1989) to test our first three hypotheses. Metabolic rate (MR) was temperature-independent within the TNZ (26° to 32°C), but increased significantly below and above that range (Fig. 13.3). SMR (0.0209 KJg^{-1} h^{-1}) was significantly lower than SMRs reported or predicted (0.0227 $KJg^{-1}\,h^{-1}$) for other sea birds of the same body mass (Ellis 1974). Patterns of body temperature (T_B) vs. ambient temperature (T_A), which indicate thermoregulation between 14° and 29°C, hypothermia below 14°C and hyperthermia above 29°C (Fig. 13.4), accord with T_As experienced by gray gull young (see Guerra et al 1988a; Fitzpatrick et al. 1989). Hyperthermia reduces both water and energy loss by lowering the T_B-T_A gradient before

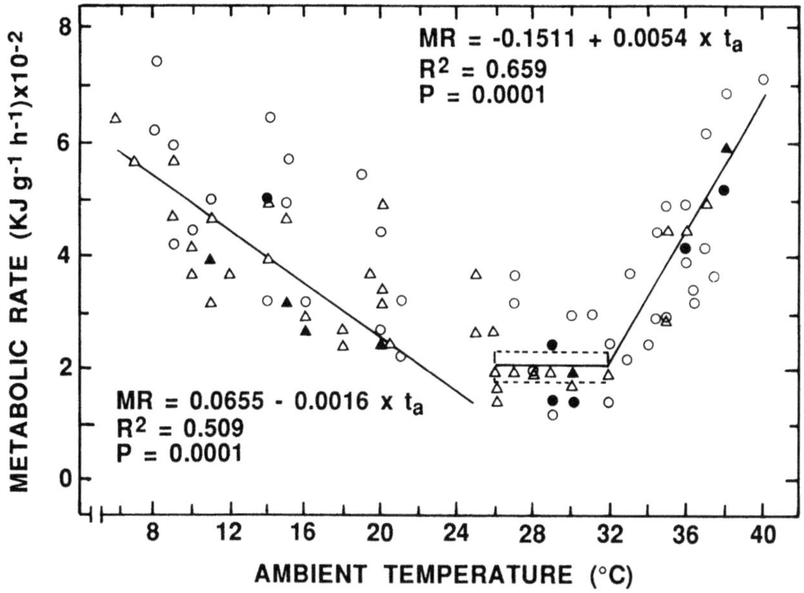

FIGURE 13.3. Metabolic rate (KJ g^{-1} h^{-1}) of Gray Gulls, *Larus modestus*, (218 to 380 g) measured at Air Temperature T_A 6 to 40°C. Regression models for metabolic rates (MR) are given above and below the thermoneutral zone. Mean and 95 percent confidence limits of standard metabolic rate are indicated in the thermoneutral zone, respectively, by continuous and dashed horizontal lines. Circles = 218 to 300 g chicks; triangles = 300 to 380 g gulls; closed figures = more than one identical determination. R^2 = Coefficient of determination and P = Probability.

gray gulls reach the upper TNZ limit (32°C). Gray gulls become hyperthermic at lower T_As than the chicks of desert-nesting California and ring-billed, *L. delawarenesis*, gulls (Dawson et al. 1976; Chappell et al. 1984) and the western plumed pigeon, *Lophopnaps ferruguinea*, (Dawson and Bennett 1973). The major difference between the desert environments of gray gulls and these other desert nesters is the latters' close proximity to water and food. Hyperthermia at relatively lower T_A reduces dependence on evaporative cooling, allowing incubating adult gray gulls to stand above their nest during the hottest hours of the day for convective thermoregulation. Desert-nesting Heermann's gulls sit tightly on their eggs, cooling evaporatively by panting (Bartholomew and Dawson 1979). Since Heermann's gulls, "ecological equivalents" of gray gulls, nest close to water and food at Isla Raza, they can afford evaporative cooling for thermoregulation.

To test the hypothesis that gray gull eggs, because of frequent exposure, should have lower water vapor conductance (G-H_2O) than that of other gulls, including *L. heermanni*, we measured water loss rate (M-H_2O) in the eggs (see Guerra et al. 1988b). Mean M-H_2O and G-H_2O for gray gull eggs ($\bar{x}$ =49 g) were 0.91 mg d^{-1} and 3.70 mg d^{-1} $torr^{-1}$, respectively. The G-H_2O is well below the 95% confidence limit for the model (G-

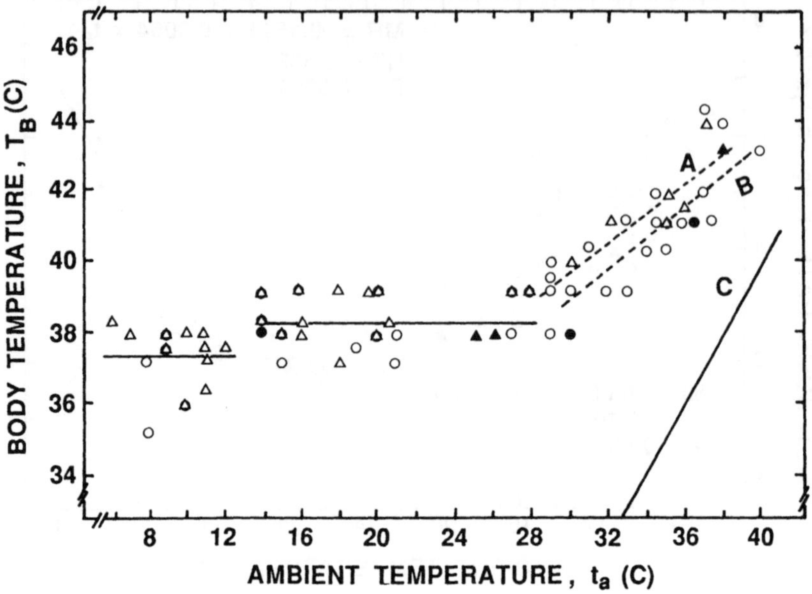

Figure 13.4. Multiple body temperature determinations of 22 Gray gulls, *Larus modestus*, (218 to 380 g) at T_A from 6° to 40°C in a metabolic chamber. Horizontal lines represent mean Body Temperature (T_B) values at $T_A = 6°$ to 14° and 14° to 28°C. (A) and (B) represent regression lines of gulls of 218 to 300 g (circles) and 300 to 380 g (triangles), respectively. Closed figures represent more than one identical determination. (C) shows the relation where $T_B = T_A$.

$H_2O = 0.760\ W^{0.688}$) which we developed from published G-H_2Os for nine larids (Fig. 13.5 Ar et al. 1974; Ar and Rahn 1978; Morgan et al. 1978; Rahn and Dawson 1979). The empirical G-H_2O is nearly one-third lower than that (11.058 mg d^{-1} $torr^{-1}$) predicted by the model for a 49 g egg and for the slightly larger (53.4 g) eggs of Heermann's gulls (Ar and Rahn 1978; Rahn and Dawson 1979). Low G-H_2O in gray gulls is an adaptation to nesting far from water in the Atacama.

We tested the hypothesis that gray gulls should have less costly thermoregulatory behavior than conspecifics by comparing a thermoregulatory index (TRI), similar to the one developed by Bartholomew and Dawson (1979) for *L. heermanni*, between gray gulls and sympatric coastal-nesting band-tailed gulls, *L. belcheri* (see Fitzpatrick et al. 1988c). The thermoregulatory index indirectly reflects potential metabolic energy and/or water costs for a suite of thermoregulatory behaviors (TRB) associated with changing T_A. Although both species increased thermoregulatory commitments above a T_A of 29°C (Fig. 13.6), gray gulls used less costly TRBs than band-tailed gulls. Their thermoregulatory behavior is consistent with the pattern of hyperthermia, which together minimize both water and energy loss.

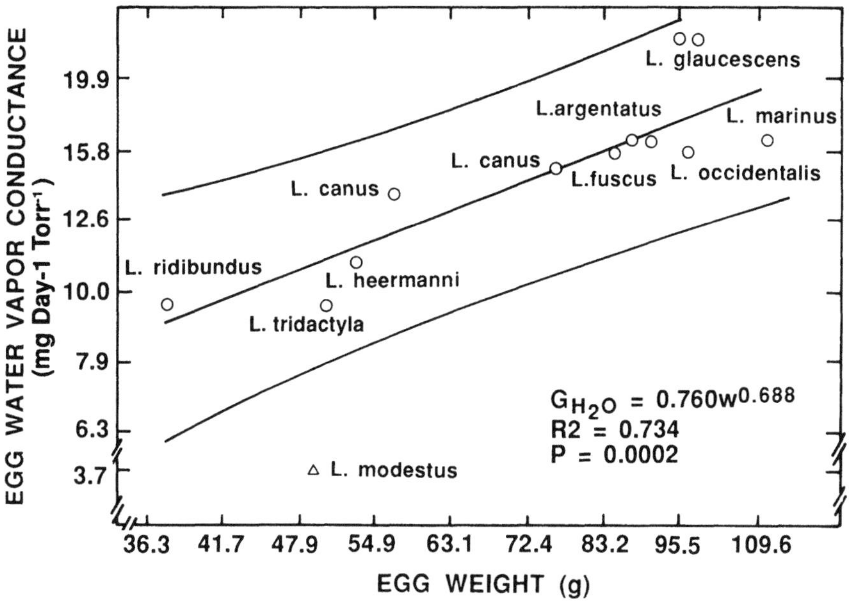

Figure 13.5. Regression model relating water vapor conductance ($G\text{-}H_2O$ = mg day^{-1} torr^{-1}) for eggs of 10 gull species. Gray gull (Larus modestus) eggs (Δ) are excluded from the model for comparative purposes. Lines above and below the regression define 95 percent confidence limits. Sources of data are given in text. R^2 = Coefficient of determination; P = Probability and W = Weight of gulls.

Reproductive Consequences and Trade-Offs of Desert Nesting

Although gray gulls are well adapted for nesting in the Atacama, there are several trade-offs affecting their reproductive biology relative to most other seagulls: (1) relatively long incubation time; (2) slow growth rate; (3) reduced clutch size; and (4) sensitivity to marine productivity. Howell et al. (1974) reported that the ca. 30-day incubation for gray gulls, which exceeds predictions for gulls with similar size eggs, resulted from egg exposure to low nightly temperatures. In contrast, we believe gray gulls incubate eggs constantly at night and that prolonged embryonic development is a consequence of the shells having reduced functional pore areas, implied by low $G\text{-}H_2O$s. Since a low functional pore area would also reduce O_2/CO exchange, overall metabolism of the developing embryos should be affected, a trade-off of nesting in the Atacama.

Other trade-offs are intimately associated with energy and time constraints of nesting far from food, requiring daily round-trip foraging flights of up to 200 km. As Wiens (1984) pointed out, "the balance or imbalance of intake and demands determines the fitness of individuals through effects

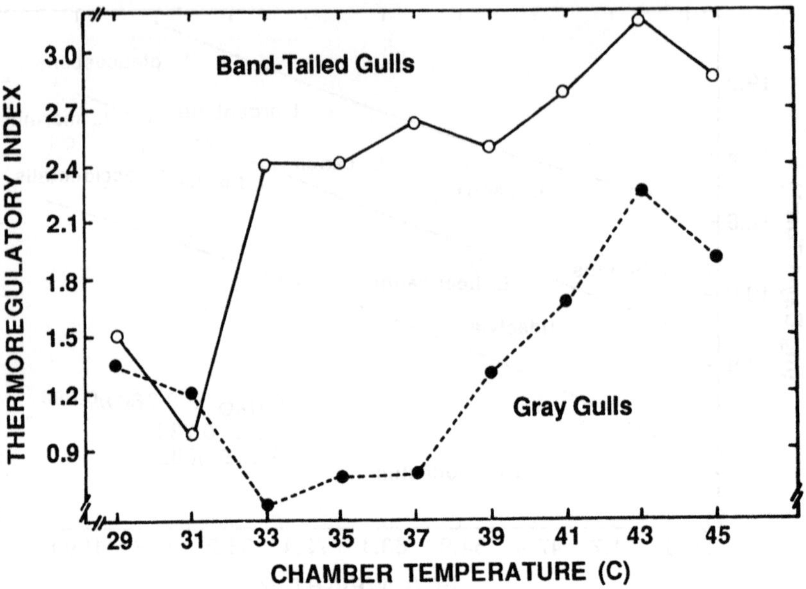

Figure 13.6. Comparative thermoregulatory costs between band-tailed gulls, *Larus belcheri*, (open circles) and gray gulls, *Larus modestus*, (closed circles). Thermoregulatory index reflects costs in terms of energy expenditure and water loss.

on reproduction." Since ecological energy budgets (Congdon et al. 1984) are necessary in understanding the life history traits which affect that balance, we estimated a time-activity energy budget for gray gulls nesting at Cerro Negro during their ca. six-month reproductive season (see Fitzpatrick et al. 1988a). The reproductive period (ca. August to February) consists of: (1) prenesting (August to October) when both gulls of a breeding pair fly daily to the desert to establish territories; (2) five weeks for incubation/brooding when pairs alternate foraging flights; (3) two months when both parents forage; and (4) a short prewinter migration period when adults supplement the diets of their fledged offspring on the beaches. During incubation and brooding, when gulls alternate daily foraging flights, each member of a pair should make 18 to 19 trips; afterwards, both make 50 to 60 flights together.

Gull activities can be partitioned into six time-energy segments: (1) nine hours in the desert interacting with mate and/or chick(s) at 83 KJ; (2) two hours flying to the coast with NE-E tailwinds at 110 KJ; (3) five hours offshore foraging for fish at 426 KJ; (4) 8.3 hours on the beach resting, occasionally interacting with other gulls or feeding on invertebrates at 172 KJ; (5) 2.5 hours congregating and spiraling before returning to the desert at 138 KJ; and (6) 2.2 hours flying into the desert, without winds, at 187 KJ. Energy (1116 KJ) expended in foraging is three times that of nest-tending (338 KJ). Integrating nest-tending and foraging activities over a 48-hour period, each member of a breeding pair expends 1454 or 727 KJ per day during incubation and brooding. Based on an empirically-determined net metabolizable energy efficiency (0.81), each adult must consume ca. 1800 KJ during a foraging trip. If intake is entirely anchovies (Cerro Negro feed almost exclusively on them), each gull would require at least 78 anchovies (101 g dry mass; 367 g wet mess; 17.8 KJ g^{-1} dry mass) or nearly its entire mass (342 and 382 g for females and males, respectively; see Fitzpatrick et al. 1988b) per trip. During the entire reproductive period, each gull expends ca. 190,000 KJ, a cost not including gamete production, corresponding to an intake of ca. 240, 000 KJ or 48 kg of anchovies. Since the overall cost of foraging flights is nearly 73, 000 KJ or 38.5% total net metabolizable energy (NME), food availability (i.e., marine productivity) must be an important regulator in the reproductive biology of gray gulls, perhaps the single most important factor determining clutch size or whether they breed at all during and immediately after ENSOs (see below). Also, since chicks depend on adults for food and water, the amount of food the adults carry should limit the number of chicks they can support and their chicks' growth rates. Each adult carries about 67 g (317 KJ) of anchovies per day, about 17.5 and 19.6% of the body mass of males and females, respectively, to its chick(s). Although somewhat higher than the "feed size" of 14–18% body mass reported for other long distance foraging seabirds (Ricklefs 1983), it matches the consumption requirements (362 KJ d^{-1}) of gray gull chicks during their maximal growth. Nonetheless, overall growth rate of gray gull chicks is significantly lower than other semiprecocial congeners (Fig. 13.7; see Guerra et al. 1988d), except for the long-distance foraging *L. furcatus*.

Although both adults share incubation, brooding, and foraging duties, females expend additional energy producing eggs. They begin yolking six follicles, but only one or two are laid, representing ca. 790 KJ and bringing the typical female's reproductive effort (RE; see Congdon et al. 1984) to ca. 122,000 KJ or 52% of her six-month NME. Compared to most gulls which have larger clutches, energy devoted to egg production is lower in gray gulls. However, after factoring in time and energy costs of foraging and thermoregulation during nesting and brooding, gray gulls exhibit a high RE (sensu Hirshfield and Tinkle 1974) per offspring. As such, their overall reproductive biology is more similar to pelagic seabirds (Ellis 1984) than other gulls. Thus, as in other long-foraging seabirds, gray gulls should be highly sensitive to food availability during breeding and nesting (Rahn and Whitlow 1984). Consequently, we believe that the combination of high RE per offspring, a long adult life span, and an environment in which marine productivity is influenced by relatively unpredictable ENSOs should have selected for an adaptive life history strategy of "bet-hedging" (Stearns 1976) in gray gulls.

During the 1982–83 ENSO and the following year (1983–84), the com-

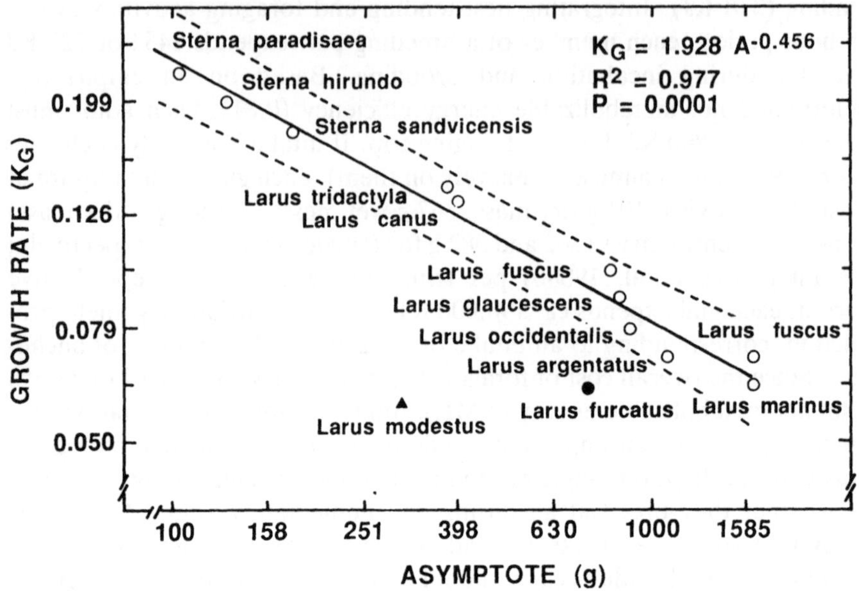

Figure 13.7. Regression (log-log) model of Gompertz growth constants (K_G = growth rate) of several Laridae species, excluding Larus modestus and *L. furcatus*. Data are from Ricklefs (1973), except for *L. modestus*). Dashed lines represent upper and lower 95% confidence limits of the regression model. Empirical values for *L. modestus* and *L. furcatus* are significantly lower than predicted values from the model (indicated as closed symbols). A = Asymptotic mass, R^2 = Coefficient of determination and P = Probability.

mercial catch of anchovies declined from over 80,000 tons in 1981 to under 500 and did not return to normal until after 1985 (SERNAP 1981–85). Although gray gulls were abundant in northern Chile, we did not observe courtship, mating, flights into the desert, or nesting in the pampas during the 1982–83 and 1983–84 breeding seasons (see Guerra et al. 1988c). Laying resumed in 1984–85 but was delayed until late December to early January (see Fig. 13.8 for normal reproductive cycle; Cikutovic et al. 1988). Laying began normally in 1985–86, coincident with the recovery of anchovies. Whereas other sea birds suffered severe mortality throughout their ranks during the 1982–83 ENSO, gray gulls apparently incurred no reproductive loss or ENSO-related adult mortality during 1982–84.

After nesting in 1985–86, we counted nearly 15,000 gray gulls on the beaches, half of which were new fledgling recruits, or ca. two per breeding pair. That conflicts with Howell et al. (1974) who reported an average clutch size at Colupo of 1.55 with relative frequencies of one, two, and three egg clutches of 0.47, 0.52 and 0.01, respectively, and argued that gray gulls are evolving from an ancestral three-egg to a single-egg clutch in response to inland nesting and long daily foraging flights. The average gray gull egg mass is less than that of the similar-sized desert-nesting Heer-

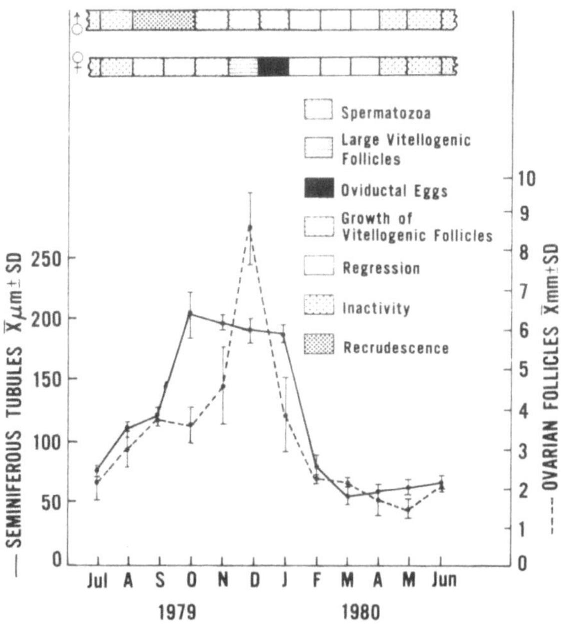

FIGURE 13.8. Normal annual reproductive cycles of male (solid line) and female (dashed line) gray gulls, *Larus modestus*.

mann's gull and represents ca. 14% of an adult female's body mass, a low value for birds (Lack 1968). Reduced egg mass and clutch size also occurs in the long-distance foraging *L. furcatus*. Our 1985–86 and 1986–87 data show a higher average (1.81) and frequency of two-egg (0.81) vs. single-egg (0.19) clutches from El Tigre-Kimurku-Valenzuela (Fig. 13.1) than Howell et al. (1974) reported for the nearby Colupo site. Our data from Cerro Negro averaged lower (1.66), but two-egg clutches were still more frequent than single-egg clutches (0.66 vs. 0.34). We speculate that the lower average and frequency of two-egg clutches at Cerro Negro than at Tigre-Kimurku-Valenzuela result from the greater costs of commuting 200 vs. 70 km daily to provision offspring. Combining all 1985–87 data, the frequency of two-egg clutches (0.71) and average clutch size (1.78) exceed those reported by Howell et al. (1974). During the first reproductive recovery year post-ENSO (1984–85), when anchovies were just returning, the frequency of two-egg clutches (0.44) was lower than that reported in 1974 for Colupo and significantly lower than that reported in 1985–86–87 when anchovies were abundant in coastal waters. We believe gray gull females "bet-hedge," altering clutch size from zero to one to two according to the probability of offspring survival which depends on marine productivity. Both sand crabs and anchovies, over 90% of gray gulls' diet, are reduced by ENSOs.

Although gray gulls are well adapted for nesting in the Atacama, reasons for it are unresolved. Three tentative hypotheses have been proposed (see Howell et al. 1974; Howell 1982): (1) gray gulls originally nested in prehistoric inland lakes, which have since dried up, forming the nitrate pampas, and simply have continued to nest at those ancestral sites; (2) being smaller and less aggressive than sympatric coastal-nesting kelp (*L. dominicanus*) and band-tailed gulls, gray gulls nest inland to avoid competition for nest sites; and (3) similarly, since they are less able to defend their nests from predators, gray gulls nest where turkey vultures are absent. Although we have no clear evidence to support or refute these hypotheses, nest predation, principally by turkey vultures and peregrine falcons, appears to be an important factor in the biology of gray gulls. First, desert flights begin after dark, even though wind conditions are favorable earlier (see Fitzpatrick et al. 1988a). Departure in the dark precludes being followed by turkey vultures. According to O'Connor (1984), predators commonly follow adult birds when returning from foraging to feed their chicks. We have observed numerous turkey vultures "searching" the pampas and significant egg predation in the Kimurku-Velenzuela area (see Guerra et al. 1988a). Apparently, turkey vultures only recently have discovered nest sites close to the coast. Although Howell et al. (1974) did not mention nest site predation as being important, local residents reported to us that turkey vultures and condors, *Vultur gryphus*, began visiting the Colupo site a few years prior to its abandonment. Predation and egg exploitation by humans may have caused the colony to leave. We have not observed turkey vultures at sites farther inland, suggesting that random searches of the inner desert are not sufficiently profitable for them. We believe the wide spatial dispersion of subcolonies and nests within them and asynchronous laying among females are adaptive responses to increasing turkey vulture predation.

Apart from *L. furcatus*, the reproductive biology of gray gulls is more similar to long-distance foraging pelagic sea birds than other sea gulls. Their unique adaptations and behavior are evolutionary responses to the complex interplay among severe conditions of the Atacama, long-distance foraging flights, marine productivity, and predation. Nonetheless, gray gulls are literally balancing on the edge of their overall resource budget, a balance easily threatened by humans through intensifying anchovy fishing and pollution of the coastal waters from mining activities.

Acknowledgments

The following assisted in the work summarized herein: R.E. Aguilar, G.S. Luna, B.J. Venables, T.L. King, G. Portflitt, O. Cayo, Y. Mendez, E. Hernandez, M. Silva, C. Guerra, A. Marin and R. Borquez. Partial support was from two NSF U.S.—Latin America Research Travel Grants and University of North Texas Research grants to Lloyd C. Fitzpatrick, UDA

(DIEXAT) Research Grants to Carlos G. Guerra, and a FONDECYT Grant (88-070) to Carlos G. Cuerra and Lloyd C. Fitzpatrick.

References

Ar A, Paganelli CV, Reeves RB, Greene DG, Rahn H (1974) The avian egg: Water vapor conductance, shell thickness and functional pore area. Condor 76:153–158

Ar A, Rahn H (1978) Interdependence of gas conductance, incubation length and weight of the avian egg. In Piiper RA (Ed.). Respiratory Function in Birds, Adult and Embryonic. Springer-Verlag, Berlin. pp. 227–236

Arntz WE (1986) The two faces of El Niño 1982–83. Mesereforsch 31:1–46

Bartholomew GA, Dawson WR (1979) Thermoregulatory behavior during incubation in Heermann's gulls. Physiol. Zool. 52(4):422–437

Cawkell E, Hamilton J (1961) The birds of Falkland Islands. Ibis. 102a:1

Chapman SE (1973) The grey gull. Sea Swallow 22:7–10

Chappell MA, Goldstein DL, Winkler DW (1984) Oxygen consumption, evaporative water loss, and temperature regulation of California gull chicks (*Larus californicus*) in a desert rookery. Physiol. Zool. 57(2):204–214

Cikutovic MA, Guerra CG, Fitzpatrick LC (1988) Gonadal cycle of gray gulls *Larus modestus* in northern Chile. Le Gerfaut 78:209–216

Congdon JD, Dunham AE, Tinkle DW (1982) Energy budgets and life histories of reptiles. In Gans C, and Pough H (Eds.) Biology of the Reptilia, Vol. 13. Academic Press, New York. pp. 233–271

Congdon JD, Dunham AE, Tinkle DW (1982) Energy budgets and life histories of reptiles. In Gans C, and Pough H (Eds.) Biology of the Reptilia, Vol. 13. Academic Press. New York. pp. 233–271

Dawson WR, Bennett AF (1973) Roles of metabolic level and temperature regulation in the adjustment of western plumed pigeons (*Lophophaps ferruginea*) to desert conditions. Comp. Biochem. Physiol. 44A:249–266

Dawson WR, Bennett AF, Hudson JW (1976) Metabolism and thermoregulation in hatching ring-billed gulls. Condor 78(1):49–60

Devillers P, Terschuren J (1976) Desertion of the gray gull (*Larus modestus*) Colony of Pedro de Valdiuia (Chile) Le Gerfaut 66:132–137

Ellis HI (1984) Energetics of free-ranging seabirds. In Whittow GC and Rahn H (Eds.). Seabirds Energetics. Plenum Press, New York. pp. 203–234

Fitzpatrick LC, Guerra CG (1988) Microclimatic features of gray gull (*Larus modestus*) nests in the Atacama Desert. Le Gerfaut 78:421–428

Fitzpatrick LC, Guerra CG, Aguilar RE (1988a) Energetics of reproduction in the desert-nesting seagull *Larus modestus*. Estud. Oceanol. 7:33–39

Fitzpatrick LC, Guerra CG, King TL (1988b) Sex determination in gray gulls *Larus modestus* using external measurements and discriminant analysis. Estud. Oceanol. 7:71–74

Fitzpatrick LC, Guerra CG, Luna GS (1989) Thermoregulatory behavior in desert-nesting gray gull (*Larus modestus*) Le Gerfaut 78:167–175

Goodall JD, Philippi RA, Johnson AW (1945) Nesting habits of the Peruvian gray gull. The Auk 62:450–451

Guerra CG, Aguilar RE, Fitzpatrick LC (1988) Water vapor conductance of gray gull (*Larus modestus*) eggs: Adaptation to desert nesting. Colonial Waterbirds 11:107–109

Guerra CG, Cikutovic MA (1983a) Un nuevo sitio de nidificacion para la garuma *Larus modestus* (Aves: Charadriiformes: Laridae). Estud. Oceanol. 3(1):13–20

Guerra CG, Cikutovic MA (1983b) Alimentacion y morfologia comparada entre adultos y juveniles de la garuma *Larus modestus*. Mems. Asoc. Lat. Am. Acuicult. A.L.A. 5(1):41

Guerra CG, Fitzpatrick LC, Aguilar RE (1988a) Growth rates in gray gulls *Larus modestus*: Influence of desert nesting and foraging distance. The Auk. 105:779–783

Guerra CG, Fitzpatrick LC, Aguilar RE, Luna GS (1988b) Location and characterization of new nesting sites for gray gulls *Larus modestus* in the Atacama Desert, northern Chile. Le Gerfaut. 78:121–129

Guerra CG, Fitzpatrick LC, Aguilar RE, Venables BJ (1988c) Reproductive consequences of El Niño-Southern Oscillation in gray gulls *Larus modestus*. Col. Waterbirds 11:170–175

Guerra CG, Fitzpatrick LC, Luna GS (1989) Physiological adaptations to desert nesting in the sea gull *Larus modestus*. Estud. Oceanol. 8:1–7

Harrison P (1983) Sea birds, An Identification Guide. Houghton Mifflin Company, Boston

Hirshfield MF, Tinkle DW (1974) Natural selection and the evolution of reproductive effort. Proceedings of the National Academy of Sciences, USA, 72:227–2231

Howell TR (1982) Desert-nesting sea gulls. Nat. Hist. 91:52–59

Howell TR, Araya B, Millie WR (1974) Breeding biology of the gray gull, *Larus modestus*. Univ. Calif. Pub. Zool. 104, Berkeley

King J (1974) Seasonal allocation of time and energy resources in birds. In Paynter R (Ed.). Avian Energetics. Publ. Nuttall Ornithol. Club No. 15

Lack D (1968) Ecological Adaptations for Breeding in Birds. Methkuen, London

Moffett GM Jr (1969) The garuma-gull of the desert. Sea Frontiers 15:330–338

Morgan KR, Paganelli CV, Rahn H (1978) Egg weight loss and nest humidity during incubation in two Alaskan gulls. Condor 80:272–275

Murphy RC (1936) Oceanic Birds of South America, Vol. 1. Am. Mus. Nat. Hist., New York

O'Connor R (1984) The Growth and Development of Birds. A Wiley-Interscience Publ., New York

Philippi RA (1964) Catalogo de las aves de Chile con su distribucion geografica. Bol. M.N.H.N. No. 45

Rahn H, Dawson WR (1979) Incubation water loss in eggs of Heermann's and western gulls. Physiol. Zool. 52:451–460

Rahn H, Whittow GC (1984) Introduction. In Rahn H, Whittow GC (Eds.). Seabird Energetics. Plenum Press, New York

Ricklefs RE (1973) Patterns of growth in birds. II. Growth rate and mode of development. Ibis 115:117–201

Ricklefs RE (1983) Some considerations on the reproductive energetics of pelagic sea birds. Stud. Avian Biol. 8:84–94

Schmidt-Nielsen K (1985) Desert Animals: Physiological Problems of Heat and Water. Clarendon Press, Oxford

SERNAP Annual Reports (1981–85) Servicio Nacional de Pesca, Santiago, Chile

Shreiber RW, Schreiber EA (1983) Reproductive failure of marine birds on Christmas Island, Fall 1982. Tropical Ocean-Atmosphere Newsletter 16:10–12
Shreiber RW, Schreiber EA (1984) Central Pacific sea birds and the El Niño Southern Oscillation 1982 to 1983 perspectives. Science 225:713–716
Stearns SC (1976) Life history tactics: A review of the ideas. Quart. Rev. Biol. 51:3–47
Tovar H (1968) Areas de reproduccion y distrilbucion de las aves marilnas en el litoral peruano. Bol. Ins. Mar Peru-Callao (10):526–540
Tovar H, Cabrera D (1985) Las aves guaneras y el Fenomeno El Niño. Bol. Ins. Mar Peru-Callao. Volumen estraordinario:181–186
Walsberg GE, Campbell GS, King JR (1978) Animal coat color and radiative heat gain: A re-evaluation. J. Comp. Physiol. 126:211–222
Wiens JA (1984) Modelling the energy requirements of a seabird population. In Whittow GC and Rahn H (Eds.). Seabird Energetics. Plenum Press, New York. pp. 255–284
Yanez PA (1949a) Vertebrados marinos chilenos. Aves, I parte. Rev. Biol. Mar. I(3):184–261
Yanez PA (1949b) Vertebrados marinos chilenos. Aves, II parte. Rev. Biol. Mar. II(1):2–40

14

Gonadal Cycles of *Pygoscelis* Penguins of the South Shetland Islands

José Valencia and Víctor Leyton

Penguins are an important component of the Antarctic ecosystem because they contribute 93% of the total biomass of antarctic birds and consume 83% of the total food required by birds in the area, which is mainly krill (Laws 1977). *Pygoscelids* are pelagic animals with a reproductive season starting in early October and ending in February; after landing and selecting a site, courtship and reproductive behavior take place.

Periodic changes of biotic and abiotic conditions contribute to synchronize the reproductive cycles of wild vertebrates, enhancing the species survival (Dawson 1986). Favorable conditions prevailing in lower latitudes induce, in many cases, continuous reproduction (Lofts and Murton 1973), whereas in high latitudes birds show a marked seasonality of breeding that affects the sequence and duration of migration, nest building, mating, egg-laying, incubation, and parental care.

Reproductive activity of males and females of seasonal breeders is restricted to definite periods when mature gametes are produced. For male birds, completion of spermatogenesis takes place within a few weeks or months. Afterwards, the testes enter a regression to a morphological condition similar to prepuberty. From that stage, a new phase of development of gametes proceeds before the beginning of a new breeding season. These changes between the gametes production and gonad involution are frequently extended over a whole year and are known as the annual cycle of reproduction (Johnston 1956).

Adult female birds have but one ovary and one fully developed oviduct. The female gonad has marked morphological variations, and only a few ova ever reach maturity so that the normal clutch can be three or four large eggs for most sea birds.

On Ardley Island, South Shetland, Antarctica (62° S latitude), three species of closely related penguins overlap their nesting distribution. *Pygoscelis papua*, *P. antarctica*, and *P. adeliae* form mixed nesting colonies. Nests of *P. papua* are predominant, followed by *P. adeliae*, and *P. antarctica* having the smallest number. The nesting area is under the climatic conditions prevalent in the maritime Antarctic and, during the spring and summer, daylight is more than 12 hours (Schlatter et al. 1968).

The three penguin species are characterized by their longevity (16 years in *P. adeliae*), delayed sexual maturity (three to four years in *P. adeliae* [Ainley 1985], two years in *P. papua* [Croxall 1984]), reduced egg clutch, parental care, and rapid chick growth.

As nesting sites and food are required, more or less at the same time, by breeding *Pygoscelis* penguins, we proposed to describe and characterize the annual cycle of gonads and the temporal sequence of sperm and egg production for the species found at Ardley Island, South Shetland. We identified differences between species to evaluate the importance of these factors in promoting or maintaining their ecological isolation.

Since the volume, weight, shape, and color of the testis do not always coincide with the developmental stages of the germ line, we studied the main histological changes during the reproductive cycle.

To obtain information on the reproductive and non-reproductive condition of penguins, gonads were obtained from 81 adult males and 78 females collected at east beaches of Ardley Island (62°13′S, 58°55′W). Samples were taken between September and February 1982–85. Afterward, gonads were weighed and measured from the time the adults returned to the colony after the winter season until the chicks had left the breeding sites. Testes and ovaries were fixed in Bouin's solution, embedded in wax, sectioned at 5 to 7 μm, and stained with routine techniques. Samples were used for histological and morphometric study.

The parameters selected to describe the macroscopic gonad changes were weight and gonadosomatic index, which relates gonad weight to total body mass. For the study of the microscopic changes of the testes, we recorded tubule diameter, seminiferous epithelium thickness and relative number, and size of interstitial Leydig cells as quantitative descriptors of the testis cycle. To determine the events of sperm generation and correct classification of each sample, we used the four stages of the testis cycle, used for *P. papua* (Fuenzalida et al. 1989), that are summarized in Table 14.1. These stages are quite different from others proposed for gulls by Johnston (1956) and the mute swan by Breucker (1982).

For the ovaries, we recorded the condition of vitellogenic and non-

TABLE 14.1. Description of stages of the testis cycle.

Resting:	Recrudescence:
Presence of Sertoli cells, spermatogonia, and a few peritubular cells. Tubule diameter is minimum and interstitial tissue is wide.	Sertoli cells, spermatogonia, spermatocytes, and spermatids present. Expanding seminiferous tubules with reduction of peritubular cells and interstitial space.
Active:	Regression:
Germ cells strain complete, including spermatozoa inside the tubule lumen and epididymis. Leydig cells in clusters.	Seminiferous epithelium disorganized. Immature cell clusters in degenerative process, disintegrated cell remains inside the lumen and epididymis.

vitellogenic eggs. Simultaneously, we recorded, for each bird, the reproductive behavioral elements such as courtship, nest building, egg incubation, and chick hatching to study the temporal synchronization of reproductive events for each of the three species.

Testicular Organization and Testicular Cycles

The testes of penguins are located within the abdomen, surrounded by a tunica albuginea which varies in thickness according to the testis cycle. Seminiferous tubules are branched, as in other birds (King 1975), encircled by a basal lamina, and two to four layers of peritubular cells. Externally the interstitial tissue is found where blood and lymphatic vessels are present with connective and Leydig cells.

During the testicular cycle, both the testis and epididymis undergo macroscopic and microscopic changes. Weight changes are dramatic when comparisons of resting and active testes are made. Average testis weight during the resting stage is 0.80 (±0.36 g) in *P. adeliae*, 0.55 (±0.48 g) in *P. antarctica*, and 1.10 (±0.06 g) in *P. papua*. During the active stage, testis weight increases to 12.6 (±3.26 g) in *P. adeliae*, 7.2 (±3.44 g) in *P. antarctica*, and 9.10 (±2.27 g) in *P. papua*. This physical change is due to the intense cell proliferation in the seminiferous tubules (Fig. 14.1).

When testis weight is related to the male body mass in a gonadosomatic index, it is found that during the resting state the value is 0.016 in *P. adeliae*, 0.004 in *P. antarctica*, and 0.018 in *P. papua*. When comparisons are made in the active testis, the gonadosomatic indices are 0.27 in *P. adeliae*, 0.19 in *P. antarctica*, and 0.22 in *P. papua*. These differences are a function of the body mass of each species. Changes are mainly influenced by body weight losses due to fasting during the initial stages of breeding. Minimum values of tubule diameter and thickness of seminiferous epithelium are found for all penguins during rest. (Table 14.2).

In the case of interstitial Leydig cells, the changes in relative number are mainly due to their spatial distribution within the connective tissue. Volume modifications occur when the expansion of tubules is higher during the change of the testis from resting to active phase. The relative changes of nuclear diameter are usually taken as indicative of physiological activity and are related to androgen production regulating the testis cycle and reproductive behavior. For all species studied, the average nuclear diameter of Leydig cells is maximum in the active testis (4.8 ± 0.6 μm) and minimum when the testis regresses to the resting stage (3.5 ± .05 μm).

To improve our understanding of the significance of testis macroscopic and microscopic changes, we include the consideration of reproductive events and temporal synchronization in three species of penguins over four consecutive years. Figure 14.2 shows the duration of each stage of the testis

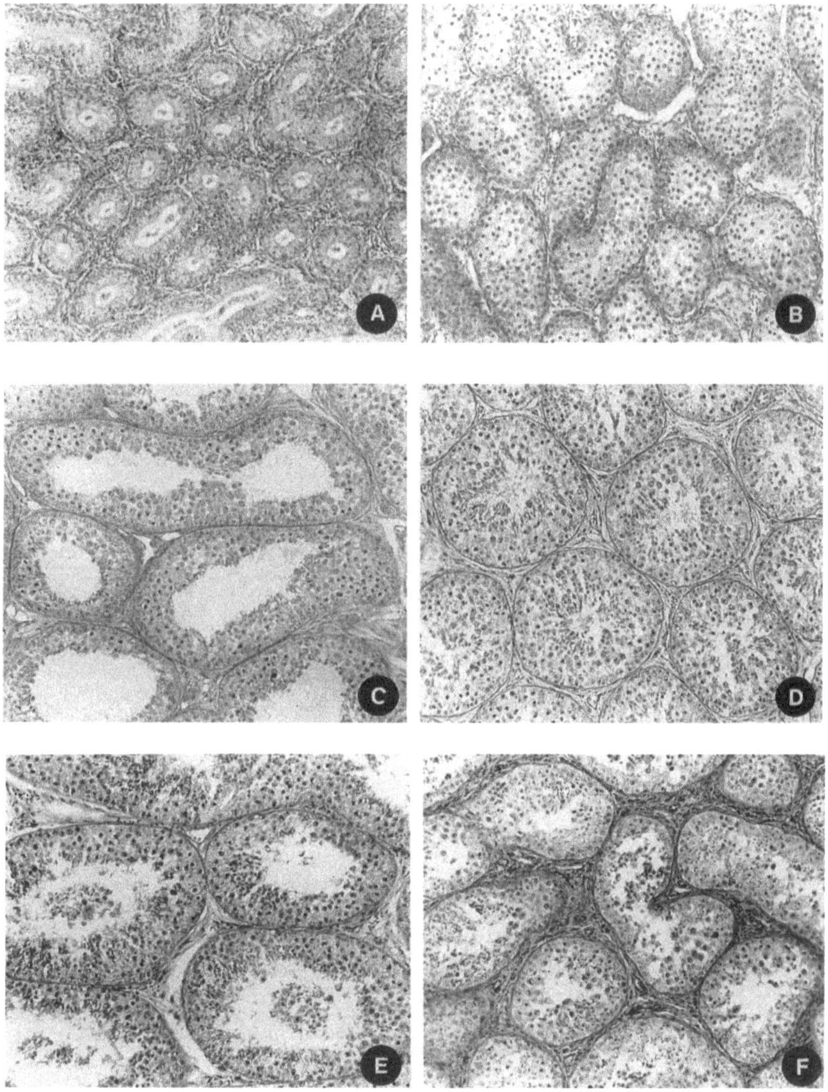

FIGURE 14.1. Testis sections of *Pygoscelis* penguin stained with hematoxylin-eosin and photographed under 20x magnification. (A) resting; spermatogonia inside seminiferous tubules; (B–C) recrudescence; proliferating gonia, spermatocytes, and spermatids lining the seminiferous epithelium; (D) active; complete germ line, including spermatozoa; (E–F) regression; degenerated cells inside the lumen.

TABLE 14.2. Tubule diameter and thickness of seminiferous epithelium in cross-section ($\bar{x} \pm SD$) during the testis cycle of *Pygoscelis* penguins.

Cycle stage	*P. adeliae*	*P. antarctica*	*P. papua*
	Tubule Diameter (μm)		
Resting	60 ± 1.0	65 ± 24	52 ± 2.4
Recrudescence	–	223 ± 9.3	–
Active	202 ± 30	117 ± 21	240 ± 5.0
Regression	126 ± 81	128 ± 21	160 ± 5.5
	Epithelium height (μm)		
Resting	18 ± 1.1	12 ± 3.4	18 ± 0.3
Recrudescence	–	66 ± 2.9	–
Active	60 ± 1.7	52 ± 11	68 ± 2.4
Regression	25 ± 12	28 ± 5.6	45 ± 1.7

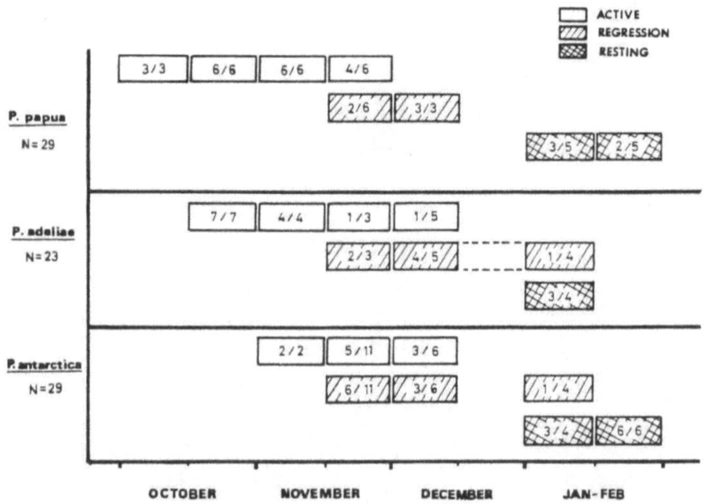

FIGURE 14.2. Temporal changes of testis cycle for three species of *Pygoscelis* penguins.

cycle as derived from our samples. The number inside the squares indicates the proportion of individuals that were found in that stage of the cycle. Data show that each species population is not absolutely synchronized for each cycle stage and that *P. antarctica* is active later during the breeding period. In general, our field observations indicate that older and experienced birds arrive earlier and begin their breeding before younger and inexperienced individuals.

Ovarian Activity

All adult females of the three species sampled during January and February possess ovaries with only translucent non-vitellogenic follicles. Initial vitellogenesis was only observed in *P. papua* in early October, whereas *P. adeliae* and *P. antarctica* arrived at the colony site in advanced vitellogenesis when yolk deposition enhances rapid follicular growth. No samples of non-vitellogenic *P. adeliae* and *P. antarctica* were found before the reproductive phase. In each ovary, approximately two or three follicles undergo vitellinization and are ovulated late in October (Fig. 14.3).

The temporal sequence of ovarian events is shown in Figure 14.4. Incubating eggs were observed from mid-October to mid-December, with the exception of *P. adeliae* where this period was principally restricted to November and December. Nesting started in early October and eggs were simultaneously incubated for the three species. When incubation finished, the ovaries contained only non-vitellogenic follicles from December onwards.

Testis Activity

Histologic and morphometric analyses of *Pygoscelis* penguin testes reveal a short and well-defined seasonal cycle. The period of reproductive activity occurs earlier in both *P. papua* and *P. adeliae* and before *P. antarctica.* The reproductive season is restricted to the spring months when gross modifications can be observed at both macroscopic and microscopic levels. Penguins land with active testes and mature gametes, begin courtship in early October, and finish reproduction by late November to early December. During this period, the females incubate eggs for about 33 to 37 days. When reproduction is complete, the gonads of the males and females rapidly regress to an inactive condition until the next season. This accelerated breeding cycle might be associated with the increased energy cost of breeding, principally supported by krill.

Modifications of male gonadosomatic indices (GSI) throughout the testis cycle show the magnitude of morphologic changes inside the seminiferous tubules. Higher GSIs found in animals with active testes are indicative of lower body weight and higher testis weight, principally due to intensive spermatogenesis. The opposite is found for the testing gonad. These changes are an expression of the modifications of reproductive behavior and food availability. High body weight at the beginning of the reproductive period is indicative of body fat storage which diminishes near mating and nesting. This aspect is especially important for the females, in whom fat reserves permit rapid follicular growth, nest building, incubation of eggs, and parental care. If energy reserves are considerably depleted, nest desertion and reproductive failure may result.

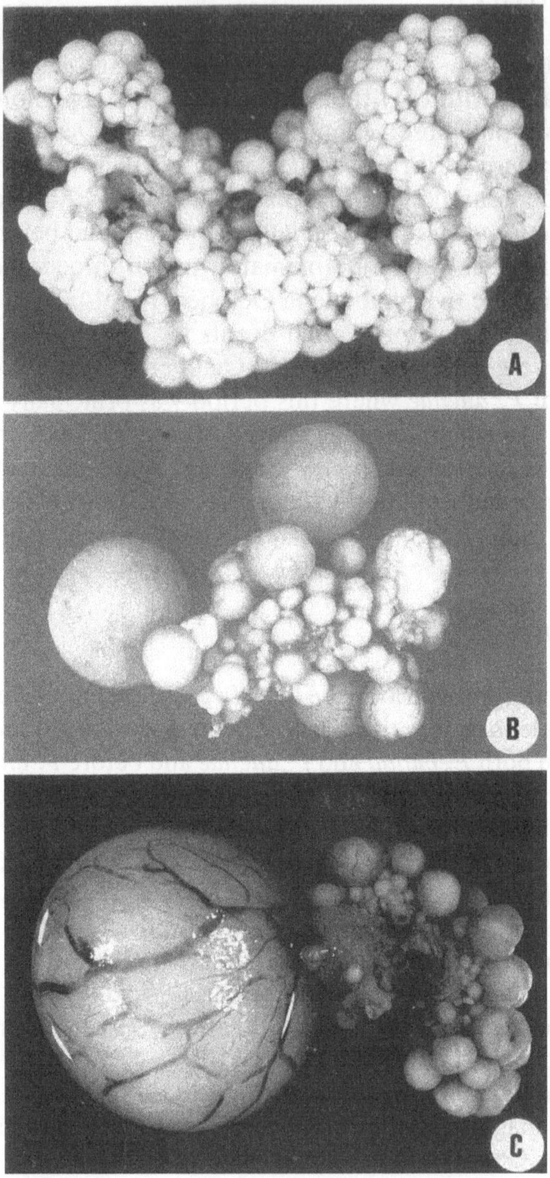

FIGURE 14.3. Ovaries of *Pygoscelis papua* in three stages of their cycle. (A) non-vitellogenic; (B) early vitellogenesis; (C) advanced vitellogenesis, prior to ovulation.

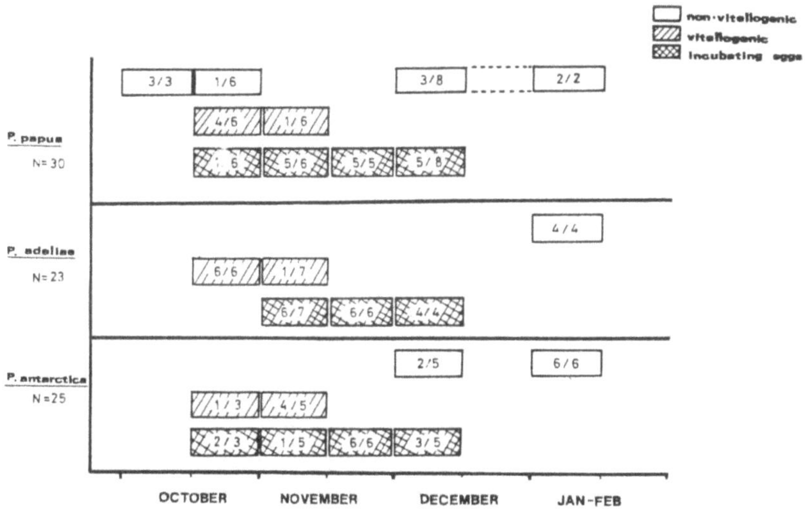

FIGURE 14.4. Temporal changes of the ovary for three species of *Pygoscelis* penguins.

Changes in interstitial Leydig cell number and nuclear diameter are indicative of high functional activity during reproductive behavior that is related to elevated levels of circulating androgens, as occurs during the breeding season of free-living species (Silverin 1984). A temporal reproductive asynchrony is clear for the three *Pgyoscelis* penguins, as described by Valencia and Sallaberry (1983). This sequence favors a temporal differentiation that contributes to a better distribution of sites for nesting and a decrease of potential food demand in the area.

Birds, as in other vertebrates, have evolved mechanisms which time both the onset and culmination of their breeding cycle and maximize the survival of parents and young (Baker 1938). The transfer of photosensitive adults from short to long daylengths initially results in gonadal growth. After a species-specific period of exposure to long days, birds become photorefractory and the gonads regress. In penguins, photorefraction is associated with resting gonads and reduced activity of the hypothalamus-hypophysis-gonad axis, whereas resumption of hormone synthesis only occurs some weeks after the birds are exposed to long days (Hector et al. 1986).

Observations of the breeding biology of *P. adeliae* and *P. antarctica* in the South Orkney Islands (Lishman 1985) have shown that *P. adeliae* arrives and breeds earlier during the spring season, as in Ardley Island. Aspects of the breeding biology of *P. papua* in the Crozet Archipielago have been shown by Bost and Jouventin (1991), where laying begins in mid-winter and continues until early austral summer.

Asynchrony in laying and incubating eggs for the three species studied at Ardley Island correspond to differences in the breeding timing. When *P.*

papua arrive, they have a maximum body weight, fat reserves, and developed gonads. They nest near the sea and spread over the east end of the island. Other species arrive later and distribute in marginal areas where competition for nesting sites with other pairs diminishes. Energy demand of the chicks and monthly food availability at sea may affect the magnitude of differences between reproductive schedules of the three penguin species.

The timing of breeding can result from differences in adaptations to exploit environmental conditions. From observations by Lishman (1985) and by us, *P. adeliae* and *P. papua* appear to be adapted to the disadvantages of poor diet in early spring and harsh environmental conditions. They commence breeding as soon as possible so that the chicks can be reared succesfully in a very short summer. *P. antarctica*, on the other hand, breed later to minimize direct competition for the nest sites of the other two species and to obtain benefits from both qualitative and quantitative food distribution for their young and themselves. Thus, a temporal partition sequence in reproduction seems to facilitate the avoidance of interference between closely related species and helps to maintain their reproductive isolation.

Acknowledgments

We are grateful for the logistical support of the Chilean Air Force and the financial support of the Chilean Antarctic Institute.

We thank Dr. Michel Sallaberry, José Yañez, and Herman Nuñez for their collaboration with obtaining field samples, and to Francisco Cortés for his technical assistance. Dr. W. Hamlett made suggestions to an earlier version of the manuscript.

References

Ainley D (1985) Development of reproductive maturity in Adeliae penguin. In Stonehouse B (ed.). Biology of Penguins. The McMillan Press, London. pp. 139–157

Baker J (1938) The relation between latitude and breeding season in birds. Proc. Zool. Soc. Lond. 108:557–582

Bost C, Jouventin P (1991). The breeding performance of the Gentoo penguin *Pygoscelis papua* at the northern edge of its range. Ibis 133:14–25

Breucker H (1982) Seasonal spermatogenesis in the mute swan (*Cignus olor*). In Beck F (ed.). Anatomy, Embryology and Cell Biology, Vol. 72. Springer-Verlag, New York, pp. 1–90

Croxall J (1984) Seabirds. In Laws RM (ed.), Antarctic Ecology, Vol. 2 Academic Press, London. pp. 533–619

Dawson A (1986) The effect of restricting the daily period of food availability on testicular growth of starlings *Sturnus vulgeris*. Ibis 128:572–575

Fuenzalida H, Leyton V, Valencia J, Blánquez M, González E (1989) Morfología del testiculo de *Pygoscelis papua* (Forster) durante el período de actividad sexual. Arch. Anat. y Embriol. 20:79–92

Hector J, Croxall J, Follet B (1986) Reproductive endocrinology of the wandering albatross *Diomedea exulans* in relation to biennial breeding and defined sexual maturity. Ibis 128:9–22

Johnston D (1956) The annual reproductive cycle of the California gull: Criteria of age and the testis cycle. Condor 58:134–162

King A (1975) Aves urogenital system. In Getty A (ed.). The Male Genital Organs Saunders, Philadelphia-London-Toronto. pp. 1927–1935

Laws R (1977) The significance of vertebrates in the Antarctic marine ecosystem. In Llano G (ed.). Adaptations within Antarctic Ecosystems. 3 SCAR Symposium. pp. 411–438

Lishman G (1985) The comparative breeding biology of Adeliae and chinstrap penguins *Pygoscelis adeliae* and *P. antarctica* at Sygny Island, South Orkney Islands. Ibis 127:84–99

Lofts B, Murton R (1973) Observation on the organization of the interstitial tissue of the testes and on the occluding junctions in the seminiferous epithelium. Adv. Biosci. 10:83–89

Schlatter R, Hermosilla W, di Castri F (1968) Distribución altitudinal de los artrópodos terrestres. Ser. Cient. INACH 15:1–26

Silverin B (1984) Annual gonadotropin and testosterone cycles in free-living male birds. J. Exp. Zool. 232:581–587

Valencia J, Sallaberry M (1983) Censo de pingüinos en isla Ardley (Shetland del Sur). Ser. Cient. INACH 30:93–96

Marsupilia

15

Reproductive Biology of South American Marsupials

John D. Harder

Introduction

Although South America is not well known for its marsupials, it rivals Australia as an important place of evolution and diversity for this group. Marsupialia is the third largest order of South American mammals, after rodents and bats. Approximately 78 marsupial species inhabit South America, representing roughly 10% of its terrestrial mammalian diversity (Kirsch 1977; Streilein 1982), and, although only three of the 15 marsupial families are found in South America, together they represent about 30% of the world's total of 250 species. In fact, marsupials are found only in the Australian and Neotropical regions with a single species, the Virginia opossum *Didelphis virginiana*, in the Nearctic (temperate North America).

Both North and South America have been regarded as the place of origin for marsupials and by the late Cretaceous the two continents shared an ancestral didelphid fauna (Marshall et al. 1990) that was similar in dental morphology to the opossums of today. Thus, the reproductive adaptations and patterns revealed in modern members of Didelphidae are viewed as primitive and are of considerable interest in comparative reproductive biology. Didelphidae also contain more species than any other marsupial family (Table 15.1), and their present distribution and ecological roles represent a unique legacy of nearly 100 million years of evolution. Unfortunately, most of these species have not been studied beyond taxonomic description; even the most basic features of their natural history are unknown. There is an urgent need to study this group before it is diminished by the onslaught of human activity. The purpose of this chapter, therefore, is to highlight the reproductive processes in this important group of mammals and to suggest ways in which their biology can be studied to the benefit of biological conservation in South America.

The Marsupial Mode of Reproduction

Modern marsupials and eutherian mammals represent two major evolutionary lines that diverged from a common ancestor in the Cretaceous,

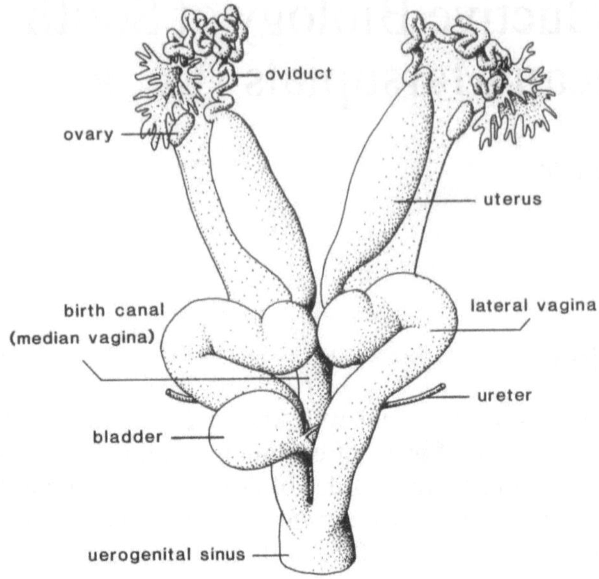

FIGURE 15.1. The didelphid reproductive tract as seen in a ventral view of a female *Monodelphis domestica.* Following copulation, spermatozoa travel up both lateral vaginae, through the uteri, and into the upper oviducts where fertilization occurs. At parturition embryos pass from the uteri, through the median vagina, and out the opening of the urogenital sinus (Drawing by David M. Dennis).

nearly 100 million years ago (Clemens 1977). The distinctions between the mammalian Infraclasses Metatheria (marsupials) and Eutheria (all other viviparous mammals) are primarily reproductive, reflecting the evolution of alternative strategies for viviparity and lactation. The marsupial-eutherian dichotomy is seen in numerous reproductive characteristics but with absolute distinction in the arrangement of the urogenital ducts (Tyndale-Biscoe and Renfree 1987; Fig. 15.1). Of greater biological importance, however, is the difference in maternal investment in gestation and lactation. For eutherians, the length and energy demands of gestation approximate those required of the mother between birth and weaning of her litter. Marsupials, on the other hand, give birth to very small, embryonic neonates following a short gestation and negligible investment in energy. For example, gestation in *D. virginiana* lasts only 13 days but the young are not weaned until about 100 days later (Fig. 15.2). Thus, for the marsupial, lactation is the dominant mode of energy transfer to the young.

One of the most remarkable features of marsupial reproduction is that gestation does not alter or interrupt the normal course of the estrous cycle, leading Sharman (1970) to suggest that gestation is physiologically equivalent to the non-pregnant cycle. For example, all temporal, morphometric, physiological, and hormonal aspects of the estrous cycle studied thus far in

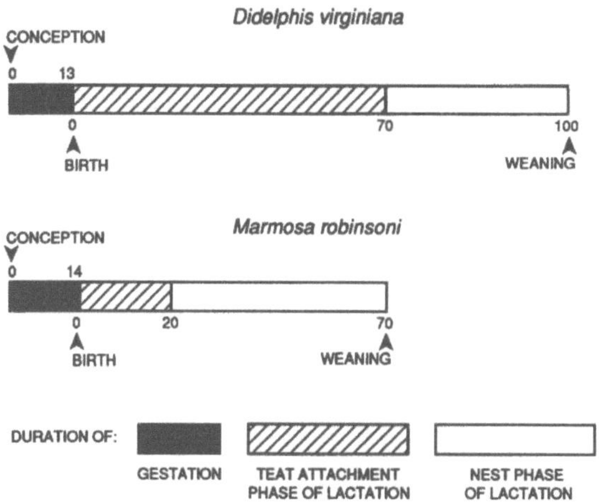

Figure 15.2. Relative duration (in days) of gestation and the phases of lactation in a large, pouched opossum, *Didelphis virginiana*, and a small, pouchless opossum, *Marmosa robinsoni*. The teat attachment phase shown for *D. virginiana* includes a period (days 50 to 70) during which young occasionally release the teat but remain in the pouch.

D. virginiana remain unaltered during gestation (Hartman 1923, Renfree 1975; Fleming and Harder 1981a, 1981b; Harder and Fleming 1981). Parturition coincides with regression of the corpus luteum (CL) and a decline in progesterone at the end of the luteal phase of the cycle. This pattern contrasts markedly with that in eutherian mammals in which pregnancy extends the life of the CL, interrupts the estrous cycle, and imposes major alterations in the physiology of the mother. Observations on other marsupial species tend to support Sharman's hypothesis, the major exceptions being recorded in studies of kangaroos and wallabies (Tyndale-Biscoe and Renfree 1987).

The foregoing introduction only serves to highlight some of the major features of marsupial reproduction. A systematic, detailed review of reproductive physiology is beyond the scope of this chapter. However, several excellent reviews of this subject are available and are cited below. Foremost among them is Tyndale-Biscoe and Renfree (1987), a thorough and engaging analysis of the world's literature on marsupial reproduction written by two of the leading scholars on the subject.

Reproductive Patterns of South American Marsupials

Discussion of reproduction of South American marsupials is, of necessity, limited to the small number of species of *Didelphis*, *Caluromys*, *Marmosa*, and *Monodelphis* that have been studied in any detail. The three species of *Didelphis*, varying from 1 to 5 kg in body weight, are by far the largest

TABLE 15.1. A classification of American marsupial genera (Kirsch 1977; Kirsch and Calaby 1977) with data on numbers of pouchless species and species in temperate South America.

Taxa	Total number of species	Number of temperate[1] species	Number of pouchless[2] species
Order Polyprotodonta			
Suborder Didelphimorphia			
Superfamily Didelphoidea			
Family Didelphidae (70 species)			
Genus *Marmosa*	44	2	44
Monodelphis	11	3	11
Lestodelphys	1	1	?
Metachirus	1	0	1
Didelphis	3	1	0
Philander	2	0	0
Lutreolina	1	1	1
Chironectes	1	0	0
Caluromys	3	0	0
Caluromysiops	1	0	?
Glironia	2	0	?
Family Microbiotheriidae (1 species)			
Genus *Dromiciops*	1	1	0
Order Paucituberculata			
Superfamily Caenolestoidea			
Family Caenolestidae (7 species)			
Genus *Caenolestes*	5	0	5
Lestoros	1	0	1
Rhyncholestes	1	1	1
Totals	78	10	64

[1] All or the majority of the known distribution of each species is south of the Tropic of Capricorn (23°S), according to the maps of Streilein (1982). *Didelphis virginiana* is found throughout much of North America south of 44°N.
[2] According to Kirsch (1977).

of the American marsupials, and each has a well-developed pouch. They are not, however, typical South American marsupials; most (64 of 78) are small, pouchless species of *Marmosa*, *Monodelphis*, or the family Caenolestidae (Table 15.1). Body weights of this group seldom exceed 100 g and are generally in the range of 40 to 80 g. Comparisons among these taxa are instructive because they represent two natural groups of contrasting body size and pouch development, which have important consequences for postnatal care of the young.

DIDELPHIS

The three species of *Didelphis* range from northern United States to southern Argentina, making it the most widely distributed marsupial genus in the world. *D. virginiana* is apparently a relatively new species that evolved from an isolated stock of *D. marsupialis* and invaded temperate North America during the Pleistocene (Gardner 1973). The common opossum, *D. marsupialis*, is very much a tropical species. Its geographical limits correspond closely to the Tropic of Cancer (23°N) and the Tropic of Capricorn 23°S). Like the other two species of *Didelphis*, it is a terrestrial omnivore with arboreal capabilities. The white-eared opossum, *D. albiventris*, is found throughout much of South America except the Amazon basin, from the higher elevations of Colombia to approximately 39°S in southern Argentina. It apparently favors cooler climates and is widely distributed in the temperate regions of South America (Streilein 1982).

Each of the three species of *Didelphis* covers a wide range of latitude, climate, and habitat where energetic constraints and variation in food availability are reflected in their reproductive strategies. The northern limits of *D. virginiana* approximate the $-7°$ January isocline (Tyndale-Biscoe 1973) or about 44°N. Here, the energetic demands of late lactation (Fleming et al. 1981) and a lack of thermoregulatory adjustment to low temperatures (Hsu et al. 1988) probably contribute to the restriction of the breeding season to the months of March through July.

Less is known of the didelphids in the temperate regions of South America, but, even at a relatively low latitude (20°S) near Belo Horizonte in southeastern Brazil, photoperiod appears to be an important proximate cue for the onset of seasonal breeding. Prostate and bulbo-urethral glands of male *D. albiventris* increased markedly in weight during the first part of the breeding period, June to January (Nogueira 1988). Females are in estrus from July to November and with pouch young from August to March (Mascarenhas 1980; Rigueira et al. 1987; cited in Nogueira 1988), a seasonal period equivalent to that of *D. virginiana* in North America. The onset of breeding in both species coincides with increasing daylength, and, although that also coincides with the dry season in Belo Horizonte, young are subsequently weaned during the rainy season (September to March) when food is abundant (Nogueira 1988). In northeast Brazil, where the annual variation in daylength is less, seasonal breeding in *D. albiventris* is more directly related to rainfall (Cerqueira 1984). *D. marsupialis* begins breeding in late January in Panama, about a month after the onset of the dry season and weans two litters during the period of fruitfall (May to September) (Fleming 1973). In eastern Colombia, *D. marsupialis* also produces two litters between January and August (Tyndale-Biscoe and Mackenzie 1976).

Litter size in *D. virginiana* varies with latitude, being largest in the north (42–44°N), e.g., 8.7 in New York (Hamilton 1958) and 8.6 in Nebraska (Reynolds 1952) where females have one, or at most two, litters per year. In Texas and Florida (30–32°N), females have two or three litters per year but the litters are smaller, about 6.3 (Hartman 1928; Burns and Burns 1957). These litter sizes represent a remarkable reduction from the number of eggs released at ovulation. An average of 30 CL (range = 15–58) per ovary (Fleming and Harder 1983) indicates a release of 60 ova per cycle, which is among the highest ovulation rates known for mammals. The number of embryos recovered in utero is lower, about 22 (Hartman 1923; Reynolds 1953) and even fewer are born, crawl to the pouch, and find an unoccupied, functional teat.

Average litter size in *D. marsupialis* was 6.5 at 5°N in French Guyana (Charles-Dominique 1983) and in eastern Colombia but was 4.5 at 3°N and at higher altitudes in western Colombia, where most females have three litters per year (Tyndale-Biscoe and MacKenzie 1976). Average litter sizes were higher, 7.1 (Hill 1918) and 8.5 (Davis 1945), at 23°S in southeastern Brazil near Rio de Janeiro. These observations suggest that females invest in larger litters when opportunities for second or third litters are reduced by cold weather, dry conditions, or high mortality.

The reproductive strategies of *Didelphis* reflect the constraints of a short reproductive life and female nutritional status at the time of parturition. Numerous studies have found that very few opossums live beyond two years in the wild (Petrides 1949; Fitch and Sandidge 1953; Llewellyn and Dale 1964; Sunquist et al. 1987). Moreover, females do not breed in their first year of life and their reproductive capacity declines sharply in their third year (Reynolds 1952; Jurgelski and Porter 1974). Thus, it appears that female opossums have but one season to achieve reproductive success.

The number of functional teats sets a finite limit on litter size in marsupials. Average litter size in *Didelphis* is considerably less than the average number of teats (13), apparently because the most anterior teats are not always fully functional (Reynolds 1952). Young on these tend to be smaller than those on the posterior teats (Cutts et al. 1978) and might suffer high mortality. Litter size appears to be related to maternal nutritional status in *D. marsupialis* and *D. virginiana* (Motta et al. 1983; Eisenberg 1988). Maximum litter size also correlates with body weight in *D. albiventris*, ranging from five in the smallest mothers to 10 in the largest (Cerqueira 1984). Results of an interesting field experiment with *D. marsupialis* in Venezuela indicated that nutritional factors can also influence the sex ratio and survival of offspring. Radiocollared females that received supplemental food near their dens each evening produced litters with male-biased sex ratios, higher body weights, and better survival than control females (Austad and Sunquist 1986).

CALUROMYS

Woolly opossums, known for their thick luxurious fur, are among the most arboreal of South American marsupials; they seldom forage on the ground, depending instead upon fruits and insects in the canopy of the tropical rainforests (Eisenberg 1989). The bare-tailed opossum, *C. philander* is somewhat smaller than the other two species, 150 to 300 g, and is widely distributed in northern and eastern regions of South America. This species is one of the best known of any marsupial in South America, particularly in the balance of ecological and physiological investigations of reproduction.

C. philander feeds primarily on fruit but also on nectar, gum, and invertebrates in the canopy of multistratal forests (Charles-Dominique 1983; Eisenberg 1989), where they can be seen running swiftly along branches and vines (Emmons 1990) and bounding from limb to limb in a manner reminiscent of lemurs (Hunsaker and Shupe 1977). *Caluromys* have a relatively high encephalization quotient (Eisenberg and Wilson 1981). They are relatively long-lived and more apt to breed in more than one season than *Marmosa* or *Monodelphis* (O'Connell 1979; Charles-Dominique et al. 1981).

Estrous cycles in *C. philander* average 38.6 days (range of 28 to 45 days in successive cycles). The luteal phase of the cycle, marked by the interval of significant elevation of plasma progesterone, lasts about 20 days (Perret and Atramentowicz 1989). This (20 of 38.6 days) is less than 60% of the cycle length, as expected for a marsupial with a Type I cycle (Tyndale-Biscoe and Renfree 1987), but is a greater proportion of the estrous cycle (20 of 38.6 days) than the comparable ratio (11 of 28 days) in *D. virginiana* (Harder and Fleming 1981). Also, mid-luteal progesterone levels (30 to 40 ng/ml) were much higher than the 10 to 15 ng/ml seen in *D. virginiana*. Records on females giving birth in captivity indicate that gestation lasts 20 to 28 days, the longest known for any American marsupial. It should be noted that the addition of this single new species to the ranks of the "studied" has revealed more variation among didelphids in estrous cycle physiology than was previously expected.

The average litter of four in *C. philander* (Charles-Dominique et al. 1981; Atrametowicz 1986) is relatively small and the period of postnatal care relatively long when compared to other didelphids of similar size (Table 15.2). Young are in the pouch for 80 days, nurse in the nest for an additional 30 days, and disperse from the nest at about 130 days (Eisenberg 1989). The pouch in *Caluromys* is developed only when the female is carrying young and regresses to lateral folds at other times (Perret and Atramentowicz 1989; Emmons 1990).

Studies of *C. philander* in French Guyana (Atramentowicz 1982; Perret and Atramentowicz 1989) have shown a positive relationship between the number of fruiting trees and the proportion of females that are reproductively active, i.e., lactating or in estrus (Fig. 15.3). Furthermore, young born shortly before the end of the fruiting season did not survive to weaning. Apparently the increased energetic demands of lactation (estimated from foraging activity) could not be met when food was scarce (Atramentowicz 1982). McNab (1986) has suggested that marsupials are unable to convert resources to high reproductive rates, which has placed them at a competitive disadvantage to eutherians in certain adaptive zones (McNab 1989).

TABLE 15.2. Duration (in days) of the estrous cycle, gestation, lactation and litter size and body weights of selected South American marsupials[1].

Species	Adult female weight (g)	Geographic distibution[2]	Length of estrous cycle	Length of gestation	Litter size	Neonatal weight (mg)	Permanent pouch exit or left in nest	Weaning	References
Didelphis virginiana	1000–3000	United States	28	13	6–9	130	70	100	Hartman (1921); Reynolds (1952); Hamilton (1958)
Didelphis marsupialis	1200	Guatemala Venezuela, Brazil	–	–	4–9	–	–	100	Fleming (1973); Tyndale-Biscoe and Mackenzie (1976)
Didelphis albiventris	1000–3000	Colombia, Brazil, Argentina	–	–	4–5	–	–	–	Tyndale-Biscoe and Mackenzie (1976); Streilein (1982); Cerqueira (1984)
Caluromys philander	285–340	Venezuela, Brazil	39	20+	4	–	75–80	100	Atramentowicz (1982, 1986); Eisenberg (1989); Perret and Atramentowicz (1989)
Marmosa robinsoni	40–80	Colombia, Venezuela	26	14	10–14	–	20	60–70	Barnes and Barthold (1969); Fleming (1973); Godfrey (1975); Hunsaker and Shupe (1977); O'Connell (1979)
Monodelphis domestica	60–80	Brazil	0	15	7	100	14	56	Fadem et al. (1982); VandeBerg (1983)
Chironectes minimus	600–800	Venezuela, Brazil	–	–	2–3	–	–	–	Ender (1966); Marshall (1978)
Caenolestes obscurus	20–30	Venezuela	–	–	3–4	–	–	–	Osgood (1921); Kirsch and Waller (1979); Tyndale-Biscoe (1980)

[1] Adapted from Tyndale-Biscoe and Renfree (1987) with modifications and additions.
[2] Nations listed represent the approximate northern and southern limits of distribution.

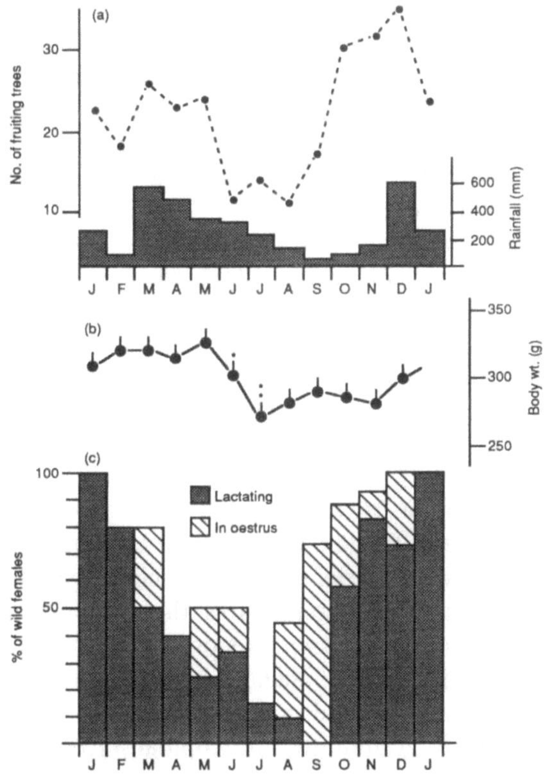

FIGURE 15.3. Relationships between seasonal variation in precipitation and fruit production (a), body weight (b), and the percentage of female woolly opossums lactating or in estrus (c) (reproduced with permission from Perret and Atramentowicz 1989).

MARMOSA

Kirsch and Callaby (1977) list 44 species of mouse opossums, making *Marmosa* the largest marsupial genus by far in South America. Known distributions of most are highly restricted, and all but two or three species are tropical in distribution (Table 15.1). Reproduction has been studied in only a few marmosids, of which *Marmosa robinsoni* (=*M. mitis*) is probably the best known. *M. robinsoni* is typical of South American marsupials, the grand majority of which are small (30 to 150 g) and pouchless, i.e., species of *Marmosa*, *Monodelphis*, and *Caenolestidae*, and it will serve as the main example of this group.

The pelage of *Marmosa robinsoni* is short and cinnamon brown on the dorsum with dark eye patches and cream to pale orange-yellow on the venter. Body weight varies from 36 to 132 g (Emmons 1990). Males are considerably larger than females and can exceed 350 mm in total length.

The relatively small range of *M. robinsoni* extends from Panama across northern Venezuela and Colombia to Ecuador (Streilein 1982) where this mouse opossum occupies a wide range of habitats including evergreen, deciduous and thorn forests, secondary growth, disturbed areas, and seasonally dry habitats (Eisenberg 1989; Emmons 1990). Like most members of the genus, *M. robinsoni* is nocturnal and forages for invertebrates and fruit in vines, trees, and undergrowth.

Marmosa robinsoni apparently is reproductively inactive from November through March in Panama (Fleming 1973), where one large litter of 10 might be the rule (Enders 1966). Lactating *M. robinsoni* were observed from May through October in the Llanos of northern Venezuela, with no clear peaks in activity during this period. Females apparently produce one or two large litters of 14 during a single breeding season in Venezuela (O'Connell 1979). The average life span in nature is probably less than one year (Hunsaker 1977).

The breeding behavior of *M. robinsoni* is similar to that observed in other didelphids, but use of the tail, as described by Barnes and Barthold (1969), is unusual. Following a brief courtship of clicking and mutual genital licking, the male grabs the shoulders of the estrous female. Simultaneously, he feels around the cage with his prehensile tail and grabs a bar or branch that will hold his weight; anchored with his tail, the male grabs the female's hind feet with his and intromission is achieved (Fig. 15.4). Copulation lasts for 15 to 20 minutes and copulatory bouts continue for more than six hours in some captive pairs (Hunsaker and Shupe 1977).

M. robinsoni is polyestrous with a cycle length of 25.5 days. The luteal phase of the cycle is the same length as gestation, about 14 days (Godfrey 1975). The mother cleans the mammary area by licking it before parturition and also cleans the young while they are firmly attached to the teats, which is only the first 20 days of the 70-day lactation period (Fig. 15.2). During this time they travel with the mother without the protection of a pouch. Infants older than 20 days of age are left in a nest while the mother forages for food. The mother will retrieve infants that are dislodged from the nest by picking them up and placing them on the mammary area (Thrasher et al. 1971; Hunsaker and Shupe 1977; Fig. 15.5). Older infants (>34 days) may again move with the mother by clinging to her back before they are weaned at 65 to 70 days (Eisenberg and Maliniak 1967; Barnes and Barthold 1969).

Conclusions Regarding Reproduction in South American Marsupials

In view of the high species diversity and wide geographical range of South American marsupials, the uniformity of reproductive patterns is remarkable. Current information suggests that all are seasonally polyestrous,

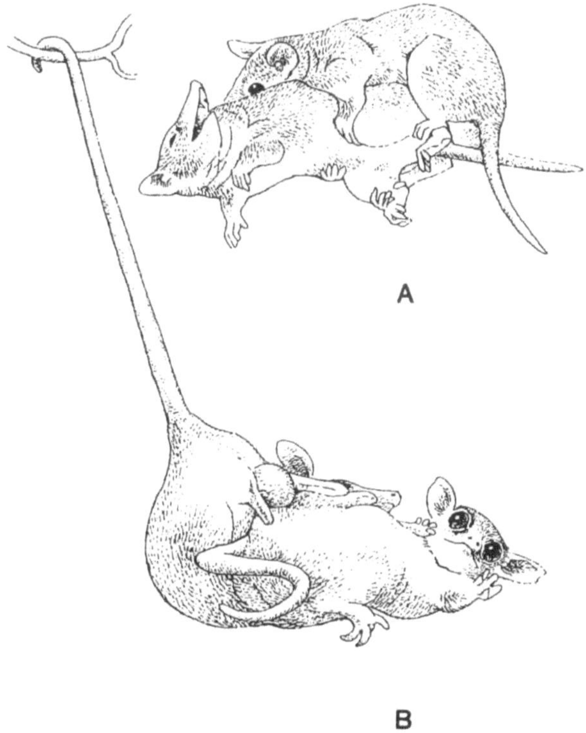

FIGURE 15.4. Typical didelphid copulatory position as seen in (A) *Monodelphis domestica* (from Trupin and Fadem 1982) and an unusual variation seen in (B) *Marmosa robinsoni*, in which the tail of the male is used to anchor and partially support the pair (from Barnes and Barthold 1969). A and B were reproduced with permission as redrawn in Tyndale-Biscoe and Renfree (1987).

although the induction of estrus by male pheromones, as seen in captive *Monodelphis domestica* (Fadem 1985), might be common. Most species attempt to produce at least two litters during a given breeding season, but there is a curious tendency toward curtailment of reproductive activity for two to four months of the year, often during October to January, regardless of latitude, climate, or habitat. Thus, the occurrence of a sexual resting or refractory period between breeding seasons, as suggested for *C. philander* by Perret and Atramentowicz (1989), might be a common feature of didelphid reproduction.

Maternal behavior is clearly influenced by presence of the pouch, which in turn is related to adult body size. At birth, neonates of all species move on their own with a wriggling motion from the opening of the urogenital sinus to the mammary area or pouch. Upon locating a teat, the neonate takes it into its well-developed oral cavity wherein structural modifications and swelling of the teat firmly affix the young to its mother. This grip is not

Figure 15.5. Maternal behavior in an opossum, *Monodelphis domestica*. The female is in the process of retrieving to her mammary area a 20-day-old young that was displaced from the nest (photograph by Nicholas B. Carter).

voluntarily released at any time during this continuous teat attachment phase. Marsupials of small (<150 g) body size often lack a pouch and the young remain firmly attached for less than one-third of the lactational period, e.g., 20 out of 70 days in *Marmosa robinsoni* (Table 15.1; Fig. 15.2). At the age such young are left in the nest, they are blind, thinly furred, and ectothermic (Wallis and Maynes 1973; Hunsaker and Shupe 1977) and the need for active maternal care is apparent. Mothers return often to the nest to nurse and warm their young and will retrieve those that become separated from littermates (Fig. 15.5). By contrast, large (>700 g) didelphids have well-developed pouches where young remain continuously attached to the teat or in the pouch for about two-thirds of the lactational period (Fig. 15.2). Mothers will clean the pouch and young by licking but will not attempt to retrieve young lost from the pouch or from the nest. However, after the young are more than 70 days of age, the mother will orient toward them and make clicking vocalizations in response to their distress calls (Hunsaker and Shupe 1977).

The pouchless condition is apparently primitive and its persistence in small marsupials might reflect strong selection for leaving the young in a nest at an early age. Litter sizes in this group tend to be large, and, by the time young are left in the nest, their combined weight is equal to about 50% of maternal body weight, apparently an unacceptable burden. By comparison, litters of larger, pouched species represent only about 20% of maternal weight by the time they are no longer carried in the pouch (Russell 1982). Thus, for larger species, the cost of carrying young for an extended period is low relative to the benefits of the protective environment provided by the pouch. Selection might also favor the pouch in larger marsupials because of the reduced availability of appropriate den sites for nursing young in many habitats. Kangaroos would be an example at the extreme. Thus, the pouch is most likely a derived condition associated with the evolution of large body size and feeding specializations (Tyndale-Biscoe and Renfree 1987).

Evolution of the pouch and elimination of the need to maintain a nest for nursing neonates might have been a significant factor in the adaptive radiation of marsupials. Transporting young in the pouch eliminates the need for frequent visits to the den, an important benefit for marsupials, such as *Didelphis*, that forage over wide areas. Such benefits are apparently realized by the water opossum, *Chironectes minimus*, the only marsupial adapted for aquatic life. A well-developed pouch opens to the rear and is closed tightly by a strong sphincter muscle (Rosenthal 1975) when the female dives to prey on fish and crustaceans. Fatty secretions in the pouch appear to aid in producing a "watertight" compartment (Marshall 1978) so that the litter of two or three (Enders 1966) is protected from drowning.

Considerations for the Future of South American Marsupials

There is a clear and pressing need for wide scale survey work on South American marsupials and the production of taxonomic keys and field guides for each country or region. The new guide for rain forest mammals by Emmons (1990) is a major contribution in this regard. The ongoing need for systematic and taxonomic work on South American marsupials is apparent in the case of *Marmosa* and Caenolestidae. Both are in need of taxonomic revision (Pine 1982) and knowledge of their reproduction is limited to only a few species. Additional information on breeding seasons, denning habits, and habitat preferences of these two pouchless groups would contribute much to our understanding of marsupial reproductive strategies.

Marsupials are particularly attractive as subjects of survey trapping because basic information on reproductive rates and seasons can be obtained from single captures in a way not possible with other mammals. Females can be inspected for number and size of young, and recapture of females with young, particularly in pouched species, will reveal patterns of postnatal mortality, estimates of number weaned, and net natality. Such observations, when coupled with climatological and ecological data, such as food availability, would contribute substantially to mammalian reproductive ecology. Because litter size appears to be closely related to the nutritional status of the mother, inspection of female marsupials during routine surveys could reveal habitat and environmental degradation before such was evident in estimates of population size or indices of abundance.

Progress in the conservation of South American marsupials can best be achieved through the cooperative efforts of citizens, institutions, and governments from throughout the Hemisphere. For example, a cooperative program between the Foundation for Nature Preservation of Suriname and the Carnegie Museum of Natural History has been successful in training biologists and establishing a scientific infrastructure for investigating and managing a system of parks and reserves throughout Suriname (Genoways et al. 1982). There is a need for more cooperative, jointly funded programs between academic institutions and South American governments to provide educational and employment opportunities for local biologists (Mares 1982). National faunal surveys focused on the distribution and abundance of vertebrates might be a natural outgrowth of these programs. Faunal surveys would not only provide critical information for the management and conservation of natural resources, but they would also provide employment opportunities for a cadre of conservation workers with leadership potential and national identities.

References

Atramentowicz M (1982) Influence du milieu sur l'activite locomotrice et la reproduction de *Caluromys Philander* (L.). Rev. Ecol. Terre Vie 36:373–395

Atramentowicz M (1986) Dynamique de population chez trois marsupiaux Didelphides de Guyane. Biotropica 18:136–149

Austad SN, Sunquist ME (1986) Sex-ratio manipulation in the common opossum. Nature 324:58–60

Barnes RD, Barthold SW (1969) Reproduction and breeding behavior in an experimental colony of *Marmosa mitis* Bangs (Didelphidae). J. Repro. Fertil. suppl. 6:477–492

Burns RK, Burns LM (1957) Observations on the breeding of the American opossum in Florida. Rev. Suis. Zool. 64:595–605

Cerqueira R (1984) Reproduction de *Didelphis albiventris* dans le nord-est du Bresil (Polyprotodontia, Didelphidae). Mammalia 48:95–104

Charles-Dominique P (1983) Ecology and social adaptations in didelphid marsupials: Comparisons with eutherians of similar ecology. In Eisenberg JF, Klieman DG (eds.). Advances in the Study of Mammalian Behavior. Amer. Soc. Mammal., Philadelphia, Philadelphia. pp. 395–422

Charles-Dominique P, Atramentowicz M, Charles-Dominique M, Gerald H, Hladik A, Hladik CM, Prevost MF (1981) Les mammiferes frugivores arboricoles nocturnes d'une foret guyanaise: Interrelations plantes-animaux. Rev. Ecol. Terre Vie 35:341–435

Clemens WA (1977) Phylogeny of the marsupials. In Stonehouse B, Gilmore D (eds.). The Biology of Marsupials. Macmillan Press, London. pp. 51–68

Cutts JH, Krause WJ, Leeson CR (1978) General observations on the growth and development of the young pouch opossum, *Didelphis virginiana*. Biology of the Neonate 33:264–272

Davis DE (1945) The annual cycle of plants, mosquitoes, birds, and mammals in two Brazilian forests. Ecol. Mono. 15:245–295

Eisenberg JF (1988) Reproduction in polyprotodont marsupials and similar-sized eutherians with a speculation concerning the evolution of litter size in mammals. In Boyce MS (ed.). Evolution of Life Histories of Mammals. Yale University Press, New Haven, Connecticut. pp. 291–311

Eisenberg JF (1989) Mammals of the Neotropics. The University of Chicago Press, Chicago, Illinois

Eisenberg JF, Maliniak E (1967) Breeding the murine opossum *Marmosa* in captivity. International Zoo Yearbook 7:78–79

Eisenberg JF, Wilson DE (1981) Relative brain size and demographic strategies in didelphid marsupials. Amer. Nat. 118:1–15

Emmons LH (1990) Neotropical rainforest mammals, a field guide. The University of Chicago Press, Chicago and London

Enders RK (1966) Attachment, nursing, and survival of young in some didelphids. In Rowlands JW (ed.). Comparative Biology of Reproduction in Mammals. Academic Press, New York. pp. 195–203

Fadem BH (1985) Evidence for the activation of female reproduction by males in a marsupial, the gray short-tailed opossum (*Monodelphis domestica*). Biol. Repro. 33:112–116

Fadem BH, Trupin GL, Maliniak E, VandeBerg JL, Hayssen V (1982) Care and breeding of the gray, short-tailed opossum (*Monodelphis domestica*). Lab Ani. Sci. 32:405–409

Fitch HS, Sandidge LL (1953) Ecology of the opossum on a natural area in northeastern Kansas. Univ. Kan. Mus. Nat. Hist. 7:305–338

Fleming MW, Harder JD (1981a) Effect of pregnancy on uterine constituents of the Virginia opossum. Comp. Biochem. Physiol. 69A:337–339

Fleming MW, Harder JD (1981b) Uterine histology and reproductive cycles in pregnant and non-pregnant opossums, *Didelphis virginiana*. J. Repro. Fertil. 63:21–24

Fleming MW, Harder JD (1983) Luteal and follicular populations in the ovary of the opossum (*Didelphis virginiana*) after ovulation. J. Repro. Fertil. 67:29–34

Fleming MW, Harder JD, Wukie JJ (1981) Reproductive energetics of the Virginia opossum compared with some eutherians. Comp. Biochem. Physiol. 70B:645–648

Fleming TH (1973) The reproductive cycles of three species of opossums and other mammals in the Panama Canal Zone. J. Mammal. 54:439–453

Gardner AL (1973) The systematics of the genus *Didelphis* (Marsupialia: Didelphidae) in North and Middle America. Texas Tech University Press, Lubbock, Texas

Genoways HH, Reichart HA, Williams SL (1982) The Suriname small mammal survey: A case study of the cooperation between research and national conservation needs. In Mares MA, Genoways HH (eds.). Mammalian Biology in South America. Pymatuning Laboratory of Ecology, University of Pittsburgh, Pittsburgh, Pennsylvania. pp. 491–504

Godfrey GK (1975) A study of oestrus and fecundity in a laboratory colony of

mouse opossums *Marmosa robinsoni*. J. Zool. London 175:541–555
Hamilton WJ (1958) Life history and economic relations of the opossum (*Didelphis marsupialis virginiana*) in New York State. Cornell University Agriculture Experiment Station Memoir 354:1–48
Harder JD, Fleming MW (1981) Estradiol and progesterone profiles indicate a lack of endocrine recognition of pregnancy in the opossum. Science 212:1400–1402
Hartman CG (1921) Breeding habits, development, and birth of the opossum. Annual Report of the Smithsonian Institution 1921:347–364
Hartman CG (1923) The oestrous cycle in the opossum. Am. J. Anat. 32:353–421
Hartman CG (1928) The breeding season of the opossum (*Didelphis virginia*) and the rate of intrauterine and postnatal development. J. Morphol. 46:143–215
Hill JP (1918) Some observations on the early development of *Didelphis aurita*. Quart. J. Microsc. Sci. 63:91–139
Hsu M, Harder JD, Lustick SI (1988) Seasonal energetics of opossums (*Didelphis virginiana*) in Ohio. Comp. Biochem. Physiol. 90A:441–443
Hunsaker D (1977) Ecology of new world marsupials. In Hunsaker D (ed.). The Biology of Marsupials. Academic Press, New York. pp. 95–156
Hunsaker D, Shupe D (1977) Behavior of new world marsupials. In Hunsaker D (ed.). The Biology of Marsupials. Academic Press, New York. pp. 279–347
Jurgelski W Jr, Porter ME (1974) The opossum (*Didelphis virginiana*) as a biomedical model. III. Breeding in captivity: Methods. Lab Ani. Sci. 24:412–425
Kirsch JAW (1977) The classification of marsupials with special reference to karyotypes and serum proteins. In Hunsaker D (ed.). The Biology of Marsupials. Academic Press, New York. pp. 1–50
Kirsch JAW, Calaby JH (1977) The species of living marsupials: An annotated list. In Stonehouse B, Gilmore D (eds.). The Biology of Marsupials. Macmillan, London. pp. 9–26
Kirsch JAW, Waller PF (1979) Notes on the trapping and behavior of the Caenolestidae. J. Mammal. 60:390–395
Llewellyn LM, Dale EH (1964) Notes on the ecology of the opossums in Maryland. J. Mammal. 45:113–122
McNab BK (1986) Food habits, energetics, and the reproduction of marsupials. J. Zool. London 208:595–614
McNab BK (1989) On the selective persistence of mammals in South America. In Redford KH, Eisenberg JF (eds.), Advances in Neotropical. Mammalogy. The Sandhill Crane Press, Inc., Gainsville, FL, pp. 605–614
Mares MA (1982) The scope of South American mammalian biology: Perspectives on a decade of research. In Mares MA, Genoways HH (eds.). Mammalian Biology in South American. Pymatuning Laboratory of Ecology, University of Pittsburgh, Pittsburgh, Pennsylvania. pp. 1–26
Marshall LG (1978) *Chironectes minimus*. Mammalian Species 109:1–6
Marshall LG, Case JA, Woodburne MO (1990) Phylogenetic relationships of the families of marsupials. In Genoways HG (ed.). Current Mammalogy. University of Nebraska, Plenum Press, New York. pp. 433–481
Mascarenhas AEAS (1980) Histologia e inervacao do sistema genital feminino na fase de anestro no gamba Didelphis albiventris (Lund 1841) (Didelphidae Marsupialia) Tese de Mestrado em Morfologia, ICB/UFMG, Belo Horizonte, Brasil

Motta M, Dezonna JC, Franco AMR (1983) A note on the reproduction of *Didelphis marsupialis* in captivity. Memorias do Instituto Oswaldo Cruz Rio de Janeiro 78:507–509

Nogueira JC (1988) Anatomical aspects and biometry of the male genital system of the white-belly opossum *Didelphis albiventris* (Lund 1841) during the annual reproductive cycle. Mammalia 52:2:233–242

O'Connell M (1979) Ecology of didelphid marsupials from northern Venezuela. In Eisenberg JF (ed.). Vertebrate Ecology in the Northern Neotropics. Smithsonian Institution Press, Washington. pp. 73–87

Osgood WH (1921) A monographic study of the American marsupial *Caenolestes*. Field Museum of Natural History Zoological Series 14:1–162

Perret M, Atramentowicz M (1989) Plasma concentrations of progesterone and testosterone in captive woolly opossums (*Caluromys philander*). J. Repro. Fertil. 85:31–41

Petrides GA (1949) Sex and age determination in the opossum. Journal of Mammalogy 30:364–378

Pine RH (1982) Current status of South American mammalogy. In Mares MA, Genoways HH (eds.). Mammalian Biology in South America. Pymatuning Laboratory of Ecology, University of Pittsburgh, Pittsburgh, Pennsylvania. pp. 27–37

Renfree MB (1975) Uterine proteins in the marsupial, *Didelphis marsupialis virginiana*, during gestation. J. Repro. Fertil. 42:163–166

Reynolds HC (1952) Studies on reproduction in the opossum (*Didelphis virginiana virginiana*). Univ. Calif. Publ. Zool. 52:223–284

Reynolds HC (1953) The opossum. Sci. Amer. 188:88–94

Rigueira SE, Valle CMC, Varejao JBM, Albuquerque PV, Nogueira JC (1987) Algumas observacoes sobre o ciclo reprodutivo anual de femeas do gamba *Didelphis albiventris* (Lund 1841) (Didelphidae, Marsupialia). Em populacoes naturais no Estado de Minas Gerais, Brasil. Revista Brasileira de Zoologia 4:129–137

Rosenthal MA (1975) Observations on the water opossum or yapok *Chironectes minimus* in captivity. International Zoo Yearbook 15:4–6

Russell EM (1982) Patterns of parental care and parental investment in marsupials. Biol. Rev. 57:423–486

Sharman GB (1970) Reproductive physiology of marsupials. Science 167:1221–1228

Streilein KE (1982) Behavior, ecology, and distribution of South American marsupials. In Mares MA, Genoways HH (eds.). Mammalian Biology in South America. Pymatuning Laboratory of Ecology, University of Pittsburgh, Pittsburgh, Pennsylvania. pp. 231–250

Sunquist ME, Austad SN, Sunquist F (1987) Movement patterns and home range patterns in the common oossum, *Didelphis marsupialis*. J. Mammal. 68:173–176

Thrasher JD, Barenfus M, Rich ST, Shupe DB (1971) The colony management of *Marmosa mitis*, the pouchless opossum. Lab Ani. Sci. 21:526–536

Trupin GL, Fadem BH (1982) Sexual behavior of the gray short-tailed opossum (*Monodelphis domestica*). J. Mammal. 63:409–414

Tyndale-Biscoe CH (1973) Life of Marsupials. American Elsevier Publishing Co., New York

Tyndale-Biscoe CH (1980) Observations on the biology of marsupials in Colombia and Venezuala. National Geographic Society Research Report 12:711–720

Tyndale-Biscoe CH, MacKenzie RB (1976) Reproduction in *Didelphis marsupialis* and *Didelphis albiventris* in Columbia. J. Mammal. 57:249–265

Tyndale-Biscoe CH, Renfree M (1987) Reproductive physiology of marsupials. Cambridge University Press, Cambridge

VandeBerg JL (1983) The gray short-tailed opossum: A new laboratory animal. Inst. Lab Ani. Res. News 26:9–12

Wallis RL, Maynes G (1973) Ontogeny of thermoregulation in *Macropus parma* (Marsupialia: Macropodidae). J. Mammal. 54:278–281

16
Reproduction in the Gray Short-Tailed Opossum, *Monodelphis domestica*

H.D.M. Moore

The gray short-tailed opossum, *Monodelphis domestica*, is a member of the Didelphidae family of opossums. These animals are of particular interest because they are the most ancient family of marsupials known and may represent the ancestors of all extant species in both the New World and Australia (Richardson 1988). However, this does not necessarily mean that they are a relic species, for the 60 American opossums have radiated to successfully fill a variety of habitat niches (Hunsaker 1977). The genus *Monodelphis* is composed of 11 species distributed from eastern Panama to central Argentina. *Monodelphis domestica* is a small nocturnal opossum found mainly in eastern and central Brazil and is thought to have a solitary, semi-arboreal lifestyle, using its partially prehensile tail for climbing. It builds nests in hollow logs and, as its name implies, may move into households where it is encouraged to feed on rodent and insect pests (Collins 1973). In the wild, *M. domestica* breeds throughout the year producing litters of 8 to 14 offspring. Females do not have a pouch but, in common with other marsupials, produce extremely altricial neonates (Tyndale-Biscoe and Renfree 1987).

Until recently, very little detail was known about the reproduction in any South American marsupial. Starting with Selenka (1887), there have been a number of classical publications on the reproductive biology of the opossum (Hartman 1923; McCrady 1938). But these studies were restricted to *Didelphis virginiana*, whose range is confined to North and Central America. In 1978, the National Zoological Park, Washington, D.C., U.S.A., obtained four male and five female short-tailed opossums from the Caatinga region of southeastern Brazil. Fortunately, these animals thrived and bred successfully over many generations so that their offspring were made available to a number of zoos and research institutes. Because *M. domestica* can be maintained under simple laboratory conditions and are small and easy to handle, they have rapidly become established as the animal of choice for the study of New World marsupial biology (Fadem et al. 1982; Baggott 1987). The main feature that distinguishes marsupials from eutherian mammals is their reproduction. It is not surprising, therefore, that most studies of *M. domestica* have focused on this aspect of their physiology.

Male Anatomy and Sperm Development

In captivity, males reach sexual maturity at 5 to 6 months of age (weight 90 to 110 g), although they may not attain full size (150 to 180 g) until over one year old. A number of secondary sexual characteristics accompany maturation. The underlying tunica vaginalis of the scrotum becomes densely pigmented with melanin at puberty, while the suprasternum sebaceous scent gland (found only in males) is more prominent and conspicuously stained with waxy secretion. Fully adult males also exhibit long canine teeth which may extend below the lower jaw line (Fig. 16.1).

The gross anatomy of the reproductive tract is similar to that described for *D. virginiana* (Hartman 1923) and consistent with the general anatomy for marsupials (Tyndale-Biscoe and Renfree 1987). Externally, the most striking feature is the dark prepenial scrotum. The reason for the pigmentation of the tunica vaginalis is unknown. In other mammals, scrotal color is used for sexual display, but this does not seem to be the case for *Monodelphis*. Biggers (1966) has suggested it may play a role in controlling testis temperature by acting as a 'black box' radiator, although, in a variety

FIGURE 16.1. Adult male gray short-tailed opossum *M. domestica*.

of opossum species, pigmentation is not obviously correlated with climatic conditions (Tyndale-Biscoe and Renfree 1987).

The oval testes are suspended in a serous fluid within the scrotal sac. The tough skin along with this fluid probably serve to protect the otherwise vulnerable pendulous testes, which lie permanently within the scrotum due to the narrow neck of the inguinal canal. In addition, a well-developed cremaster muscle can draw the scrotum close to the body cavity.

The genital tract consists of paired testes and epididymides, vasa deferentia leading to a large segmented prostate gland, a narrow urethra, and two pairs of Cowper's glands (bulbourethral glands). As in all American marsupials, the glans penis is bifurcate. The significance of this feature is unclear, but it may have evolved to help direct the ejaculate into each lateral vaginal canal during copulation (Biggers 1966).

Perhaps the most unusual feature of male reproduction in American marsupials is the development and maturation of spermatozoa. In *M. domestica*, proliferation of primary germ cells within the seminiferous tubules of the testis is unremarkable and consistent with the pattern in Eutheria. However, the morphogenic transformation to form the testicular spermatozoon during post-meiotic spermiogenesis and the continuation of this process within the epididymis involves a highly complex reorientation of the nucleus and redistribution of cytoplasm. This gamut of morphological differentiation is completed in the mid-corpus epididymidis by the unique phenomenom of sperm pairing (Fig. 16.2). This association is effected by the intimate contact of the plasma membrane over the acrosomes of opposed spermatozoa with the formation of a junctional complex at the peripheral borders (Baggott 1987). Such sperm pairing occurs in all American marsupials with the possible exception of *Dromiciops australis* (Temple-Smith and Grant 1986), but is unknown in any other vertebrate. Only in insects has another form of sperm pairing been described (Bawa 1978).

The reason for sperm pairing remains obscure but several hypotheses have been put forward. Since the flagella of paired spermatozoa normally beat in an alternate but synchronous manner, it has been suggested that sperm associations may be mutually beneficial in marsupials for passage through the lateral vagina and uterus to the ampulla of the oviduct before fertilization (Phillips 1970). However, sperm pairing has not evolved (or has not been conserved) in Australian marsupials with similar anatomy, nor does it confer on spermatozoa any greater progressive motility. In modified Tyrode's medium, *M. domestica*, spermatozoa paired or unpaired exhibit progressive motility of about 150 μm/sec at 37°C (Baggott 1987). Although this motility is relatively fast for mammalian sperm, it is slower than the mean motility for the separate spermatozoa of the wallaby (*Macropus rufogriseus*) which have a velocity of 240 μm/sec. Another argument for pairing is that it protects the acrosome and the overall survival of spermatozoa in the female tract prior to fertilization (Bedford et al. 1984).

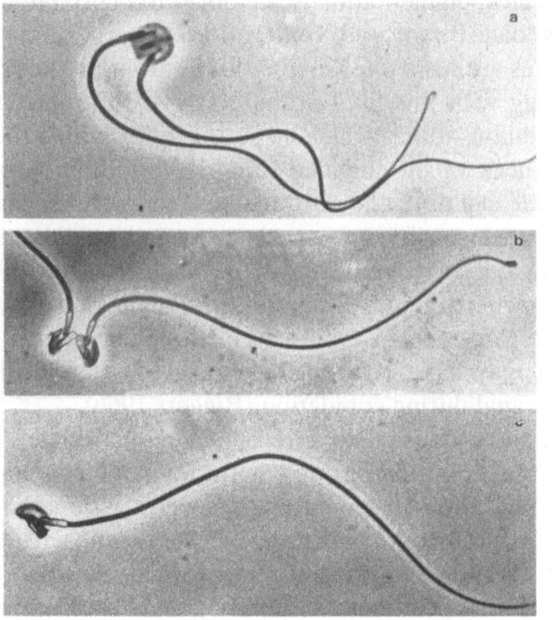

Figure 16.2. Spermatozoa of *M. domestica* (×450). (A) paired spermatozoa from the cauda epididymidis; (B) spermatozoa unpairing during in vitro incubation; (C) single spermatozoon after unpairing.

Certainly, the number of spermatozoa produced by the testes of American marsupials is low compared with many (but not all) of their Australian counterparts. In *M. domestica* the mean number of sperm in the proximal and distal epididymidis is 1.07 ± 0.3 million and 1.47 ± 0.5 million, respectively; from experiments involving the ligation of the epididymis, it has been estimated that the daily testicular output is only about 130,000 sperm. This compares with a daily sperm production rate of 120 million in the rabbit (Moore and Bedford 1978). Bedford et al. (1984) have surmised that opossums can afford such low sperm numbers because the acrosomes are protected by pairing and sperm are sequestered in oviductal crypts. Low sperm numbers may also be due to the lack of sperm competition due to the mating system adopted by these animals.

Female Tract and Follicular Development

The female gray opossum reaches maturity slightly earlier than the male, at 4 to 5 months of age (50 to 60 g) in captivity. However, young adults produce small litters of three or four offspring and full fecundity (litter size up to 14 offspring) is not reached until 8 to 12 months of age. Females are

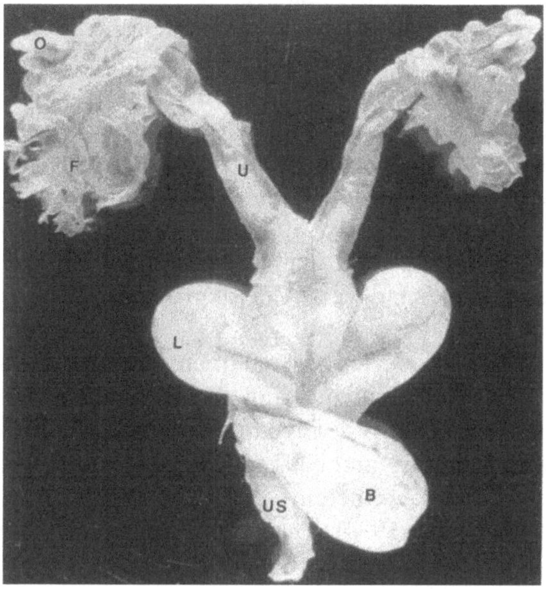

FIGURE 16.3. Female reproductive tract from *M. domestica* (×5). F = Fimbria; O = Oviduct; U = uterine horn; L = lateral vagina; B = bladder; US = urogenital sinus.

pouchless and a distinct arrangement of 14–16 teats develop from nipple anlagen formed during sexual differentiation. Nipple development is most pronounced just before parturition. Most females have little or no fur surrounding teats and tend to have loose flaps of skin on either flank which may serve to protect offspring in the absence of a pouch. Whether these features are hormonally induced or are a consequence of nursing large litters is not known.

The genital tract is characterized by a vaginal complex and urogenital sinus, as observed in various forms in all marsupials. Two lateral vaginae open posteriorly into a urogenital sinus which also receives the urethra (Fig. 16.3). Anteriorly, each lateral vagina, separated by a common median septum, forms vaginal cul de sacs into which the separate uterine cervices open (Baggott 1987; Moore and Thurstan 1990). At the posterior end of the lateral vaginae and the anterior end of the urogenital sinus, there is fibrous connective tissue termed the urogenital strand. This tissue breaks down before parturition to form the pseudo-vaginal birth canal (Tyndale-Biscoe 1966). Unlike some marsupials, where the birth canal remains after first parturition, in *M. domestica* the canal closes soon after birth (Moore unpubl. obs.).

The two uteri are entirely separate and each is served by a cervix. The uterine endometrium exhibits a pseudo-stratified epithelium with a complex arrangement of secretory glands. Following estrus and ovulation, there is a marked and progressive enlargement of the uteri due to hypertrophy and secretory activity of the endometrial epithelium. Epithelial cells display abundant endoplasmic reticula and secretory vesicles which enter the uterine lumen (Moore and Smith unpubl. obs.). This increased cell activity corresponds to high levels of progesterone in the circulation during the luteal phase. The ovaries in *Monodelphis* have a similar histology to those of other didelphids. Behavioral and hormonal studies indicate that, in captivity, *Monodelphis* does not display an ovarian cycle as such, but follicular development is induced by close proximity of an unfamiliar adult male (see below).

Following ovulation, oocytes (devoid of any cumulus mass) are captured by a large ciliated fimbria surrounding each ovary and subsequently pass into the oviduct. The latter is divided into an ampullary region with a wide lumen lined with secretory and ciliated cells and a narrow isthmus region with a greatly infolded epithelum containing crypts or glands (Baggott 1987; Baggott and Moore 1990). The isthmus region may function, as in *D. virginiana* (Rodger and Bedford 1982), as a reservoir for spermatozoa prior to ovulation and fertilization.

Sexual Behavior, Induction of Estrus, and Copulation

Early reports suggested the female gray opossum exhibited an ovarian cycle of about 32 days (Fadem et al. 1982; Fadem and Rayve 1985). However, the observation that swapping males between females maintained breeding in captive colonies of *M. domestica* led to the surprising but crucial finding that mature females display little, if any, spontaneous mating behavior or ovarian cyclicity. Instead, estrus is induced by the presence of an unfamiliar male (Fadem and Rayve 1985; Baggott et al. 1987). Following estrus (and possibly mating), a female enters an obligatory luteal phase leading to parturition after 14 days or, if not pregnant, to a quiescent state unless activated once more by a new male. Since this latter situation, leading to the next estrus, is reached in about 30 to 35 days, it may have caused the initial confusion over an ovarian cycle.

Using closed-circuit video recording, it has been possible to accurately follow mating behavior in *M. domestica.* Quiescent anestrus females are considerably aggressive towards a new male and will show antagonistic behavior such as open mouth threats, hissing, and attack leaps. However, within 3 to 5 days, pro-estrous behavior begins, leading to typical courtship such as naso-cloacal contact, soliciting chases, and rump dragging. Full estrus and mating then ensues over the following 24 to 72 hours; the median time between pairing to full estrus (as determined by mating) being about $8\frac{1}{2}$ days (Baggott et al. 1987). Castrated males usually fail to elicit an estrous response, suggesting that androgens are at least indirectly involved, and there is now cogent evidence that male primer pheromones play a role (Fadem 1987). Cage switching experiments have indicated that soiled bedding alone can induce estrous behavior, and more detailed investigations of various scent markings by the male suggest that urine (rather than feces or suprasternum gland secretion) provides the major stimulatory odor (Moore unpubl. obs.). The functional pathways for pheromones are the main olfactory system and an accessory olfactory system, the vomeronasal organ, both of which are well developed in *M. domestica.* A recent investigation involving ablation of these organs in adult females demonstrated that the main olfactory pathway appeared to be the primary route for estrous induction (Pelengaris et al. 1992). This finding is in contrast to the situation in rodents where the vomeronasal organ mediates the pheromonal signal (Keverne and Rosser 1986).

Copulation involves the male grasping the female's ankles with his hind feet while clasping her just anterior to the pelvis with his forepaws (Trupin and Fadem 1982). The male usually bites the female by the scruff of the neck and it is this action, together with a kneading of the forepaws, that leads to a characteristic reflex posture with the female invariably on her right side fully immobilized and rigid (Fig. 16.4). Intromission is accompanied by a penis-urogenital sinus lock and the female is usually dragged backwards before disengagement. Copulation takes four to five minutes and females are receptive to mating for 24 to 36 hours after full behavioral estrus. Ovulation occurs 18 to 24 hours after the first mating, but it still

FIGURE 16.4. Mating and intromission in *M. domestica.*

remains unclear whether *M. domestica* females are reflex ovulators. It is known that receptive females that are prevented from copulating proceed to ovulate (Moore unpubl. obs.); therefore, intromission is not an essential requirement. However, such females are in close contact with males and their physical presence may be sufficient to promote ovulation. Either way, hormonal profiles and direct observation of ovaries suggest a close association between behavioral estrus and ovulation. This is in contrast to such marsupial species as Australian dasyurids in which mating may precede ovulation by more than a week (Tyndale-Biscoe and Renfree 1987).

Fertilization and Embryonic Development

Since ovulation can be predicted from the time of first mating, *M. domestica* is an excellent laboratory marsupial model to investigate fertilization and early embryonic development. After mating, spermatozoa are deposited in the lateral vaginae in a seminal plasma gel. They subsequently pass through the uteri to colonize the isthmus region of the oviduct, and then, near the time of ovulation, migrate to the ampulla (Baggott 1987). Fertilization must take place almost immediately after ovulation, as oocytes recovered from the ampulla of mated females are invariably fertilized while few spermatozoa are recovered (Baggott and Moore 1990). A glycoprotein secretion produced by non-ciliated oviducal epithelial cells forms a mucin layer around the zona pellucida (Fig. 16.5) which traps any super-

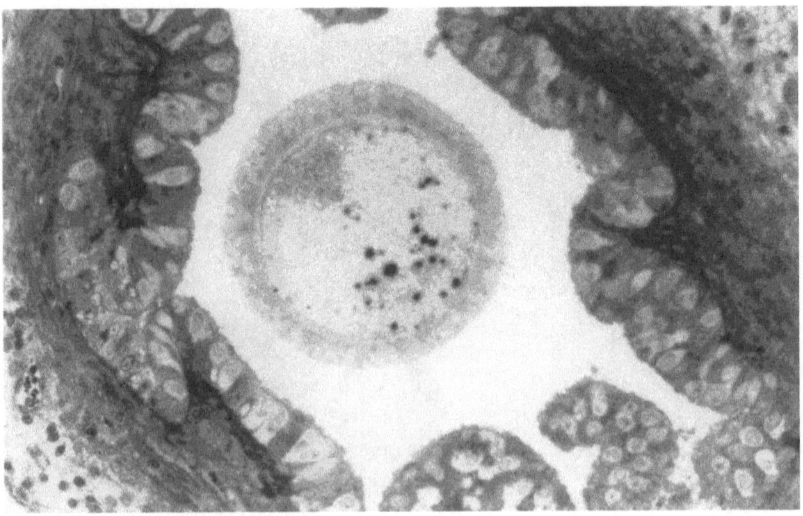

FIGURE 16.5. A fertilized oocyte in the oviductal ampulla. Note the mucopolysaccharide mucin coat formed from secretions of the oviducal epithelial cells (×180).

numerary sperm and forms an effective block to polyspermy (Phillips and Fadem 1987; Baggott and Moore 1990).

The *M. domestica* oocytes are similar to those described for *D. virginiana* (Baggott and Moore 1990) and display a characteristic peripheral granular cytoplasm containing many mitochondria and a yolky lipid occupying the central region. Following fertilization, one pole of the zygote (possibly at the site of sperm entry) becomes more refractive under phase contrast microscopy as the ooplasm undergoes further partitioning and polarization. The first cleavage occurs after passage into the uterus and is accompanied by a complete extrusion and vesiculation of the vitellus contents. This process, deutoplasmolysis, is nearly identical to that described for *D. virginiana* (Hartman 1916), although the resulting blastomeres of *M. domestica* may be smaller. The appearance of a shell membrane investing coat also corresponds with entry into the uterus and, hence, it is likely that the secretory cells responsible for shell formation reside in the endometrium rather than the oviduct. Generation of the shell membrane continues for several days, reaching maximum thickness prior to blastocyst expansion.

A unilaminar blastocyst forms five days after fertilization (Fig. 16.6). The development of the mesoderm and embryonic disc, and organogenesis occur over the next four days, followed by the breakdown of the shell membrane and a superficial implantation $10\frac{1}{2}$ days after fertilization. Parturition occurs $13\frac{1}{2}$ to 14 days after fertilization when ovulation is accurately determined by laparotomy (Moore unpubl.). Small differences in the gestation length may be due to a variable delay in development at the 4 to 8 cell stage.

Parturition and Development of Young

Nesting behavior of the mother increases the day before parturition, which usually takes place in the first few hours after dusk with the animal in a hunched position. A litter may consist of up to 16 young but only those attached to a teat survive (Fig. 16.7). At birth, neonates have a crown-rump length of about 10 mm and weigh 90 to 120 mg (Baker et al. 1990). Teat position, litter size, or gender has no obvious influence on the initial weight of the young, which remain continuously attached to the nipple for at least 14 days. At weaning (50 to 70 days after birth), however, there are significant differences in the weight of male and female offspring due to the dimorphic growth curves (Moore and Thurstan 1990).

Sexual differentiation in marsupials is unusual in that it takes place mainly after birth rather than in utero, as for eutherian mammals. *M. domestica* pups are externally sexually undifferentiated at birth and our initial studies indicated that the gonadal anlagen were indifferent (Moore and Thurstan 1990). However, a recent quantitative investigation has revealed that developing testes and ovaries can be distinguished on the day of birth by the

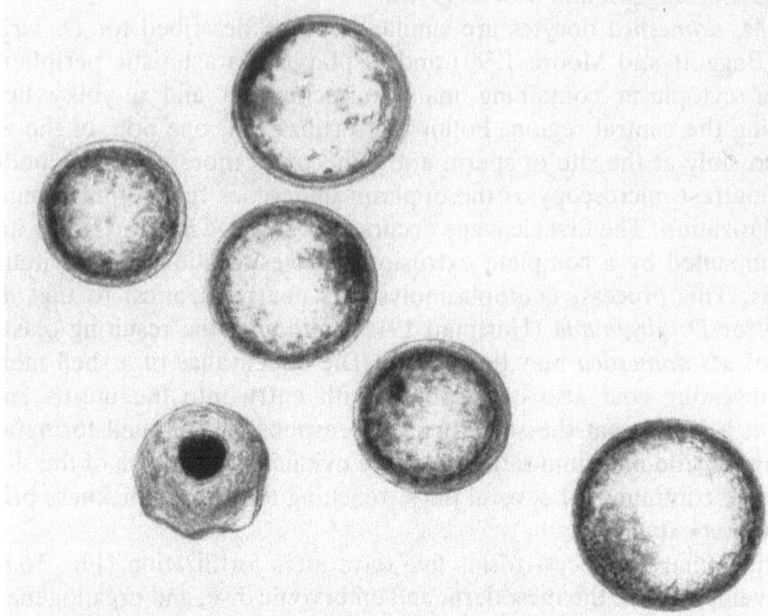

Figure 16.6. Unilaminar blastocysts recovered from the uterus five days after fertilization. The blastocysts display a variation in size. An unfertilized oocyte is present (×60).

size and histological differences in the sex cords and tunica albuginea (Baker et al. 1990). These findings suggest, therefore, that *M. domestica* gonadal differentiation may be initiated in utero, similar to the wallaby (O et al. 1988), although scrotal development, which may be under genetic control, begins two days after birth. Male neonates treated with estrogen show severe disruption of sexual differentiation with the formation of streak gonads and development of a female-like genital tract (Fadem and Tesoriero 1986; Moore and Thurstan 1990). The growth rates and subsequent sexual behavior of these treated males are also feminized. Sex reversal, with the formation of ovarian tissue, as reported by Burns (1956) for *D. virginiana*, does not occur.

Conclusion

Aside from the intrinsic interest, the marsupial mode of reproduction in *M. domestica* provides a fascinating comparison with that occurring in eutherian mammals. Fundamental differences are apparent in the gross anatomy

Figure 16.7. Neonates attached to the nipples of the mother on the day of birth (×3).

and physiology and in gamete and embryo development. Such features provide important insights into how reproductive systems may have evolved in mammals. Studies in *M. domestica* are particularly relevant in this respect, since fossil records suggest that the didelphids represent the oldest marsupial family and, therefore, may have retained reproductive attributes of their proto-mammalian ancestors.

Perhaps the most intriguing characteristic of reproduction in *M. domestica* is the induction of estrus brought about by male pheromones. In a number of eutherian mammals, such as the mouse and sheep, the presence of an adult male or his pheromones accelerates the onset of estrus (Vandenburgh and Coppola 1986), but these animals normally enter estrus spontaneously. By contrast, in *M. domestica*, spontaneous estrus rarely occurs in females housed singly and hence the influence of male pheromones is rather exceptional, although recently a similar effect has also been observed in the woolly opossum, *C. philander*, (Perret and Ben M'Barek

1991). Over the next few years it will be important to observe such marsupials in their natural habitat in order to understand the significance of such a mating strategy and the other features of their reproduction which have contributed to their long success.

References

Baggott LM (1987) Comparative aspects of the reproductive biology of the grey short-tailed opossum, *Monodelphis domestica wagner*. Ph.D. thesis, University of London

Baggott LM, Moore HDM (1990) Early embryonic development of the grey short-tailed opossum, *Monodelphis domestica*, in vivo and in vitro. J. Zool. Lond. 222, 623–639

Baggott LM, Davis-Butler S, Moore HDM (1987) Characterisation of oestrus and timed collection of oocytes in the grey short-tailed opossum, *Monodelphis domestica*. J. Reprod. Fert. 79:105–114

Baker PJ, Moore HDM, Penfold LM, Burgess AMC, Mittovoch U (1990) Gonadal sex differentiation in the neonatal marsupial, *Monodelphis domestica*. Development 109:699–704

Bawa SR (1978) Joined spermatozoa. In Afzelius B (ed.). The Functional Anatomy of the Spermatozoon. Pergamon Press, Oxford. pp. 259–266

Bedford JM, Rodger JC, Breed WG (1984). Why so many mammalian spermatozoa—a clue from marsupials? Proc. Roy. Soc. London. B. 221:221–233

Biggers JD (1966) Reproduction in male marsupials. Symp. Zool. Soc. Lond. 15:251–280

Burns RK (1956) Hormones versus constitional factors in the growth of embryonic sex primordia in the opossum. Am. J. Anat. 98:35–67

Collins LR (1973) Monotremes and marsupials. A reference for zoological institutions. Smithsonian Institution Press, Washington

Fadem BH (1987) Activation of estrus by pheromones in a marsupial: stimulus control and endocrine factors. Biol. Reprod. 36:328–332

Fadem BH, Rayve RS (1985) Characteristics of the oestrous cycle and influence of social factors in grey short-tailed opossums (*Monodelphis domestica*). J. Reprod. Fert. 73:337–342

Fadem BH, Tesoriero JV (1986) Inhibition of testicular development and feminization of the male genitalia by neonatal oestrogen treatment in a marsupial. Biol. Reprod. 34:771–776

Fadem BH, Trupin GL, Maliniak E, Vande Berg JL, Hayssen V (1982) Care and breeding of the grey short-tailed opossum (*Monodelphis domestica*). Lab. Anim. sci. 32:405–409

Hartman CG (1916) Studies in the development of the opossum, *Didelphis virginiana*. I. History of early cleavage; II. Formation of the blastocyst. J. Morph. 27:1–83

Hartman CG (1923) The oestrous cycle in the opossum. Amer. J. Anat. 32:353–421

Hunsaker D II (1977) Ecology of new world marsupials. In Hunsaker D II (ed.). The Biology of Marsupials. Academic Press, New York. pp. 95–156

Keverne EB, Rosser AE (1986) The evolutionary significance of the olfactory block to pregnancy. In Duvall D, Muller-Schwarze D, Silverstein RM (eds.). Chemical Signals in Vertebrate Ecology, Evolution and Comparative Biology. Plenum Press, New York. pp. 433–439

McCrady E (1938) The embryology of the opossum. Amer. Anat. Memoirs 16:1–233

Moore HDM, Bedford JM (1978) Fate of spermatozoa in the male. I. Quantitation of sperm accumulation after vasectomy in the rabbit. Biol. Reprod. 17:784–790

Moore HDM, Thurstan SM (1990) Sexual differentiation in the grey short-tailed opossum, *Monodelphis domestica*, and the effect of oestradiol benzoate on development in the male. J. Zool. Lond. 221:639–658

O NS, Short RV, Renfree MB, Shaw G (1988) Primary genetic control of somatic sexual differentiation in a mammal. Nature, Lond. 331:716–717

Pelengaris SA, Abbott DH, Barrett J, and Moore HDM (1992) Induction of estrus and ovulation in female grey short-tailed opossums monodelphis domestica, involves the main olfactory epithelium. Proc. Chemical Signals in Vertebrates VII (in press).

Perret M, Ben M'Barek S (1991) Male influence on estrus cycles in female woolly opossum (*Calaromys philands*). J. Reprod. Fert. 91:557–566

Phillips DM (1970) Development of spermatozoa in the woolly opossum with special reference to the shaping of the sperm head. J. Ultrastruct. Res. 33:369–380

Phillips DM, Fadem BH (1987) The oocyte of a new world marsupial, *Monodelphis domestica*: Structure, formation and function of the enveloping mucoid layer. J. Exp. Zool. 242:363–371

Richardson BJ (1988) A new view of the relationships of Australian and American marsupials. Aust. Mammol. 11:71–73

Rodger JC, Bedford JM (1982) Separation of sperm pairs. Sperm/egg interaction in the opossum, *Didelphidae virginiana*. J. Reprod. Fert. 64:171–179

Selenka E (1887) Studien uber Entwickelungsgeschichte der Thiere. 4. Das Opossum, *Didelphis virginiana*. C. W. Kreidel, Wiesbaden

Temple-Smith PD, Grant TR (1986) Sperm structure and marsupial phylogeny. In Acher M (ed.). Possums and Opossums: Studies in Evolution. Australian Mammal Society, Sydney

Trupin GL, Fadem BH (1982) Sexual behaviour of the grey short-tailed opossum (*Monodelphis domestica*). J. Mammol. 63:409–414

Tyndale-Biscoe LH (1966) The marsupial birth canal. Symp. Zool. Soc. Lond. 15:233–250

Tyndale-Biscoe H, Renfree M (1987) Reproductive physiology of marsupials. Cambridge University Press, Cambridge

Vandenburgh JG, Coppola DM (1986) The physiology and ecology of puberty modulation by primer pheromones. Adv. Stud. Behav. 16:71–107

Eutheria

17

Reproduction in Female Marmoset Monkeys, *Callithrix jacchus*

David H. Abbott

Introduction

Marmoset and tamarin monkeys, frequently referred to simply as "marmosets," are small (100 to 800 g), arboreal, anthropoid primates of the family Callitrichidae. The common marmoset, *Callithrix jacchus*, the focus of this chapter, belongs to one of the two "short-tusked" genera of callitrichid monkeys, both of which are anatomically, ecologically, behaviorally, and often physiologically distinct from the two tamarin genera of *Saguinus* and *Leontopithecus*. Thus, the aspects of female reproductive biology covered by this chapter may not always be typical of female callitrichid monkeys in general.

The common marmoset inhabits the coastal and inland forests of northeastern Brazil and lives in groups of 3 to 15 individuals. A single breeding female usually gives birth to two infants twice a year. While captive studies show no evidence of seasonality in breeding, there is preliminary evidence from field studies for two possible birthing seasons in each year (Stevenson and Rylands 1988). Common marmosets in captivity are usually kept as male-female pairs or in social groups comprising either an adult pair and their offspring or five to eight unrelated adult animals. When adult females are housed together in a group, only one female breeds (Abbott 1984a; Hearn 1987). Laboratory studies have established that the breeding female of a group is the most socially dominant female. In most instances in captivity, and possibly in the wild, this female effectively suppresses the fertility of other females of reproductive age (Abbott and Hearn 1978; Stevenson and Rylands 1988). Social contraception is therefore a part of everyday life for female marmosets. Male marmosets, however, are not always so constrained as females. While mating can be an exclusive behavior of the dominant breeding pair, subordinate males in some groups can mate with the dominant female and ejaculate apparently normal spermatozoa. In subordinate males, the production of mature gametes and the display of copulatory behavior, therefore, can still occur despite their suppressed plasma testosterone and luteinizing hormone (LH) levels and impaired testis endocrine function (Abbott 1986; Sheffield et al. 1989).

This chapter will now focus on female reproductive biology in the common marmoset, the most widely used callitrichid monkey in reproductive research. It will highlight some of the unusual and fascinating aspects of female reproduction in this primate and the major contributions the marmoset has made to our understanding of female reproductive function. Almost all of our knowledge of marmoset reproductive biology comes from studies of captive animals. There is little field work on reproduction from which to draw. This chapter will, therefore, provide a physiological focus rather than an ecological one.

Sexual Differentiation

The normal development of female marmoset monkeys provides a challenge to our current perception of the sexual differentiation process in mammals (Goy and McEwen 1980; Baum et al. 1990). Marmosets commonly produce dizygotic twin embryos, which establish vascular anastomoses between their placental circulations and develop a common chorionic cavity early in pregnancy (Wislocki 1939; Benirschke and Layton 1969). Consequently, females share their blood circulation and most of their fetal development with a male co-twin in 50% of pregnancies. Sexual differentiation of the gonads occurs relatively late in pregnancy, at about day 80 (Chambers and Hearn 1985), yet females emerge at birth (144 days of gestation) with only XX/XY haemapoietic chimerisim (Benirschke and Brownhill 1962) to reflect their intimate contact with a developing male; there is no apparent masculinization of behavior or morphology (Abbott 1984b). Females co-joined to males in other mammalian species became "intersex freemartins" (Marcum 1974) with partial sex reversal of the gonads (Short et al. 1969). To circumvent this problem, marmosets may have developed protective mechanisms in female embryos to prevent their masculinization by the testicular secretions of their male co-twins, such as developing delayed sexual differentiation of certain biological processes, e.g., neural mechanisms controlling reproductive behavior, until after birth. Abbott and Hearn (1979) provided evidence of the latter when they administered testosterone to neonatal females for 50 days after birth (when males normally experience a rise in circulating testosterone levels) and masculinized components of the females' adult sexual behavior. As illustrated in Figure 17.1, the neonatally androgenized females courted and attempted to mount normal females in a manner similar to males and refused most mount attempts by males, unlike normal females. This neonatal effect of androgen did not appear to compromise the fertility of treated females or their ability to rear offspring (Abbott 1984b). Dixson (1990) reinforced the importance of this neonatal period in marmosets for the neural differentiation of sexual behavior when he investigated the importance of high circulating testosterone levels in neonatal males. Males castrated at birth, and thus deprived of experiencing the post-natal elevation

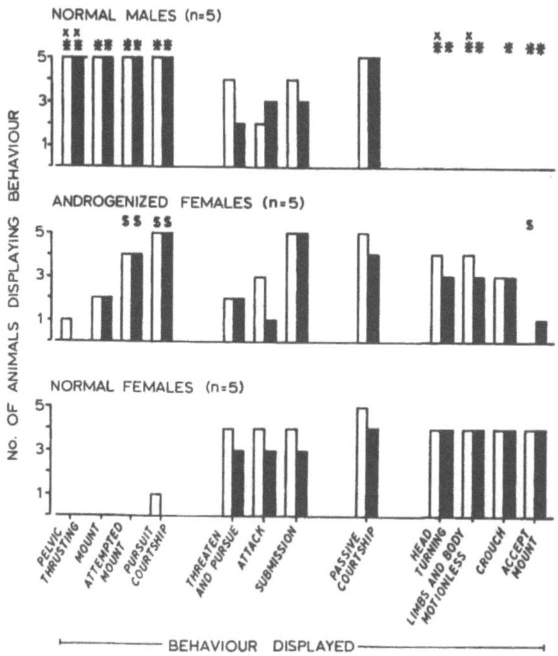

FIGURE 17.1. Sexual and aggressive behavior displayed by normal male, androgenized female (treated neonatally with testosterone), and normal female marmosets during two series of 15-min. behavioral trials of all possible pair combinations. The open bars denote the first series of all pair combinations and the closed bars denote the second series. *: $p < 0.05$, normal males versus normal females; x: $p < 0.05$, normal males versus androgenized females; $: $p < 0.05$, normal females versus androgenized females.

of testosterone, showed significant deficits in masculine components of their adult sexual behavior, even while receiving replacement testosterone therapy in adulthood.

Puberty

The timing of puberty is difficult to determine in female marmosets because breeding females generally suppress ovulation in their daughters of reproductive age (Abbott 1984a). However, from about 200 days of age onwards, females were found to exhibit elevations in plasma concentrations of estradiol extending into the preovulatory range (about 0.8 to 3.0 ng/ml; Fig. 17.2). Elevations in plasma LH concentrations were observed from about 300 days of age onwards, but elevations in plasma progesterone concentrations, indicative of ovulation (>10 ng/ml), were not observed until 400 days and then mostly in females removed from their families at about 300 days of age. While females could also conceive for the first time between 300 to 400 days, the median age range for first conception was 500

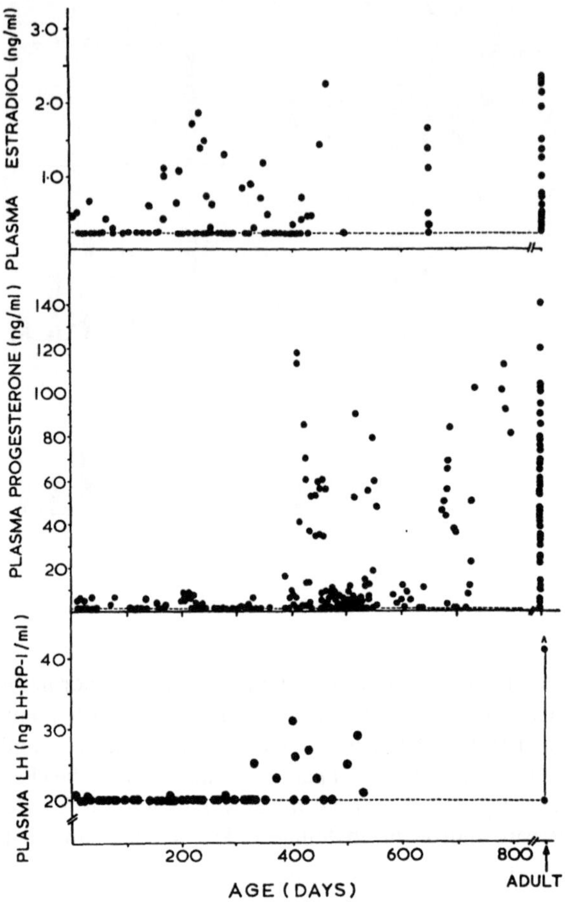

Figure 17.2. Plasma estradiol-17β, progesterone and immunoreactive LH concentrations in female marmoset monkeys. Individual sample values are plotted, except for the adult range of LH where "^" denotes values up to 120 ng/ml at the mid-cycle LH peak. The dotted line marks the sensitivity limit of the assays.

to 600 days, when the females reached mature body weight. Therefore, the onset of puberty probably occurs at about eight months of age in female marmosets and full maturity is attained at about 18 months (Abbott and Hearn 1978).

Ovarian Function

Female callitrichids, unlike most anthropoid primates, regularly ovulate two to three ova. The ovarian cycle of the common marmoset is 28.6 ± 1.1 days long (mean ± s.e.m.), with a preovulatory or follicular phase of

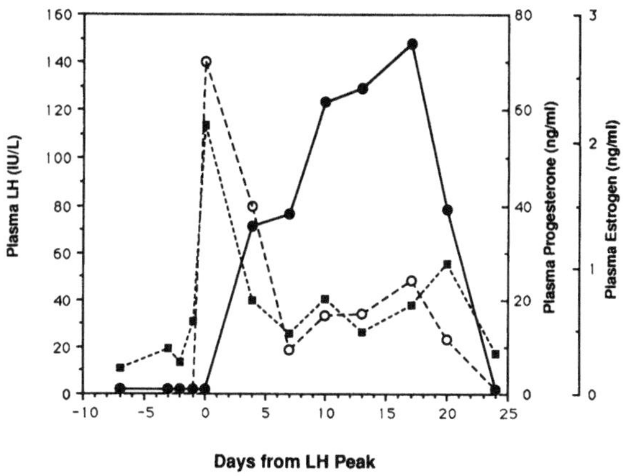

FIGURE 17.3. Plasma concentrations of progesterone (●), estrogen (■) and bioactive LH (○) in an adult female marmoset monkey (200W) across a typical ovarian cycle. Day 0 is denoted as the day of the bioactive LH peak.

8.2 ± 0.3 days and a postovulatory or luteal phase of 19.2 ± 0.6 days (Harlow et al. 1983). As marmosets do not menstruate and do not exhibit any obvious external signs of ovulation, the ovarian cycle must be monitored using measurements of changing hormone concentrations in blood or urine (Harlow et al. 1983; Eastman et al. 1984); this necessity is reflected throughout this chapter. Plasma estrogen concentrations during the follicular phase remain at 100 to 300 pg/ml until about day six to seven when they rise to reach a preovulatory peak of 1 to 3 ng/ml. Bioactive LH concentrations commonly exceed 60 IU/L at the time of the ovulatory peak. Plasma progesterone concentrations only start to exceed 10 ng/ml one day after ovulation and remain elevated at levels of 50 ng/ml and more for the duration of the luteal phase. A typical ovulatory cycle is illustrated by the hormonal profile shown by female 200W in Figure 17.3 and reflects the high circulating steroid hormone concentrations commonly found in South American primates.

A major improvement in the timing and prediction of ovulation was achieved by Summers et al. (1985) with the introduction of timed luteolysis. An analogue of prostaglandin F2α (cloprostenol) injected once at 10 days or more following ovulation, in either pregnant or non-pregnant animals, resulted in complete luteolysis, the onset of the next ovarian cycle, and ovulation after 10.7 ± 0.3 days. This treatment was effective up to day 43 of pregnancy (the approximate time of the luteo-placental shift in the maintenance of pregnancy in marmosets; Hearn 1978) and is now commonly used in studies of ovarian function, in which term pregnancies are undesirable (Hodges et al. 1987; Abbott et al. 1988; Harlow et al. 1988) or in which controlled timing of ovulation is essential, such as in studies of early pregnancy (Smith et al. 1987; Lopata et al. 1988).

A pair of ovaries from a female marmoset undergoing normal ovarian cycles commonly contains up to 100 antral follicles greater than or equal to 0.5 mm in diameter (Hillier et al. 1987). Most are not atretic. Only 6 to 9% of antral follicles reach 1.0 to 1.9 mm in diameter and usually only two to three follicles of 2 to 4 mm in diameter are present in a pair of ovaries at mid-cycle, preceding ovulation (Hillier et al. 1987; Abbott et al. 1989) No large follicles are left after ovulation apart from the post-ovulatory follicles or corpora lutea, and ovulation sites are usually found on both ovaries. Typically, there is no dominant or ovulatory ovary in a marmoset ovarian cycle.

Similar to the human and the rodent, in vitro studies of follicular granulosa cell function have shown that marmoset granulosa cells are sensitive to follicle-stimulating hormone (FSH) and respond with increased estradiol biosynthesis (an indicator of aromatase enzymatic activity) and progesterone biosynthesis (Harlow et al. 1988). Unlike the rodent, androgens, but not estrogen, enhanced the FSH responsiveness of granulosa cells in small follicles (0.5 to 1.0 mm in diameter; Hillier et al. 1988). Nevertheless, estrogen was important in the marmoset for the initiation of granulosa cell responsiveness to insulin-like growth factor-1 (IGF-1; Shaw and Hodges 1990); IGF-1 is an important regulator of intra-ovarian function. FSH was also shown to be important for the induction of LH responsiveness in granulosa cells, a crucial step for final follicular maturation and ovulation (Shaw et al. 1989).

Pregnancy

Gestation lasts 144 ± 2days in the marmoset (Chambers and Hearn 1979). Hearn et al. (1978) were the first to describe the endocrinology of pregnancy. The typical pattern of circulating levels of progesterone, estradiol, estrone, and LH/chorionic gonadotrophin (CG) is shown in Figure 17.4. Implantation is estimated to commence at 12 days following ovulation (Moore et al. 1985), relatively late for a primate species, and early embryonic development is also relatively tardy (Chambers and Hearn 1985). However, there is an acceleration in growth and development following day 80.

By 31 days following ovulation, the embryonic trophoblast extends laterally over the entire uterine surface and, in this way, probably permits close enough proximity to the maternal circulation to provide adequate nutrition for the embryo; chorionic villus development is poor at this stage (Smith et al. 1987). The various stages of implantation are illustrated diagrammatically in Figure 17.5. Smith and Moore (1988) have also shown that type C RNA retroviral particles are present in marmoset placentae at an early stage of implantation and that they appear to be an endogenous part of the host genome. As C-type viral particles suppress the immune response and cause cell fusion, their expression at the syncytiotrophoblast-

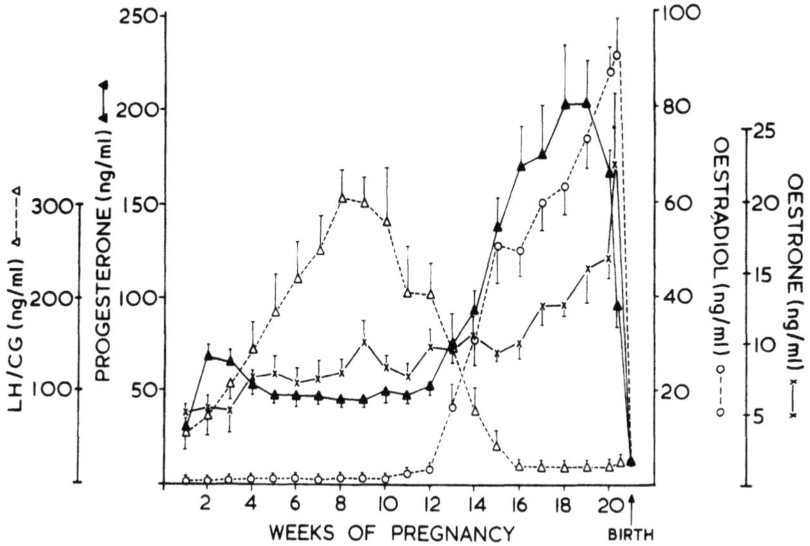

Figure 17.4. Mean (±s.e.m.) plasma concentrations of progesterone (▲), estradiol-17β (○), estrone (x), and immunoreactive LH/CG (Δ) during pregnancy in the marmoset monkey. Reprinted from Chambers and Hearn (1979), with kind permission of the Journal of Reproduction and Fertility Ltd.

cytotrophoblast interface may be highly important for implantation and early pregnancy development.

Summers and colleagues (Lopata et al. 1988; Summers et al. 1988) have also demonstrated that the marmoset may be valuable for investigations related to in vitro fertilization (IVF) research and infertility in women. Improved methods of collecting, freezing, and transferring of embryos were established using this species. In addition, a procedure was developed to remove trophectoderm cells from day-8 blastocysts which did not compromise an embryo's further development yet which provided sufficient cells to establish a pre-implantation diagnosis of a genetic disorder. Oocytes were also recovered during timed collections just prior to human chorionic gonadotrophin (hCG)-induced ovulation. They were fertilized in vitro and transferred to the uteri of three females with their ovarian cycles synchronized to that of the donors. This resulted in two of the recipient females giving birth to live infants.

Ultrasound Monitoring of Ovarian Function and Pregnancy

du Boulay and Wilson (1988) were the first to demonstrate the usefulness of non-invasive ultrasound monitoring of reproduction in female marmosets without sedation. A 5 mHz linear probe was used with a "waterbath

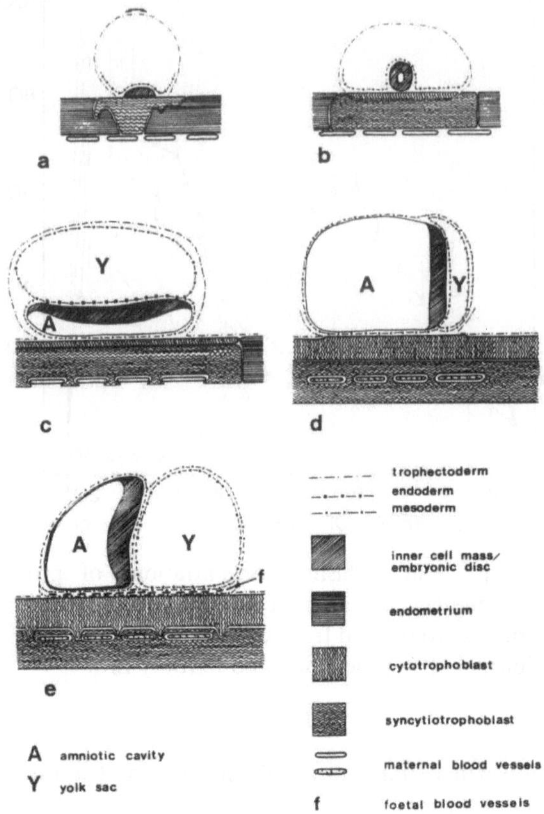

Figure 17.5. Diagrammatic representation of the various stages of implantation in the marmoset monkey: (a) 13 days after ovulation. The arrow denotes the proposed migration of trophoblast around the blastocoel; (b) 16 days; (c) 19 days; (d) 23 days; and (e) 31 days after ovulation. See the *key* for explanation of shading and labeling. Reprinted from Smith et al. (1987), with kind permission of Springer-Verlag.

stand-off" attachment to bring the reproductive tract to an optimal focal distance. The animals were restrained on a padded base with velcro straps for 5 to 10 min. Two to three preovulatory follicles of 2 to 4 mm in diameter were reliably observed by ultrasound, only one to two days prior to ovulation; Figure 17.6 provides a typical example (du Boulay and Wilson 1988; Abbott et al. 1989). Corpora lutea were observed by ultrasound from about two to three days following ovulation onwards. However, the ultrasound observations only accurately predicted the location of all ovulation sites on the two ovaries (observed at laparotomy) 60% of the time. Pregnancy was reliably determined by day 17 to 18 following ovulation; a typical visualization of a gestation sac within the uterus is illustrated in Figure 17.7. Ultrasound monitoring of pregnancy has permitted non-

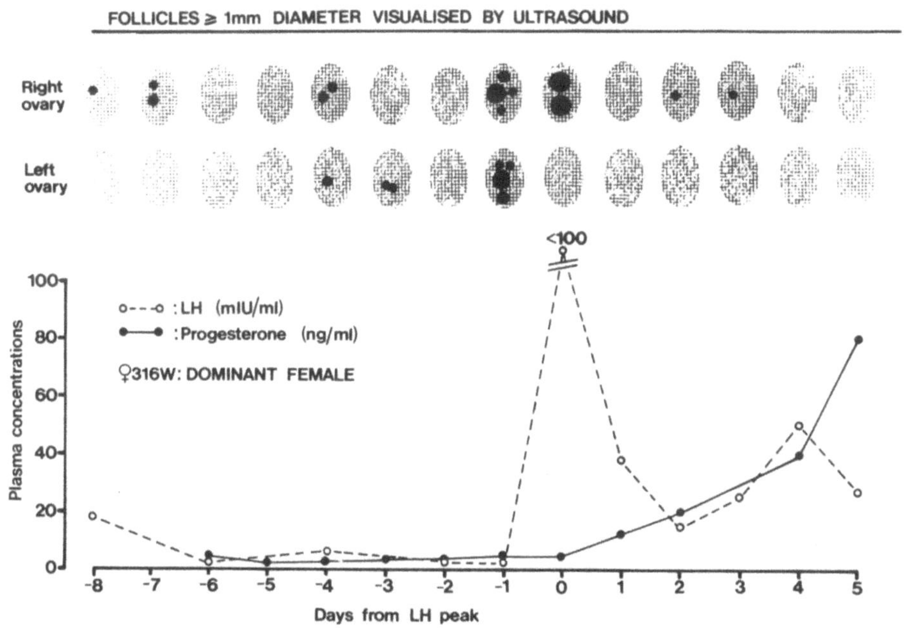

Figure 17.6. Diagrammatic representation of ovarian follicles of ≥ 1 mm in diameter across a typical ovarian cycle in female 316W. Follicles were visualized by a Siemens Sonoline SL real-time ultrasound scanner, using a 5mHZ linear probe with a "waterbath stand-off" provided by the manufacturer. Note that large follicles of 1.5 mm in diameter and more were only visualized on Days −1 and 0 (Day-1: right ovary, 2.2, 1.5, 1.2 and 1.0 mm, left ovary, 1.8, 1.5, 1.0 and 1.0 mm; Day 0: right ovary, 3.3 and 3.3 mm, left ovary, none). No such large follicles were seen after Day 0. Two corpora lutea were found in the right ovary and one on the left ovary from both ultrasound and laparotomy procedures. Plasma concentrations of progesterone (•) and bioactive LH (○) are displayed relative to the day of the LH peak (Day 0).

invasive, quantitative assessment of normal fetal development and identification of abnormal development, such as hydrocephaly and small gestation sacs indicating impending fetal death (du Boulay and Wilson 1988).

Social Suppression of Reproduction

The restriction of reproduction to one female per social group is a well-known characteristic of callitrichid monkeys. In the common marmoset this is achieved by the total suppression of ovulation in many daughters of reproductive age and in most unrelated subordinate females (Abbott 1984a). The ovarian suppression is attributable to subordinate social status. If a subordinate female is removed from her dominant cage mate, then

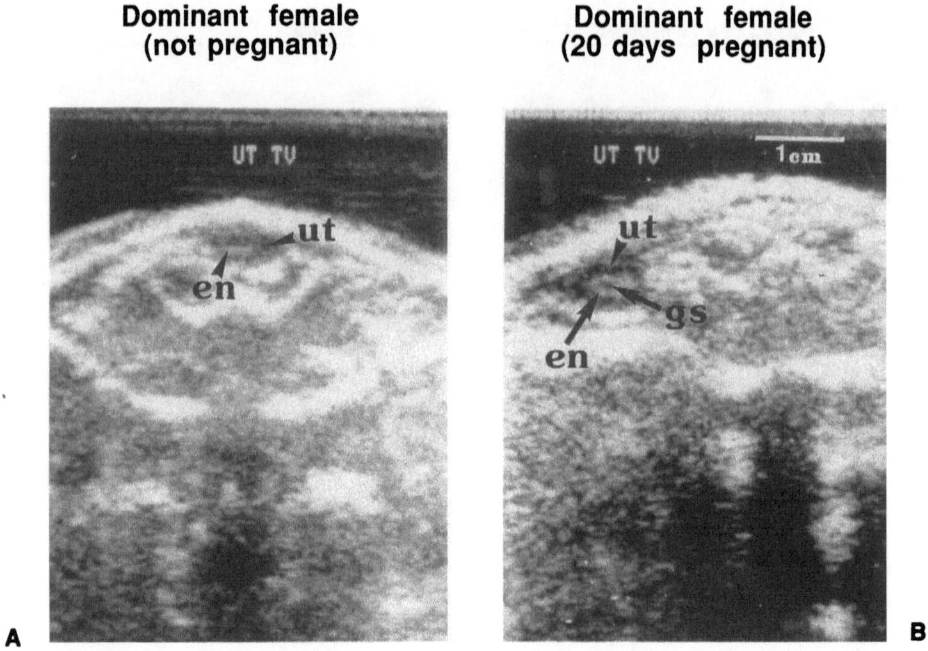

FIGURE 17.7. Ultrasonograms of the uterus and abdominal cavity of two dominant female marmosets in transverse view (UT TV). In (A) the uterus (ut) of a non-pregnant, luteal phase female is illustrated (seven days after ovulation), showing the highly echogenic interface between adjacent endometrial layers (en). In (B) the enlarged uterus (ut) of a pregnant female, 20 days following ovulation is shown, illustrating the gestation sac (gs) between the highly echogenic endometrial layers (en). The white bar line in (b) gives an indication of the scale in both ultrasonograms. The ultrasound details are given in the legend to Figure 17.6.

normal ovulatory cycles commence within 13.5 ± 3.8 days. A typical example of this is illustrated by female 224W in Figure 17.8; as a social subordinate in a group, this female was anovulatory until removed from the group and housed singly. Plasma LH levels then rose rapidly, inducing ovulation after nine days. Upon return of the female to subordinate status in a new social group, plasma LH levels fell below 2 IU/L within 24 hours, resulting in the premature demise of the luteal phase of her second ovulatory cycle and the subsequent anovulatory condition. Removal of only the dominant female from the social group will achieve the same ovulation-inducing effect in the highest-ranking subordinate female (Abbott 1989).

The physiological mechanisms responsible for ovarian suppression do not involve elevated blood levels of cortisol or prolactin, altered patterns of melatonin secretion, or low body weight (Abbott et al. 1981; Abbott 1986; Webley et al. 1989). As illustrated in Figure 17.8, reduced plasma LH concentrations accompany the anovulatory condition. These lowered

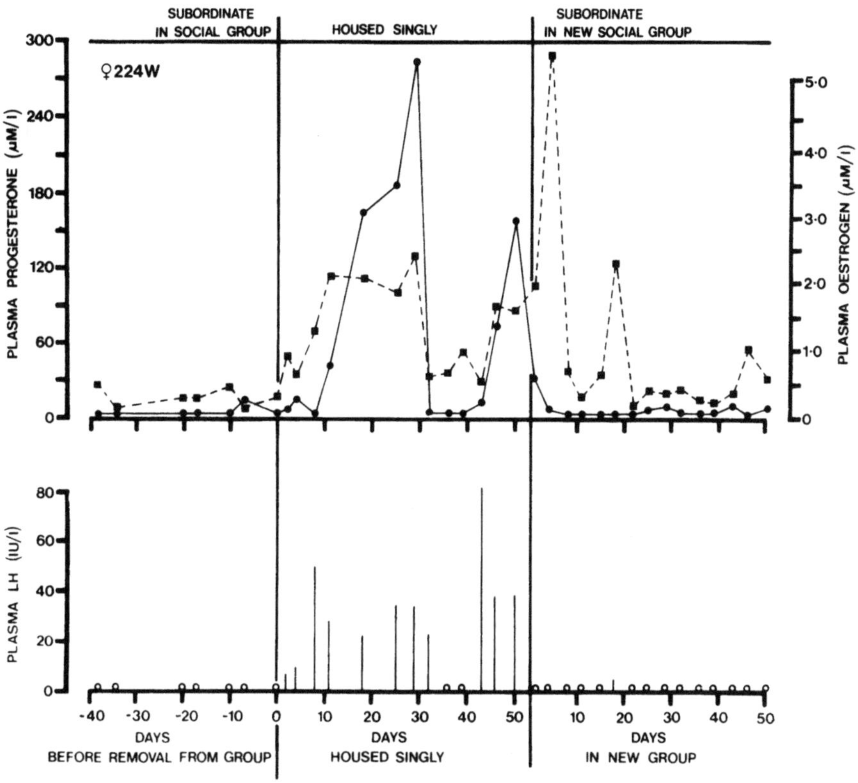

FIGURE 17.8. Plasma concentrations of progesterone (●), estrogen (■) and bioactive LH (vertical bars) in an adult female marmoset monkey (224W) while (A) subordinate in a social group (established 9 months previously), (B) housed singly (for 53 days), and (C) subordinate in a new social group. Open circles = below the sensitivity of the LH bioassay (<2 IU/L). Reprinted from Abbott et al (1988), with kind permission of Journal of Endocrinology Ltd.

LH levels can be increased by exogenous gonadotropin-releasing hormone (GnRH) treatment of subordinates remaining in their groups via a mini-infusion pump strapped to the female's back. Such GnRH treatment resulted in ovarian follicular growth, ovulation, and one short-term pregnancy (Abbott 1987, 1989). This finding suggests a suppression of endogenous GnRH secretion in subordinate females. Two components of the neuroendocrine mechanism inhibiting GnRH secretion have been identified so far: one is manifest as an increased sensitivity to estradiol negative feedback, and the other appears to be independent of ovarian steroid feedback and involves the endogenous opioid peptides (Abbott 1988; Abbott et al. 1990). The actual cues used by the dominant female to activate and maintain this neuroendocrine mechanism in subordinate females involve a combination of pheromonal, visual, and behavioral signals (Barrett et al. 1990, in press). However, in the activation of this neuroendocrine suppression, pheromones from the dominant female may be the single most important cue.

Such a highly developed socioendocrine mechanism controlling fertility in female marmosets may be due to the need for non-reproductive "helpers" to help care for and carry the offspring of the breeding female in the wild, so as to maximize the chances of the dominant's offspring surviving. This situation is to the obvious reproductive benefit of the dominant female, but it may also be of benefit to non-breeding female subordinates to (a) maintain their membership of a group, which in itself may be crucial for the survival of individual marmosets (lone marmosets are rarely observed in the wild and may be prone to predation or harassment from territory-holding marmoset groups), (b) gain experience in infant care (this is an acquired skill in the marmoset), and (c) wait and attain dominant breeding status at an appropriate time, such as when the breeding female dies in their group or an adjacent one (Sussman and Garber 1987). Such a breeding system could severely limit gene flow from one generation to the next in marmosets unless the turnover of breeding females or female migration is high in the wild. Both of these latter scenarios have been reported in a recent field study of the common marmoset (Scanlon et al. 1988). However, the occurrence of frequent migration between groups and a short reproductive tenure for any particular breeding female may not be typical for the genus or the species (Ferrari and Lopes Ferrari 1989).

Sexual Behavior

Female callitrichid monkeys were thought to exhibit no behavioral signs of estrus around the time of ovulation. Kendrick and Dixson (1983), however, showed that while copulations occurred throughout the ovarian cycle of the common marmoset, females clearly exhibited increased proceptive behavior (soliciting sexual responses from male partners) during the periovulatory period. The frequencies of ejaculatory mounts also peaked during the periovulatory period, and females refused significantly more of the males' mount attempts (i.e., showed reduced receptivity) during the luteal phase. Ovariectomy and adrenalectomy experiments suggested that neither ovarian nor adrenal hormones are important for the maintenance of sexual behavior in female marmosets (Dixson and Lloyd 1988). However, treatment with estradiol (but not testosterone) significantly increased proceptivity and receptivity of females and the frequency of ejaculatory mounts by males. The anteromedial hypothalamus was implicated as a major neural site involved in the expression of female proceptivity, while lesions to the more medial regions of the hypothalamus inhibited both proceptive and receptive displays in female marmosets, suggesting a possible distinction between hypothalamic control of proceptive and receptive sexual behavior in this species (Dixson and Lloyd 1988). In the wild, sexual behavior is rarely observed (Stevenson and Rylands 1988). The few copulations that have been observed occurred briefly and with no observable preliminaries.

Overview

The success of reproductive research on the common marmoset monkey owes much to the prolific fecundity of this primate in captivity, its relatively short maturation time of 18 months, and the ability to easily maintain captive social groups because of the marmoset's small body size and "easy-going" temperament. The marmoset has repeatedly proven to be particularly useful at all levels of reproductive research, from differentiation and development to neural controls of physiology and behavior. The direct applicability of this research to improvements in human infertility are immediately obvious in IVF procedures and the preimplantation diagnosis of genetic disorders in embryos. The potential of this female South American primate in reproduction research is still growing. It certainly provides an interesting and stimulating alternative to the commonly studied Old World primates, such as the macaques.

The common marmoset in the wild has not suffered from the same catastrophic decline in population numbers, as has certainly been the case for some of the tamarin species, such as the golden lion tamarin (*Leontopithecus rosalia*). There has, as yet, been no need to supplement the wild population of common marmosets with re-introduced animals bred in captivity. This particular marmoset seems to adapt fairly well to human encroachment into its natural environment. The common marmoset has also been successfully introduced into the state and city of Rio de Janeiro and several other parts of southeastern Brazil, illustrating the versatility of this species in adapting to environments far from its usual northeastern location. Northeastern Brazil provides the coolest and most arid habitat occupied by the genus *Callithrix*. Perhaps the adaptation of the common marmoset to this rather xeric region has resulted in a hardy and resilient callitrichid monkey, capable of meeting the new demands required by man's devastation of Brazil's natural forests. However, the fortunate state of affairs for the common marmoset is far from typical of that experienced by many marmoset and tamarin species in South and Central America, where numbers in the wild are becoming extremely low.

ACKNOWLEDGMENTS

The author would like to thank his colleagues, including J. Barrett, O.L. Wilson, L.M. George, J.K. Hodges, D.J. Cheesman, E. Wilkinson, and S.E. Aldridge for their collaboration in some of the work reported here; I.S. Sutherland and G.R. Chambers (National Institute of Medical Research, Mill Hill, London, UK) for designing and building the mini-infusion device; J.P. Hearn, W. Saltzman and N. Schultz-Darken for criticism of the manuscript; T.J. Dennett, M.J. Walton and R. Dodsworth for preparation of the illustrations and A. McCormick for typing the manuscript. The studies reported in this chapter were supported in part by the Wellcome Trust (UK), the Medical Research Council and the Agricultural and Food Research Council (UK), and NIH grant RR00167 to the Wisconsin Regional Primate Research Center (WRPRC). This is publication number 30-046 of the WRPRC.

References

Abbott DH (1984a) Behavioral and physiological suppression of fertility in subordinate marmoset monkeys. Am. J. Primatol. 6:169–186

Abbott DH (1984b) Differentiation of sexual behavior in female marmoset monkeys: Effects of neonatal testosterone or a male co-twin. Prog. Brain Res. 61: 349–35

Abbott DH (1986) Social suppression of reproduction in subordinate marmoset monkeys (*Callithrix jacchus jacchus*). In De Mello MT (ed.). A Primatologia No Brasil. Sociedade Brasileira de Primatologia, Brasilia 2:1–16

Abbott DH (1987) Behaviorally mediated suppression of reproduction in female primates. J. Zool. Lond. 213:455–470

Abbott DH (1988) Natural suppression of fertility. In Hearn JP, Smith GR (eds.). Reproduction and Disease in Captive and Wild Animals. Symposia of the Zoological Society of London, Vol 60. Oxford University Press, Oxford. pp. 7–28

Abbott DH (1989) Social suppression of reproduction in primates. In Standen V, Foley RA (eds.). Comparative Socioecology: The Behavioral Ecology of Humans and Other Mammals. Blackwells Scientific Publishers, Oxford. pp. 285–304

Abbott DH, George LM, Barrett J, Hodges JK, O'Byrne KT, Sheffield JW, Sutherland IA, Chambers GR, Lunn SF, Ruiz de Elvira M-C (1990) Social control of ovulation in marmoset monkeys: A neuroendocrine basis for the study of infertility. In Bercovitch FB, Ziegler TE (eds.). Socioendocrinology of Primate Reproduction. Alan R. Liss, Inc., New York. pp. 135–158

Abbott DH, Hearn JP (1978) Physical, hormonal and behavioral aspects of social development in the marmoset monkeys, *Callithrix jacchus*. J. Reprod. Fertil. 53:155–166

Abbott DH, Hearn JP (1979) The effects of neonatal exposure to testosterone on the development of behavior in female marmoset monkeys. In Sex, Hormones and Behavior. Ciba Foundation, Symposium 62 (New Series), Excerpta Medica, Amsterdam. pp. 299–316

Abbott DH, Hodges JK, George LM (1988) Social status controls LH secretion and ovulation in female marmoset monkeys (*Callithrix jacchus*). J. Endocrinol. 117:329–339

Abbott DH, McNeilly AS, Lunn SF, Hulme MJ, Burden FJ (1981) Inhibition of ovarian function in subordinate female marmoset monkeys (*Callithrix jacchus jacchus*). J. Reprod. Fertil. 63:335–345

Abbott DH, Wilson OL, George LM, Aldridge SE, Wilkinson E, Barrett J (1989)

Ultrasound examination of ovarian activity in dominant and subordinate female marmoset monkeys. J. Reprod. Fertil., Abstract Series No. 3:22 (Abstract No. 34)

Barrett J, Abbott DH, George LM (1990) Extension of reproductive suppression by pheromonal cues in subordinate female marmoset monkeys, *Callithrix jacchus*. J. Reprod. Fertil. 90:411–418

Barrett J, Abbott DH, George LM (in press) The involvement of sensory cues in the suppression of reproduction in subordinate female marmoset monkeys, *Callithrix jacchus*. J. Reprod. Fertil.

Baum MJ, Carroll RS, Cherry JA, Tobet SA (1990) Steroidal control of behavioral, neuroendocrine and brain sexual differentiation: Studies in a carnivore, the ferret. J. Neuroendocrinol. 2:401–418

Benirschke K, Brownhill LE (1962) Further observations on marrow chimerism in marmosets. Cytogenetics 1:245–257

Benirschke K, Layton W (1969) An early twin blastocyst of the golden lion marmoset, *Leontocebus rosalia L.* Folia Primatol. 10:131–138

Chambers PL, Hearn JP (1979) Peripheral plasma levels of progesterone, estradiol-17β, estrone, testosterone, androstenedione and chorionic gonadotrophin during pregnancy in the marmoset monkey, *Callithrix jacchus*. J. Reprod. Fertil. 56:23–32

Chambers PL, Hearn JP (1985) Embryonic, fetal and placental development in the common marmoset monkey (*Callithrix jacchus*). J. Zool. Lond. (A) 207:545–561

Dixson AF (1990) Effects of castration neonatally, prepubertally or in adulthood upon sexual and aggressive behavior in the male marmoset (*Callithrix jacchus*). Abstract presented at the XIIIth Congress of the International Primatological Congress, Nagoya and Kyoto, Japan

Dixson AF, Lloyd SAC (1988) The hormonal and hypothalamic control of primate sexual behavior. In Hearn JP, Smith GR (eds.). Reproduction and Disease in Captive and Wild Animals. Symposia of the Zoological Society of London, Vol 60. Oxford University Press, Oxford. pp. 81–117

du Boulay GH, Wilson OL (1988) Diagnosis of pregnancy and disease by ultrasound in exotic species. In Hearn JP, Smith GR (eds.). Reproduction and Disease in Captive and Wild Animals. Symposia of the Zoological Society of London, Vol 60. Oxford University Press, Oxford. pp. 135–150

Eastman S-AK, Makawiti DW, Collins WP, Hodges JK (1984) Pattern of excretion of urinary steroid metabolites during the ovarian cycle and pregnancy in the marmoset monkey. J. Endocrinol. 102:19–26

Ferrari SF, Lopes Ferrari MA (1989) A reevaluation of the social organization of the Callitrichidae, with reference to the ecological differences between genera. Folia Primatol. 52:132–147

Goy RW, McEwen BS (1980) Sexual Differentiation of the Brain. MIT Press, Cambridge, Massachusetts

Harlow CR, Gems S, Hodges JK, Hearn JP (1983) The relationship between plasma progesterone and the timing of ovulation and early embryonic development in the marmoset monkey (*Callithrix jacchus*). J. Zool. Lond. 201:273–282

Harlow CR, Shaw HJ, Hillier SG, Hodges JK (1988) Factors influencing follicle-stimulating hormone-responsive steroidogenesis in marmoset granulosa cells: Effects of androgens and the stage of follicular maturity. Endocrinology 122: 2780–2787

Hearn JP (1978) The endocrinology of reproduction in the common marmoset, *Callithrix jacchus*. In Kleiman DG (ed.). The Biology and Conservation of Marmosets. Smithsonian Institute Press, Washington, D.C. pp. 163–171
Hearn JP (1987) Marmosets and tamarins. In Poole TB (ed.). The UFAW Handbook on the Care and Management of Laboratory Animals, Sixth Edition. Longman Scientific and Technical. pp. 568–581
Hearn JP, Abbott DH, Chambers PL, Hodges JK, Lunn SF (1978) Use of the common marmoset, *Callithrix jacchus*, in reproductive research. In Gengozian N, Dienhardt FW (eds.). Marmosets in Experimental Medicine, Primates in Medicine, Vol 10. Karger, Basel. pp. 40–49
Hillier SG, Harlow CR, Shaw HJ, Wickings EJ, Dixson AF, Hodges JK (1987) Granulosa cell differentiation in primate ovaries: The marmoset monkey (*Callithrix jacchus*) as a laboratory model. In Stouffer RL (ed.). The Primate Ovary. Plenum Press, New York. pp. 61–73
Hillier SG, Harlow CR, Shaw HJ, Wickings EJ, Dixson AF, Hodges JK (1988) Cellular aspects of preovulatory folliculogenesis in primate ovaries. Human Reprod. 3:507–511
Hodges JK, Cottingham PG, Summers PM, Yingnan L (1987) Controlled ovulation in the marmoset monkey (*Callithrix jacchus*) with human chorionic gonadotropin following prostaglandin-induced luteal regression. Fertil. Steril. 48:299–305
Kendrick KM, Dixson AF (1983) The effect of the ovarian cycle on the sexual behavior of the common marmoset (*Callithrix jacchus*). Physiol. Behav. 30:735–742
Lopata A, Summers PM, Hearn JP (1988) Births following the transfer of cultured embryos obtained by in vitro and in vivo fertilization in the marmoset monkey (*Callithrix jacchus*). Fertil. Steril. 50:503–509
Marcum JB (1974) The freemartin syndrome. Anim. Breed. Abst. 12:227–242
Moore HDM, Gems S, Hearn JP (1985) Early implantation stages in the marmoset monkey (*Callithrix jacchus*). Am. J. Anat. 172:265–278
Scanlon CE, Chalmers NR, Monteiro Da Cruz MAO (1988) Changes in size, composition and reproductive condition of wild marmoset groups (*Callithrix jacchus jacchus*) in Northeast Brazil. Primates 29:295–305
Shaw HJ, Hodges JK (1990) Developmental changes in granulosa cell responsiveness to IGF-1 in the marmoset ovary. J. Reprod. Fertil. Abst. Series No. 5:36 (Abstract No. 62)
Shaw HJ, Hillier SG, Hodges JK (1989) Developmental changes in luteinizing hormone/human chorionic gonadotropin steroidogenic responsiveness in marmoset granulosa cells: Effects of follicle-stimulating hormone and androgens. Endocrinology 124:1669–1677
Sheffield JW, O'Shaugnessy DJ, Abbott DH (1989) Social rank effects testicular steroidogenesis in male marmoset monkeys (*Callithrix jacchus*). J. Endocrinol. 122(Suppl):83
Short RV, Smith J, Mann T, Evans EP, Hallett J, Fryer A, Hamerton JL (1969) Cytogenetic and endocrine studies of a freemartin heifer and its bull co-twin. Cytogenetics 8:369–388
Smith CA, Moore HDM (1988) Expression of C-type viral particles at implantation in the marmoset monkey. Human Reprod. 3:395–398
Smith CA, Moore HDM, Hearn JP (1987) The ultrastructure of early implantation in the marmoset monkey (*Callithrix jacchus*). Anat. Embryol. 175:399–410

Stevenson MF, Rylands AB (1988) The marmoset monkeys, genus *Callithrix*. In Mittermeier A, Rylands AB, Coimbra-Filho AF, Fonseca GAB (eds.). Ecology and Behavior of Neotropical Primates. World Wildlife Fund, Washington, D.C. pp. 131–222

Summers PM, Campbell JM, Miller MW (1988) Normal in vivo development of marmoset monkey embryos after trophectoderm biopsy. Hum. Reprod. 3:389–393

Summers PM, Wennick CJ, Hodges JK (1985) Cloprostenol-induced luteolysis in the marmoset monkey (*Callithrix jacchus*). J. Reprod. Fertil. 73:133–138

Sussman RW, Garber PA (1987) A new interpretation of the social organization and mating system of the Callitrichidae. Int. J. Primatol. 8:73–92

Webley GE, Abbott DH, George LM, Hearn JP, Mehl H (1989) The circadian pattern of plasma melatonin concentrations in the marmoset monkey (*Callithrix jacchus*). Am. J. Primatol. 17:73–79.

Wislocki GB (1939) Observations on twinning in marmosets. Am. J. Anat. 64:645–483

18

Reproductive Biology of the Female Black Mastiff Bat, *Molossus ater*

John J. Rasweiler IV

Introduction

The black mastiff bat, *Molossus ater* (Fig. 18.1), belongs to the Family Molossidae, a moderately large group of insectivorous bats (12 genera, about 82 species) represented in most tropical and warm temperate regions of the world (Koopman 1970; Freeman 1981). *M. ater* is one of the most common and widespread neotropical molossids occurring from southern Mexico to southern Brazil, Paraguay, and northern Argentina, generally at low altitudes (Handley 1976; Dolan 1989; Koopman pers. comm.). This bat is a strong, fast-flying species that usually forages well above ground level, but can also scamper about quadrapedally with great agility.

The overall impact of man's activities upon *M. ater* is presently unclear. Lumbering and forest clearing certainly must result in the elimination of many roosts, since this bat often lives in hollow trees. However, like a number of other molossid bats, *M. ater* also regularly establishes colonies in the roofs and attics of buildings (Greenhall and Stell 1960; Goodwin and Greenhall 1961). For this reason, these animals are often regarded as pests and exterminated. On balance, the molossid bats are probably beneficial to man because of their prodigious consumption of insects, and many infestations of buildings might be dealt with better by bat-proofing the structures.

Although *M. ater* and a number of other molossid bats are presently very abundant, there is some reason to be concerned about their long-term survival as species. Dramatic declines have been noted in recent years in certain populations of one species, the free-tailed bat *Tadarida brasiliensis*, in the United States and Mexico. This may be attributable both to human disturbance of their roosts and pesticide contamination of their insect diet (McCracken 1986).

Until quite recently, information on reproduction in *M. ater* was restricted to scattered, brief comments about the condition of wild-caught specimens, some histological observations on early uterine embryos (Sansom 1932), and preliminary reports that ovulation and implantation

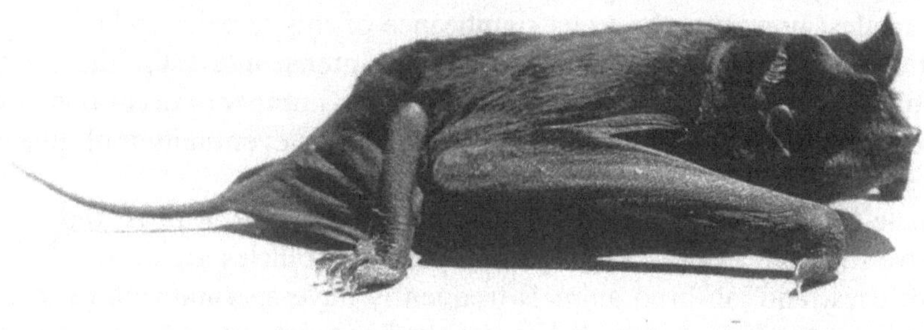

FIGURE 18.1. The black mastiff bat, *Molossus ater*.

appeared to occur only on the right side of the female tract (Wimsatt 1975, 1979). In a somewhat more extensive field study, Marques (1986) took pregnant *M. ater* during much of the year in Manaus, Brazil, but did obtain evidence for reproductive synchronization in the population.

Some fortuitous observations on a closely related species (*Molossus bondae*) suggested that further work with one of the molossid bats might contribute new insights on important, long-standing problems related to mammalian placentation (Rasweiler 1991a). A study of reproduction in laboratory-maintained *M. ater* was therefore initiated. This established that *M. ater* is not only a useful model for investigations of placental development, but also exhibits some remarkable reproductive similarities to the anthropoid primates, including true menstruation.

Reproductive Behavior

For studies of their reproductive biology, captive *M. ater* were generally housed in sexually-segregated groups for some months. A single male was then placed with a group of females for breeding purposes (Rasweiler 1987, 1988). Although bats in the all-male cages seemed to live together without serious problems, ferocious fighting quickly erupted when males used as breeders were returned to such groupings. Breeding males also occasionally exhibited aggressive behavior towards humans, but most *M. ater* proved to be very docile animals.

M. ater possesses a large, cutaneous gland at the base of its throat which is prominent in adult males but non-functional in females (Werner and Lay 1963; Horst 1966). Because of the difficulty of observing these animals in the wild, the function of this gular gland has long been unclear. In captivity, the animals frequently roosted or were active out in the open portion of their cages during daylight hours, and they sometimes exhibited behavior involving the gland. Breeding males would occasionally approach and rub it on females; however, the exact significance of this remains to be determined. Also, the females sometimes exhibited intense interest in the gular gland of the male and attempted to nuzzle it. On a number of occasions this led the females to crawl under the male, and coitus eventually took place (Rasweiler 1987, 1988).

Female *M. ater* do not exhibit a limited period of estrus. Rather, they are receptive to males even when large preovulatory follicles are absent from their ovaries, and lab-bred animals frequently have spermatozoa in their smears for extended periods. It has not yet been determined if the latter may have been due in part to the storage of spermatozoa in the female tract (Rasweiler 1987). An Australian molossid, *Mormopterus planiceps*, and a number of rhinolophid and vespertilionid bats have been found to possess this capability (Racey, 1979; Crichton and Krutzsch 1987).

Ovarian Function

Although *M. ater* has two ovaries, the right was significantly larger than the left in nearly all animals examined. Both ovaries contained primordial follicles, but follicular development was retarded in the left ovary. Large preovulatory follicles and corpora lutea were noted only in the right ovary. Similar observations have been made in a number of other molossid bats (Wimsatt 1975, 1979; Rasweiler 1988).

This restriction of advanced follicular development and ovulation to the right side in *M. ater* may be related to differences in the blood supply to the two ovaries. The left ovary had a much less prominent intraovarian blood supply, and this may serve to limit follicular development on that side (Rasweiler 1988). Normally, as follicles grow and approach ovulation, they develop a progressively richer network of blood vessels in their thecae internae (Ellinwood et al. 1978; Murakami et al. 1988). Obviously, this does not address the important issue of what underlies differential development of the right and left ovaries in the first place.

Ovulation in *M. ater* appears to be spontaneous, rather than induced by copulation, because a number of females which had been sexually isolated for prolonged periods possessed multiple corpora lutea of different ages (Rasweiler 1988). In some species (e.g., rats, mice, and hamsters), ovulation is spontaneous but the corpora lutea do not become fully functional unless copulation has occurred. This does not seem to be the case in *M. ater* because pronounced post-ovulatory endometrial development and differentiation was observed in females which had been sexually isolated (Rasweiler 1988, 1991a).

During advanced pregnancy in *M. ater*, large antral follicles were generally observed in the right ovary along with the corpora lutea of pregnancy. These may have been destined for ovulation at a post-partum estrus (Rasweiler 1988). Females which were simultaneously pregnant and lactating have been taken from a wild population of *M. ater* (Marques 1986).

The Non-Pregnant Uterine Cycle

The endometrial lining of the uterus in *M. ater* exhibited an unusually advanced degree of differentiation late in the cycle. At this time, stromal fibroblasts in the lamina functionalis were converted into plump, more tightly-packed decidual cells (Figs. 18.2, 18.3). Also, the endothelial cells lining some of the endometrial blood vessels hypertrophied and assumed a glandular appearance (the possible significance of which is discussed later in this chapter; Rasweiler 1991a). In most mammals which develop decidual reactions, the stromal fibroblasts only undergo this change upon invasion of the endometrium by trophoblast cells of the embryos or following experimental manipulation of the uterus (e.g., by subjecting the endometrium to mild mechanical trauma after appropriate hormonal stimulation). The elephant shrew, *Elephantulus myurus*, is one exceptional species, since decidual cells develop spontaneously in its uterus during the non-pregnant cycle (van der Horst and Gillman 1941). Late in the menstrual cycle of the human and chimpanzee, many endometrial stromal cells begin to differentiate into decidual cells; this is the so-called "predecidual reaction." Much less pronounced changes of a similar nature have also been observed at this time in the baboon and rhesus monkey (Dollar et al. 1972; Brenner and Maslar 1988).

At the end of the cycle in *M. ater*, this decidualized endometrial tissue becomes necrotic and is sloughed off with associated bleeding (Fig. 18.4). Although virtually all of the lamina functionalis is shed during this menstrual process, most of the desquamated endometrial tissue undergoes dissolution within the uterus and relatively little recognizable cellular debris passes into the vagina (Rasweiler 1991a).

The Adaptive Significance of Menstruation

Menstruation is relatively uncommon among mammals and has now been observed in the catarrhine and platyrrhine primates, two families of bats (the Molossidae and the Phyllostomidae), and the elephant shrew (Hamlett 1934; van der Horst and Gillman 1941; Eckstein and Zuckerman 1956b; Rasweiler 1970, 1979a, 1979b, 1991a; Quintero and Rasweiler 1973; Bonilla and Rasweiler 1974; Rasweiler and Bonilla in press). These species possess a number of common reproductive characteristics which suggest

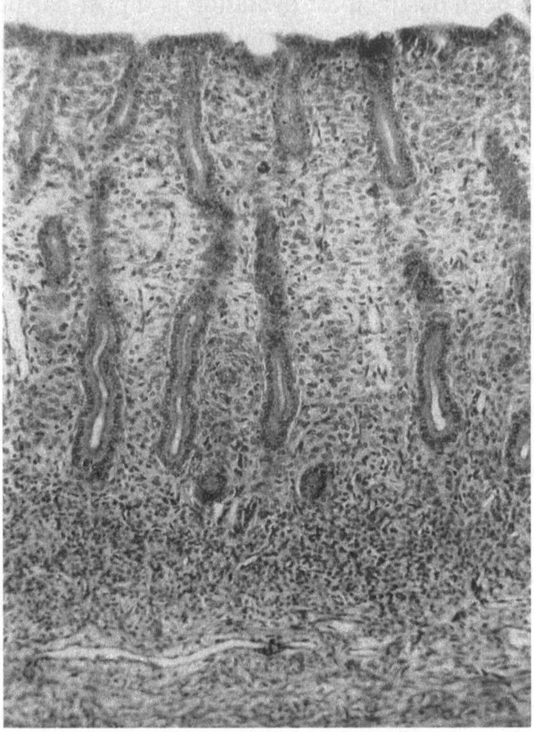

FIGURE 18.2. Section of early luteal phase endometrium in *M. ater*. This female had a luteinizing, recently ruptured follicle in her right ovary. Masson's trichrome (×145).

that they may have developed the ability to menstruate for similar reasons. The anthropoid primates and many of the phyllostomid bats (e.g., the long-tongued bat, *Glossophaga soricina*) possess simplex uteri, which have only a single body (Fig. 18.5; Eckstein and Zuckerman 1956; Rasweiler 1979; Hood and Smith 1983). All of the menstruating species which have been adequately studied exhibit substantial post-ovulatory endometrial growth and differentiation (Rasweiler 1970, 1991a; Finn 1987; Rasweiler and Bonilla in press). Since these animals possess embryos with invasive trophoblast, one must suspect that some of these endometrial changes may be related to the development of mechanisms for controlling and/or limiting trophoblastic growth. This is widely considered to be one of the functions of decidual cells. Some of the vascular changes observed late in the cycle in *M. ater* may influence trophoblastic growth and/or differentiation in the event of pregnancy as well (see below). All of these species also have low reproductive potentials, usually giving birth to only one or two

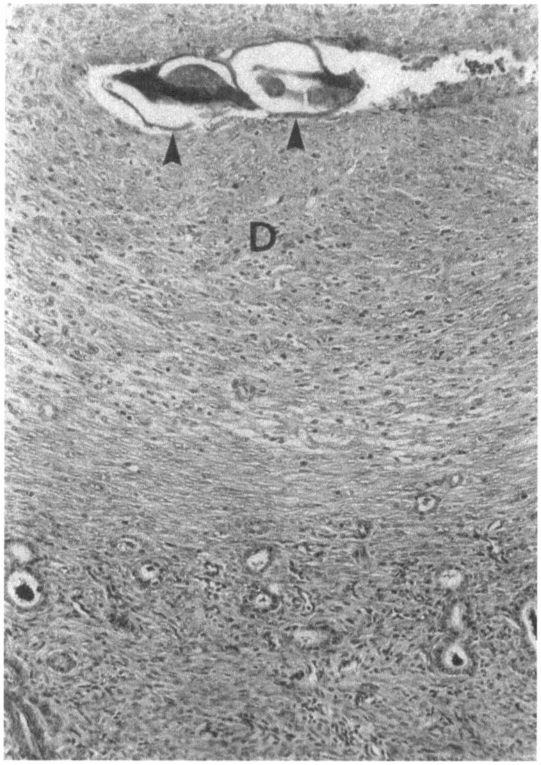

FIGURE 18.3. Section of endometrium which had undergone a spontaneous decidual reaction. This non-pregnant female had a pair of well-developed corpora lutea in her right ovary. Remnants of the unfertilized oocytes (arrowheads) are evident in the uterine lumen. D = decidua. PAS-hematoxylin. (×145).

young after relatively long gestation periods. Finally, many are seasonal breeders. Modern day humans breed relatively, but not completely, free of seasonal influences (Roenneberg and Aschoff 1990). It seems reasonable to assume, however, that our own species probably evolved from menstruating ancestors which were also seasonal breeders. The ability to menstruate would seem to provide all of these species with an efficient mechanism for eliminating highly differentiated endometria in the event of early reproductive failures (i.e., fertilization failures or early embryonic losses). In the case of seasonally breeding mammals, this may permit the females to establish new pregnancies and to produce young while environmental conditions are still reasonably favorable. This would be of obvious adaptive significance to species with low reproductive potentials (Rasweiler 1982, 1991a).

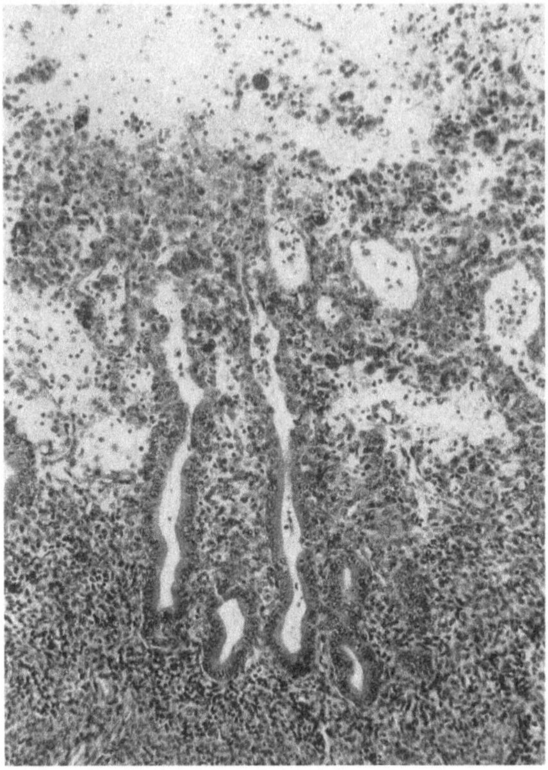

Figure 18.4. Section of endometrium in a menstruating female. Much of the lamina functionalis was necrotic and in the process of being sloughed off. Masson's trichrome (×145).

The Evolution of Menstruation in Bats

The fact that menstruation occurs in both *M. ater* and some of the phyllostomids is of further interest because these bats are not generally considered to be closely related within the Suborder Microchiroptera (Koopman and Jones 1970; Smith 1976; Luckett 1980). The phyllostomids are a large, highly diverse family (49 genera, 137 species) found primarily in tropical and subtropical regions of the New World (Jones and Carter 1976). This group includes species which are predominantly frugivorous, pollenivorous and nectarivorous, insectivorous, carnivorous, omnivorous, or sanguivorous.

The observation of menstruation in both the molossid and phyllostomid bats raises two interesting possibilities: that menstruation may be even more widespread in the Chiroptera or that the phenomenon may have

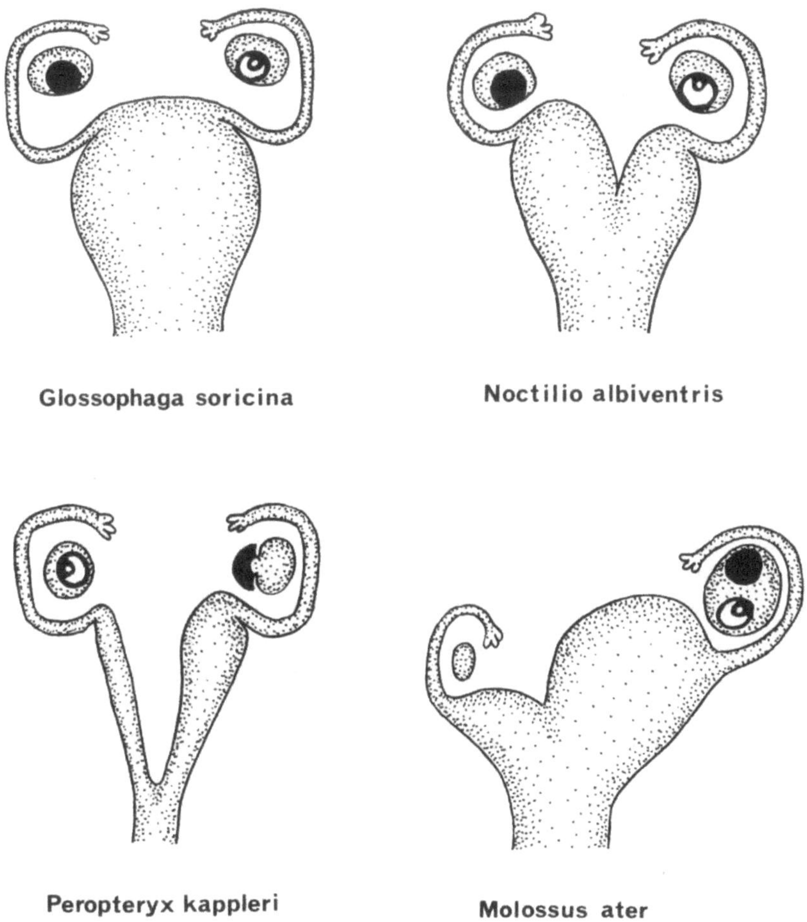

FIGURE 18.5. Diagramatic representations of the female reproductive tracts in selected microchiropteran bats. The presence of a Graafian follicle or corpus luteum (●) indicates that the ovary is capable of ovulating. The corpora lutea in *P. kappleri* are frequently everted. *M. ater* ovulates only from the right ovary. Post-ovulatory unilateral uterine reactions are shown for *N. albiventris* and *P. kappleri*. The right uterine horn of *M. ater* is depicted enlarged as is typical of parous females.

evolved independently in ancestors of these families. While too little information is available to dismiss the first of these, several lines of evidence provide some support for the second (Rasweiler 1991a). The menstrual cycle and events during early pregnancy in members of these two families are in some respects quite different. In most *M. ater*, considerable post-menstrual repair occurs prior to ovulation, as in the catarrhine primates. In the event of pregnancy in this bat, the embryo enters the uterus at the morula stage, apparently after a relatively short period of transit through the oviduct. A uterine morula was observed in one female on the fourth day after spermatozoa were first detected in her vaginal smears. Since the precise time of ovulation in this animal was unknown, tubal transit of the embryo could have been even less than four days (Rasweiler 1988). In some of the phyllostomid bats (*G. soricina*, *Carollia* sp., and *Desmodus rotundus*), on the other hand, menstruation in cycling females occurs in close temporal proximity to ovulation, and most endometrial repair is post-ovulatory. In the event of conception in these species the embryo is retained in the oviduct for an extraordinarily long time (12 to 13 days in *G. soricina* and more than 13 days in *Carollia* sp. and *D. rotundus*) and develops to the zona-free blastocyst stage there (Rasweiler 1970, 1979a, 1979b; Quintero and Rasweiler 1973, 1974; Bonilla and Rasweiler 1974; Rasweiler and Bonilla in press). In most commonly-studied mammals, the embryos do not develop to the blastocyst stage and lose their zonae pellucidae until after passing into the uterus, and the tubal journey of the embryo is much shorter, generally on the order of two to four days. Prolonged retention of the embryo in the oviduct in the phyllostomid bats provides time for endometrial repair, both in cycling animals which have menstruated and in those (e.g., *Carollia*) which can conceive at a post-partum estrus.

Finally, there is some evidence that the phyllostomid bats may have evolved from non-menstruating ancestors. *G. soricina*, *Carollia* sp., and *D. rotundus* are monovular and all exhibit tendencies to alternate successive ovulations between the two ovaries. This tendency is relatively weak in *G. soricina* (occurring in about 62% of the cycles in lab-maintained animals) but is stronger in *Carollia* (78%) and *D. rotundus* (93%). Furthermore, all three species exhibit preferential stimulation of the oviduct on the side of ovulation (Rasweiler 1972, 1979b; Bonilla and Rasweiler 1974; Quintero and Rasweiler 1974). This is noteworthy because it has generally been assumed in most mammals that ovarian hormones synthesized by the Graafian follicles or corpora lutea enter the systemic circulation before being delivered to the reproductive ducts, and this should result in equal stimulation of both sides even in monovular species. In the case of these bats, there must be a local route for preferential delivery of such hormones from the ovary containing the Graafian follicle or corpus luteum to the adjacent oviduct. It has been suggested that this local route may involve a countercurrent exchange of hormones between the ovarian venous and/or lymphatic drainage and the arterial supply of the adjacent oviduct (Bonilla and Rasweiler 1974; Rasweiler 1978, 1984). No evidence of such unilateral reactions has been observed in the simplex uteri of *G. soricina* and *Carollia* or the partially bicornuate uterus of *D. rotundus*. Such reactions do occur, however, in the partially bicornuate uterus of the noctilionid bat, *Noctilio albiventris* (Fig. 18.5; Rasweiler 1978) which, on the basis of reproductive characteristics, appears to be relatively closely related to the phyllostomids (Rasweiler 1979a; Luckett 1980). *Noctilio* also exhibits a high frequency (92%) of alternation of successive ovulations (Rasweiler 1979b).

When evidence for the alternation of successive ovulations and the occurrence of unilateral oviductal reactions were first observed in these bats, their biological significance was difficult to explain. The picture began to clear when studies were extended to another microchiropteran bat, *Peropteryx kappleri* (Family Emballonuridae). This species also tends to alternate successive ovulations (in about 76% of successive cycles) and demonstrates unilateral oviductal and uterine reactions (Fig. 18.5). The ability to alternate ovulations appears to be advantageous to females which ovulate relatively soon after giving birth. In one female which had not alternated ovulations, a fertilization failure had occurred because spermatozoa were unable to pass debris in the recently parturient uterine horn. It was also quite impressive in *P. kappleri* how little stimulation of the endometrium occurred during early pregnancy in the uterine horn opposite to the ovary containing the corpus luteum. This suggests that if *P. kappleri* suffers a fertilization failure or the early loss of an embryo and alternates ovulations, the endometrium contralateral to the first corpus luteum probably could be prepared to accommodate spermatozoa and then a conceptus within a reasonably short period of time. A highly differentiated and aging post-ovulatory endometrium would not first have to be dealt with on that side (Rasweiler 1982).

Since the phyllostomids *G. soricina* and *Carollia* sp. possess simplex uteri (Fig. 18.5), they cannot locally stimulate and utilize the right and left sides of the uterus in the same way as *P. kappleri*. The ability of *G. soricina* and *Carollia* sp. to alternate ovulations and to control their oviducts locally would seem to be of questionable biological significance and may instead represent primitive reproductive characteristics retained from ancestral bats which did not menstruate (Rasweiler 1982, 1991a).

Although *M. ater* has two uterine horns, alternate utilization of these horns is precluded because ovulation occurs only from the right ovary (Fig. 18.5). It is therefore perhaps not surprising that, in the event of a fertilization failure or an early embryonic loss, this bat can also efficiently eliminate a highly differentiated, post-ovulatory endometrium by means of menstruation (Rasweiler 1991a).

Placentation

During pregnancy in many mammals, trophoblast invades the uterine mucosa and brings the chorion (one of the fetal membranes) into an intimate morphological relationship with the maternal blood supply, in the form of a placenta, for the purposes of physiological exchange. Among the problems which have long perplexed students of placentation are (1) what controls the growth of the trophoblast when it is invasive and (2) why are prominent extracellular layers so frequently evident in the barrier separating the fetal and maternal circulations in mammalian placentas? An examination of placentation in *M. ater* has provided some valuable new insights into these questions (Rasweiler 1990, 1991b).

During implantation in *M. ater*, which is of the central or superficial type, the blastocyst expands in the lumen of the right uterine horn, and the trophoblast engulfs the most superficial maternal capillaries in the endometrium over a broad area. This is accompanied by expansion of the yolk sac. As mesoderm and vitelline blood vessels spread completely around the yolk sac, an extensive choriovitelline placenta is formed where the yolk sac is closely apposed to the trophoblast-engulfed maternal vessels (Fig. 18.6). This choriovitelline placenta has only a brief existence, as the yolk sac soon collapses and is replaced by vascularized mesoderm derived

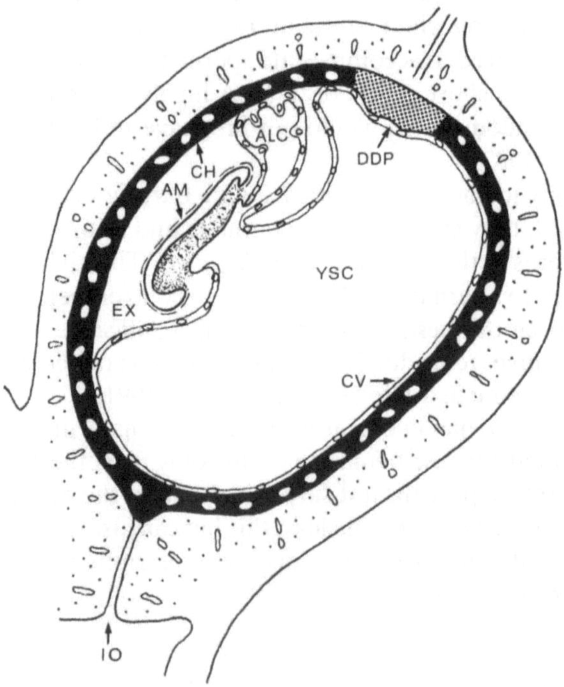

Figure 18.6. Diagram of the right uterine horn in a *M. ater* carrying a somite stage embryo. Vascularization of the yolk sac wall has created an extensive choriovitelline placents (CV). The amnion (AM) has been formed dorsal to the embryo, and the allantois has made its initial contact with the chorion (CH). ALC = allantoic cavity, DDP = developing discoidal hemochorial placenta, EX = exocoelom, IO = internal os, YSC = yolk sac cavity. From Rasweiler 1990.

from the allantois. This creates an extensive diffuse endotheliodichorial placenta (Fig. 18.7), which functions as the principal site of physiological exchange between the conceptus and mother during mid-gestation. The barrier between the two circulations is classified as being endotheliodichorial because it includes the following cellular layers: maternal endothelial cells, syncytiotrophoblast, cytotrophoblast, and fetal endothelial cells (Fig. 18.8). An unusually thick basal lamina (the "interstitial membrane") lies between the maternal endothelium and the syncytiotrophoblast.

As pregnancy progresses the diffuse endotheliodichorial placenta of *M. ater* involutes and is supplanted by a discoidal hemochorial placenta as the major site of feto-maternal exchange (Fig. 18.9). Development of the discoidal placenta is of particular interest because this always seems to occur at the cranial end of the right uterine horn. This positioning of the discoidal placenta is related to the presence of a peculiar vascular tuft which forms in the superficial endometrium of this region after ovulation (Fig. 18.10). This

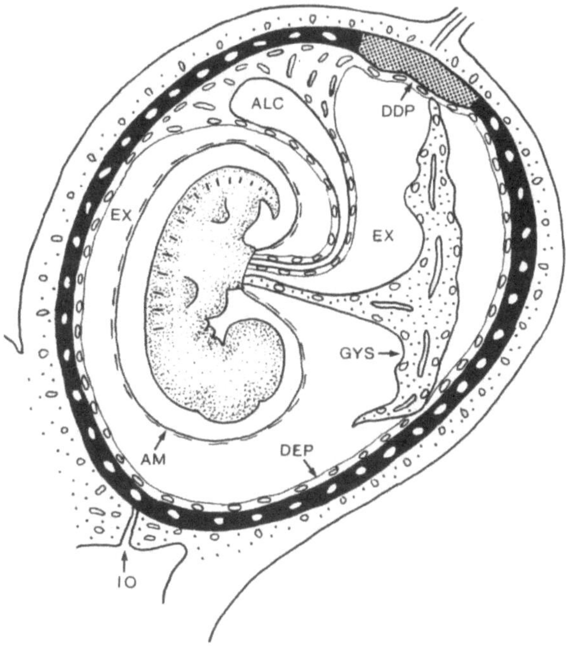

FIGURE 18.7. Diagram of the right uterine horn in a *M. ater* carrying a limb bud embryo. The yolk sac has separated almost completely from the chorion and has been converted into a glandular-appearing body (GYS). Spread of the vascularized allantoic mesoderm has created a diffuse endotheliodichorial placenta (DEP). The developing discoidal placenta (DDP) is larger than previously but still perfused by only a very small volume of blood. ALC = allantoic cavity; AM = amnion; EX = exocoelom; IO = internal os. From Rasweiler 1990.

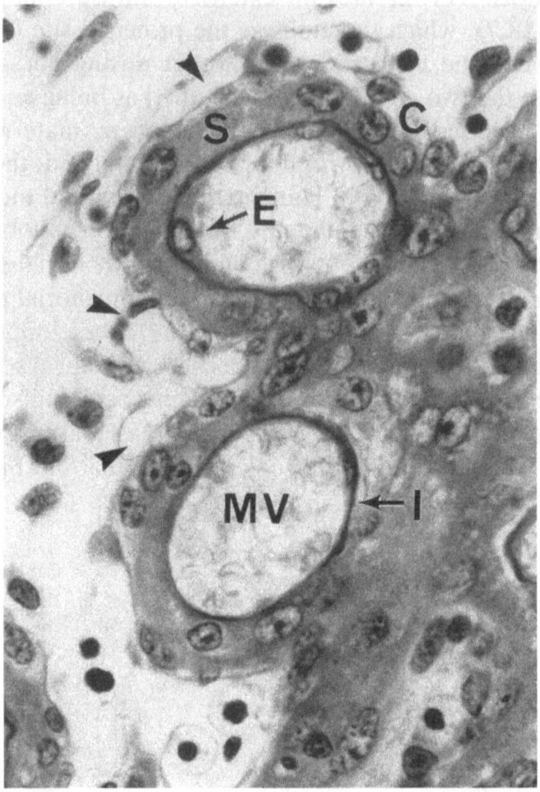

FIGURE 18.8. Interhemal barrier in the diffuse endotheliodichorial placenta of *M. ater*. The maternal vascular tubules (MV) are lined by hypertrophied endothelial cells (E). The latter are separated from the syncytiotrophoblast (S) by a thickened basal lamina, the interstitial membrane (I). Outside of the syncytiotrophoblast is a layer of cytotrophoblast (C). The arrowheads denote several fetal capillaries. Amylase-PAS-hematoxylin. (×796).

tuft is composed of a closely packed network of capillaries and possibly endothelial tubules connected to specialized arterioles and venules which run parallel to the intramural portion of the oviduct. These vessels are specialized in that they exhibit morphological and histochemical changes which are different from those occurring in other endometrial vessels. As decidual reactions spread through the endometrium late in the nonpregnant cycle or during implantation, the endothelial cells of the tuft vessels hypertrophy and their basal laminae become more prominent in histological sections stained for glycoproteins.

Although the trophoblast invades the decidua all around the embryo, much of that engulfing the maternal vessels within the choriovitelline and diffuse placentas is syncytial (Fig. 18.8), while that proliferating around the

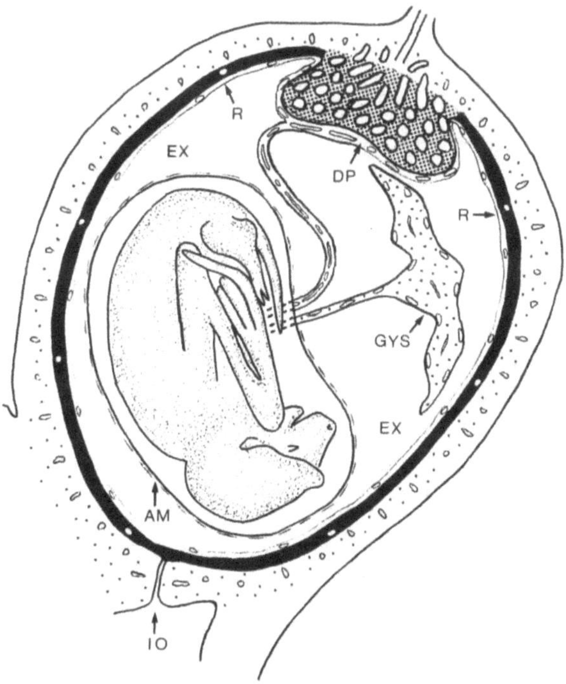

FIGURE 18.9. Diagram of the right uterine horn in a *M. ater* carrying a near-term fetus. Only remnants of the diffuse endotheliodichorial placenta (R) remain, and much of the maternal uterine blood flow has been shifted to the now fully functional discoidal hemochorial placenta (DP). AM = amnion; EX = exocoelom; GYS = glandular-appearing yolk sac; IO = internal os. From Rasweiler 1990.

vascular tuft remains cellular (Fig. 18.11). Furthermore, much of the cytotrophoblast in this latter region appears to be growing on the thickened basal laminae of the tuft vessels. These observations suggest that the hypertrophied endothelial cells of the vascular tuft may be secreting factors which control early trophoblastic growth and are incorporated, at least in part, into the endothelial cells' basal laminae. There is considerable evidence that basal laminae and components of the extracellular matrix play important morphogenetic roles in other tissues and organs (Rasweiler 1991b).

The pattern of trophoblastic development in the choriovitelline and diffuse placentas of *M. ater* is different from that seen in the discoidal placenta. It seems reasonable to speculate that this might be attributable in part to differences in the secretory activities of maternal endothelial cells in the non-tuft decidua.

These observations also raise the possibility that maternal endothelial cells may be influencing the growth and/or differentiation of trophoblast in

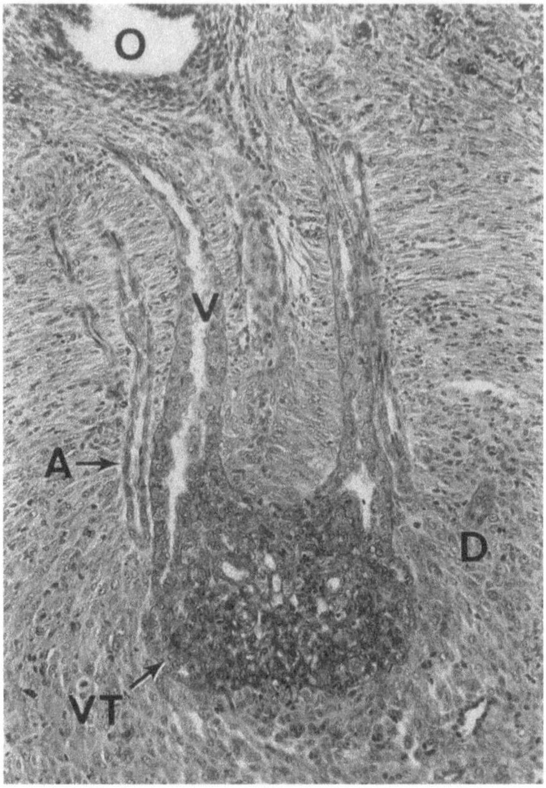

FIGURE 18.10. Frontal section through the cranial end of a spontaneously decidualized right uterine horn in *M. ater* showing the vascular tuft (VT). Note the hypertrophy of the endothelial cells lining the tuft arterioles (A) and venules (V). D = decidua; O = oviduct approaching the tuft. Amylase-PAS-hematoxylin. (×145). From Rasweiler 1991a.

other species. Thickened basal laminae are evident between the maternal endothelial cells and the trophoblast in the interhemal barrier of the endotheliochorial placentas in a number of other bats, insectivores, tupaiids, carnivores (including seals), and elephants (reviewed by Rasweiler 1991b). In some bats the maternal endothelial cells are lost during development of the chorioallantoic placenta, thereby creating a hemochorial condition. However, components of the endothelial basal lamina appear to persist as part of an interrupted layer (the "intrasyncytial lamina") that is completely engulfed by trophoblast (Enders 1982; Cukierski 1987; Bhiwgade 1990). This may also occur in the hyena (a carnivore), several sciurid rodents, and the elephant shrew (a macroscelid). It has been suggested that this peculiar morphological relationship (i.e., the engulfment of a fragmented basal lamina by adjacent epithelial cells) may have evolved as a biological com-

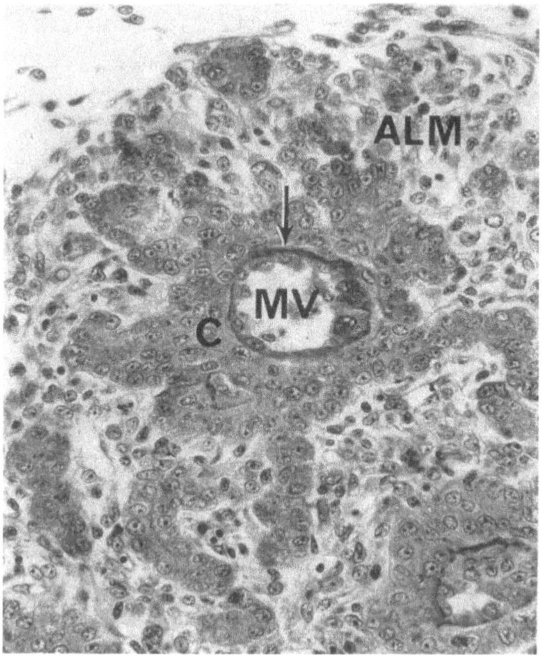

FIGURE 18.11. Portion of the developing discoidal placenta showing a maternal vessel (MV) of the vascular tuft engulfed by cytotrophoblast (C). Intense staining (arrow) is evident at the base of the endothelial cells lining this vessel. This is attributable to thickening of their basal lamina and the accumulation of glycoprotein-rich granules in their cytoplasm. ALM = allantoic mesoderm. Amylase-PAS-hematoxylin (×283).

promise which retains tissue of maternal origin in the interhemal barrier for morphogenetic reasons, but isolates potentially thrombogenic subendothelial components from exposure to the maternal blood. One of the functions of endothelial cells is to provide a nonthrombogenic lining for the vascular system, while subendothelial components of the vessel walls (including the endothelial basal laminae) can induce platelet adhesion and aggregation (Rasweiler 1991b).

At the limb bud stage, the relationship between the cytotrophoblast and maternal blood vessels within the developing discoidal placenta of *M. ater* begins to change. Solid trophoblastic tubules then start to sprout from the surface of the trophoblastic cuffs surrounding these vessels (Fig. 18.11) and from the mesodermal (embryonic) face of a band of cytotrophoblast extending across the base of the placenta. Many of these new tubules lack any maternal vascular elements. Such sprouting greatly increases the number of trophoblastic tubules and serves to fill in the center of the developing placenta. Patent tubules then form by a canalization process (Fig. 18.12).

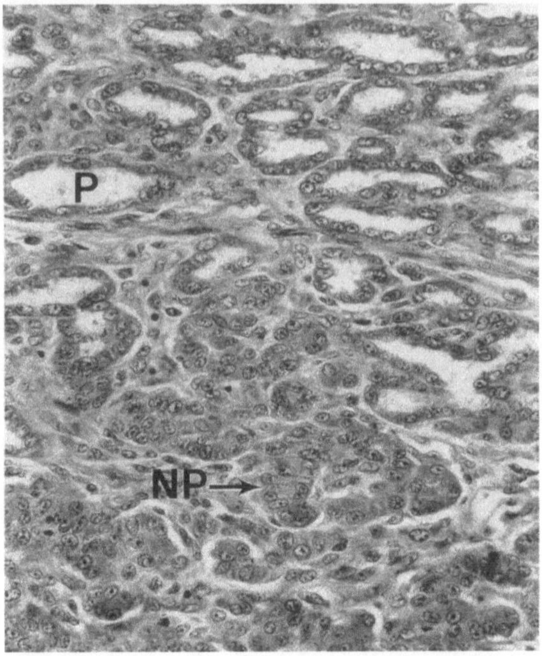

FIGURE 18.12. Portion of the developing discoidal placenta containing both patent (P) and non-patent (NP) trophoblastic tubules. The patent tubules shown here still contained relatively little maternal blood. Amylase-PAS-hematoxylin. (×283).

Endothelial cells in the small, trophoblast-engulfed maternal vessels of the vascular tuft either become necrotic or are redistributed to larger vessels at the base of the placenta. The basal laminae which had been associated with these cells lose their staining affinities at about the same time and disappear. Once maternal blood begins to flow through the tubules, a hemochorial type of placental relationship is established (Fig. 18.13) that is, one in which maternal blood directly bathes the trophoblast. The discoidal placenta of *M. ater* is quite unusual in that it is one of the few hemochorial placentas which lacks a continuous layer of syncytiotrophoblast in its interhemal membrane (Rasweiler 1991b).

Concluding Comments

Because *M. ater* is quite abundant in the wild and either ignored by humans or actually disliked due to its propensity to roost in buildings, there has been little compelling reason to study this bat's reproductive biology. It therefore came as a particular surprise that many aspects of reproduction in this poorly-known species turned out to be quite unusual and instructive. The discovery of menstruation in *M. ater* has contributed to a better under-

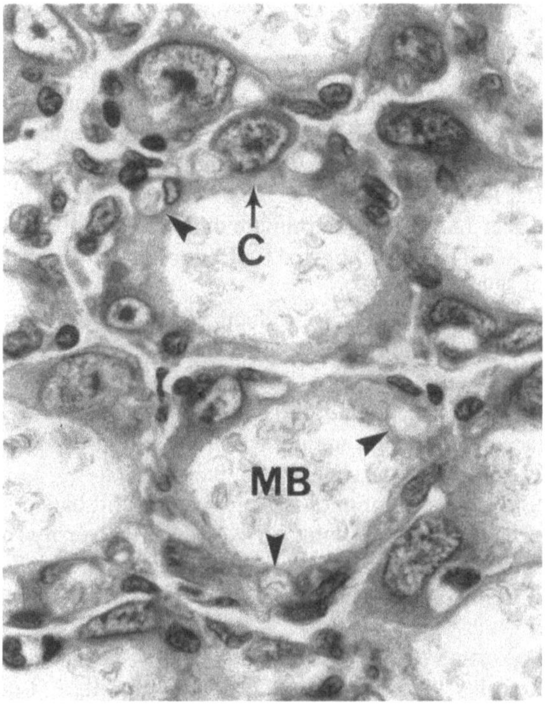

FIGURE 18.13. Portion of the fully-developed discoidal placenta serving a near-term fetus in *M. ater*. Maternal blood (MB) flows through a labyrinth lined by giant cytotrophoblast cells (C) which possess hypertrophied, euchromatic nuclei. The thickness of the interhemal barrier has been reduced in many areas because fetal capillaries (arrowheads) now occupy indentations of the trophoblast layer. Amylase-PAS-hematoxylin (×796).

standing of why this process probably evolved and indicates that this bat might provide a relevant new animal model for use in human reproductive research. Furthermore, observations on placental development in *M. ater* suggest that endometrial endothelial cells may be playing a very important role in controlling trophoblastic growth and differentiation in this bat and possibly many other mammals as well.

ACKNOWLEDGMENTS

The author's work with *M. ater* was supported by grants HD-17739 and RR-05396 from the National Institutes of Health. The generous assistance of Mr. Indar Ramnarine in photographing *M. ater* is gratefully acknowledged. This chapter is dedicated to the memory of Professor William A. Wimsatt who devoted much of his research career to studying the reproductive biology of bats.

References

Bhiwgade DA (1990) Comparative electron microscopy of chorio-allantoic placenta barrier in some Indian Chiroptera. Acta Anatomica 138:302–317

Bonilla H de, Rasweiler JJ IV (1974) Breeding activity, preimplantation development, and oviduct histology of the short-tailed fruit bat, *Carollia*, in captivity. The Anatomical Record 179:385–404

Brenner RM, Maslar IA (1988) The primate oviduct and endometrium. In Knobil E, Neill JD (eds.). The Physiology of Reproduction, Vol. 1. Raven Press, New York. pp. 303–329

Crichton EG, Krutzsch PH (1987) Reproductive biology of the female little mastiff bat, *Mormopterus planiceps* (Chiroptera: Molossidae) in southeast Australia. The American Journal of Anatomy 178:369–386

Cukierski MA (1987) Synthesis and transport studies of the intrasyncytial lamina: an unusual placental basement membrane in the little brown bat, *Myotis lucifugus*. The American Journal of Anatomy 178:387–409

Dolan PG (1989) Systematics of middle American mastiff bats of genus *Molossus*. Special Publications of the Museum at Texas Tech University No. 29:1–71

Dollar JR, Graham CE, Reyes FI (1982) Postovulatory predecidual development in the baboon, chimpanzee and human. American Journal of Primatology 3:307–313

Eckstein P, Zuckerman S (1956a) Morphology of the reproductive tract. In Parkes AS (ed.). Marshall's Physiology of Reproduction, Vol. 1. Longmans, Green, London. pp. 43–155

Eckstein P, Zuckerman S (1956b) Changes in the accessory reproductive organs of the non-pregnant female. In Parkes, AS (ed.). Marshall's Physiology of Reproduction, Vol. 1. Longmans, Green, London. pp. 543–654

Ellinwood WE, Nett TM, Niswender GD (1978) Ovarian vasculature: Structure and function. In Jones RE (ed.). The Vertebrate Ovary, Comparative Biology and Evolution. Plenum Press, New York. pp. 583–614

Enders AC (1982) Whither studies of comparative placental morphology? Journal of Reproduction and Fertility, Supplement 31:9–15

Finn CA (1987) Why do women and some other primates menstruate? Perspectives in Biology and Medicine 30:566–574

Freeman PW (1981) A multivariate study of the Family Molossidae (Mammalia, Chiroptera): Morphology, ecology, evolution. Fieldiana Zoology, New Series, No. 7, Publication 1316:1–173

Goodwin GG, Greenhall AM (1961) A review of the bats of Trinidad and Tobago. Descriptions, rabies infection and ecology. Bulletin of the American Museum of Natural History 122:187–302

Greenhall AM, Stell G (1960) Bionomics and Chemical Control of Free-tailed House Bats (*Molossus*) in Trinidad. Special Scientific Report: Wildlife No. 53, United States Department of the Interior, Fish and Wildlife Service, Washington, D.C.

Hamlett GWD (1934) Uterine bleeding in a bat (*Glossophaga soricina*). The Anatomical Record 60:9–17

Handley CO Jr (1976) Mammals of the Smithsonian Venezuelan Project. Brigham Young University Science Bulletin, Biological Series, Vol. 20, No. 5

Hood CS, Smith JD (1983) Histomorphology of the female reproductive tract in phyllostomoid bats. Occasional Papers of the Museum at Texas Tech University No. 86:1–38

Horst R (1966) Observations on the gular gland of *Molossus rufus nigricans*. The Anatomical Record 154:465

Jones JK Jr, Carter DC (1976) Annotated checklist, with keys to subfamilies and genera. In Baker RJ, Jones JK Jr, Carter DC (eds.). Biology of Bats of the New World Family Phyllostomatidae, Part I. Texas Tech University Press, Lubbock, Texas. pp. 7–38

Koopman KF (1970) Zoogeography of bats. In Slaughter BH, Walton DW (eds.). About Bats. A Chiropteran Biology Symposium. Southern Methodist University Press, Dallas. pp. 29–50

Koopman KF, Jones JK Jr (1970) Classification of bats. In Slaughter BH, Walton DW (eds.). About Bats. A Chiropteran Biology Symposium. Southern Methodist University Press, Dallas. pp. 22–28

Luckett WP (1980) The use of fetal membrane data in assessing chiropteran phylogeny. In Wilson DE, Gardner AL (eds.). Proceedings, Fifth International Bat Research Conference. Texas Tech University Press, Lubbock, Texas. pp. 245–265

McCracken GF (1986) Why are we losing our Mexican free-tailed bats? Bats. Newsletter of Bat Conservation International 3(3):1–2, 4

Marques SA (1986) Activity cycle, feeding and reproduction of *Molossus ater* (Chiroptera: Molossidae) in Brazil. Boletim do Museu Paraense Emilio Goeldi Zoologia 2:159–179

Murakami T, Ikebuchi Y, Ohtsuka A, Kikuta A, Taguchi T, Ohtani O (1988) The blood vascular wreath of rat ovarian follicle, with special reference to its changes in ovulation and luteinization: A scanning electron microscopic study of corrosion casts. Archives of Histology and Cytology 51:299–313

Quintero F, Rasweiler JJ IV (1973) The reproductive biology of the female vampire bat, *Desmodus rotundus*. American Zoologist 13:1284

Quintero F, Rasweiler JJ IV (1974) Ovulation and early embryonic development in the captive vampire bat, *Desmodus rotundus*. Journal of Reproduction and Fertility 41:265–273

Racey PA (1979) The prolonged storage and survival of spermatozoa in Chiroptera. Journal of Reproduction and Fertility 56:391–402

Rasweiler JJ IV (1970) The laboratory biology of the long-tongued bat, *Glossophaga soricina*: Maintenance procedures, estivation, the menstrual cycle, histophysiology of the oviduct and intramural implantation. Ph.D. Thesis, Cornell University, Ithaca, New York

Rasweiler JJ IV (1972) Reproduction in the long-tongued bat, *Glossophaga soricina*. I. Preimplantation development and histology of the oviduct. Journal of Reproduction and Fertility 31:249–262

Rasweiler JJ IV (1978) Unilateral oviductal and uterine reactions in the little bulldog bat, *Noctilio albiventris*. Biology of Reproduction 19:467–492

Rasweiler JJ IV (1979a) Early embryonic development and implantation in bats. Journal of Reproduction and Fertility 56:403–416

Rasweiler JJ IV (1979b) Bats as models in studies on folliculogenesis, menstruation, early pregnancy, and sperm survival. In Alexander NJ (ed.). Animal Models for Research on Contraception and Fertility. Harper and Row, Hagerstown, Maryland. pp. 437–446

Rasweiler JJ IV (1982) The contribution of observations on early pregnancy in the little sac-winged bat, *Peropteryx kappleri*, to an understanding of the evolution of reproductive mechanisms in monovular bats. Biology of Reproduction 27:681–702

Rasweiler JJ IV (1984) Reproductive failure due to the entrapment of oocytes in luteinized follicles of the little bulldog bat (*Noctilio albiventris*). Journal of Reproduction and Fertility 71:95–101

Rasweiler JJ IV (1987) Prolonged receptivity to the male and the fate of spermatozoa in the female black mastiff bat, *Molossus ater*. Journal of Reproduction and Fertility 79:643–654

Rasweiler JJ IV (1988) Ovarian function in the captive black mastiff bat, *Molossus ater*. Journal of Reproduction and Fertility 82:97–111

Rasweiler JJ IV (1990) Implantation, development of the fetal membranes, and placentation in the captive black mastiff bat, *Molossus ater*. The American Journal of Anatomy 187:109–136

Rasweiler JJ IV (1991a) Spontaneous decidual reactions and menstruation in the black mastiff bat, *Molossus ater*. The American Journal of Anatomy 191:1–22

Rasweiler JJ IV (1991b) Development of the discoidal hemochorial placenta in the black mastiff bat, *Molossus ater*. Evidence for a role of maternal endothelial cells in the control of trophoblastic growth. The American Journal of Anatomy 191:185–207

Rasweiler JJ IV, Bonilla H de (in press) Menstruation in shorttailed fruit bats (Carollia spp.) Journal of Reproduction and Fertility

Roenneberg T, Aschoff J (1990) The annual rhythm of human reproduction. I. Biology, sociology, or both? Journal of Biological Rhythms 5:195–216

Sansom GS (1932) Notes on some early blastocysts of the South American bat *Molossus*. Proceedings of the Zoological Society of London, 1932. pp. 113–118

Smith JD (1976) Chiropteran evolution. In Baker RJ, Jones JK Jr, Carter DC (eds.). Biology of Bats of the New World Family Phyllostomatidae, Part I. Texas Tech Press, Lubbock, Texas. pp. 49–69

van der Horst CJ, Gillman J (1941) The menstrual cycle in *Elephantulus*. South African Journal of Medical Science 6:27–47

Werner HJ, Lay DM (1963) Morphological aspects of the chest gland of the bat, *Molossus ater*. Journal of Mammalogy 44:552–555

Wimsatt WA (1975) Some comparative aspects of implantation. Biology of Reproduction 12:1–40

Wimsatt WA (1979) Reproductive asymmetry and unilateral pregnancy in Chiroptera. Journal of Reproduction and Fertility 56:345–357

19

Ecological Aspects of Reproductive Patterns in South American Small Rodents

Thomas E. Lacher Jr.

Introduction

The Class Mammalia is a diverse and varied taxon (over 4200 species) with a global distribution. The order Rodentia contains approximately 418 recent genera and 1749 recent species. The family Muridae is the largest of all mammalian families, containing about 1135 recent species, or one-fourth of all mammals (Anderson and Jones 1984). Murid rodents (commonly referred to as rats and mice) occur virtually everywhere that mammals are found and are only absent from Iceland, New Zealand, Antarctica, and some oceanic and arctic islands (Anderson and Jones 1984).

Classification of the Murids of South America

The family Muridae (Carleton and Musser 1989) occurs throughout South America, including recent and extinct forms on the Galapagos. They are important components of the vertebrate fauna; the subfamily Sigmodontinae contains 22 percent of the known living species of mammals of South America. Reig (1984, 1987) classifies the South American forms into two subfamilies: the Neotominae, with three species in two genera; and the Sigmodontinae, with at least 250 species and over 50 genera (but see Carleton 1980). The Sigmodontinae is divided into seven tribes by Reig (1984): Oryzomyini (with 44.2% of extant species), Icthyomyini (3.2%), Akodontini (24.9%), Scapteromyini (2.4%), Wiedomyini (0.4%), Phyllotini (18.1%), and Sigmodontini (4.0%). He considers four genera with seven species to be *incertae sedis.* Three tribes dominate the sigmodontine fauna of South America, and three genera, *Akodon*, *Thomasomys*, and *Oryzomys*, have undergone dramatic radiations (Table 19.1). All three genera are widely distributed. Many genera are monospecific (Reig 1984) and some have very limited distributions.

Reig's 1984 classification has already undergone modification by other authors (e.g., Reig 1987; Carleton and Musser 1989). There is even dis-

TABLE 19.1. Tribes and genera, with number of species in parentheses, of South American rodents of the subfamily Sigmodontinae. Number of species includes North and South American species. Numbers are composites from three sources and are rough approximations. Based upon Honacki et al. (1982), Reig (1984), and Carleton and Musser (1989).

Family Muridae	Subfamily Sigmodontinae
Tribe Oryzomyini	Tribe Wiedomyini
Oryzomys (37–38)	*Wiedomys* (1)
Microryzomys (2)	
Oligoryzomys (15)	Tribe Phyllotini
Thomasomys (23)	*Phyllotis* (12)
Oecomys (2–3)	*Calomys* (8)
Rhipidomys (7)	*Euneomys* (4)
Nectomys (2)	*Auliscomys* (4)
Neacomys (3)	*Graomys* (3)
Delomys (1)	*Andalgalomys* (3)
Aepeomys (2)	*Eligmodontia* (1)
Nesoryzomys (5)*	*Pseudoryzomys* (2)
Phaenomys (1)	*Galenomys* (1)
Chilomys (1)	*Reithrodon* (1)
Scolomys (1)	*Chinchillula* (1)
Wilfredomys (1)	*Neotomys* (1)
	Andinomys (1)
Tribe Icthyomyini	*Irenomys* (1)
Ichthyomys (3)	
Anotomys (2)	Tribe Sigmodontini
Daptomys (3)	*Sigmodon* (8)
Neusticomys (1)	*Holochilus* (4)
Tribe Akodontini	Sigmodontinae *incertae sedis*
Akodon (33)	*Zygodontomys* (3)
Oxymycterus (9)	*Rhagomys* (1)
Bolomys (6)	*Punomys* (1)
Chelemys (4)	*Abrawayaomys* (1)
Microxus (3)	
Notiomys (2)	Tribe Scapteromyini
Blarinomys (1)	*Bybimys* (3)
Podoxymys (1)	*Scapteromys* (1)
Lenoxus (1)	*Kunsia* (2)
Juscelinomys (1)	

*Several of these species are probably extinct.

agreement as to whether the family Muridae should include all rats and mice or whether it should be split into two or more distinct families. Reig (1984) places the sigmodontines in the family Cricetidae and Honacki et al. (1982) use three families (Muridae, Cricetidae, Arvicolidae) and treat sigmodontines as a subfamily of Cricetidae. This emphasizes the dynamic nature of the taxonomy of the South American forms of the family. New taxa are being described as a result of ongoing systematic and biogeographic research. It is not unreasonable to expect the number of genera and species to increase as research efforts proceed in the vast areas of the rain forests, savannas, and mountains that remain unstudied.

Evolutionary Origins and Relations

The origin of the South American rats and mice has been the subject of debate. Murid rodents were restricted to North America in the western hemisphere certainly prior to the Miocene and perhaps as late as the Pliocene. One school argues for a Pliocene origin that coincides with the closure of the Panamanian land bridge, approximately three million years ago (Mares 1975, 1976, 1985; Baskin 1978; Simpson 1980; Marshall et al. 1982). This would make the arrival of murids in South America a relatively recent event, a remarkable fact considering their dramatic radiation and diversification on the continent. Others have disputed this scenario, arguing that murids, in particular the ancestors of the subfamily Sigmodontinae, arrived during the late Oligocene or early Miocene by rafting across the water prior to the closure of the land bridge (Hershkovitz 1966, 1972; Reig 1981, 1984, 1987). The absence of fossil records prior to the Pliocene has been an argument against this scenario. The debate is currently unresolved. By either account, the South American forms are relatively recent yet have radiated throughout all of the major biomes of the continent.

Reproduction and Environment

Reproductive Patterns in Muroid Rodents

Most research on the environmental determinants of reproductive patterns in murid rodents was done in temperate zone habitats. Bronson (1989) describes murid rodents as possessing a number of characteristics that limit the degree to which they can respond to environmental factors. They (1) have exceptionally short life spans; (2) mature quickly; (3) produce large litters rapidly and repeatedly; (4) have a short gestation of three weeks or less; and (5) exhibit a post-partum estrus. Because of these aspects of their biology, murids exhibit extreme opportunism in response to favorable environmental conditions (Bronson and Perrigo 1987). The common North American genus *Peromyscus*, for example, shows a highly seasonal three-month breeding period above 60 degrees north latitude and breeds year-round in tropical Mexico (Bronson 1989). At intermediate latitudes breeding is seasonal, with a longer breeding season, or bimodal, with reproduction in the spring and fall. When winters are mild, winter breeding is observed, even in typically cold climates. Only one species (*Peromyscus maniculatus*) shows nearly all of these patterns over its range (Bronson 1985). The house mouse (*Mus domesticus*) is a common human commensal in temperate regions. Where it is feral, it breeds seasonally; in human dwellings where temperature and food are constant, it breeds continuously.

The major environmental factors that influence reproductive patterns in temperate zone species are food availability, ambient temperature, precipitation, and photoperiod (Bronson 1985). In the wet tropics, there is likely little annual variability in these parameters. Given the opportunistic nature of breeding in small rodents with short life-spans, tropical rodents should breed year-round. As Bronson (1985) stated, however, few data are available on reproduction in tropical mammals in the wild. Tropical environments do show seasonality in precipitation, and several Old World species have seasonal reproduction determined by precipitation (Delany and Neal 1969; Temme 1981). Recently published research on sigmodontine reproduction in South America now allows for a preliminary evaluation of the reproductive patterns in these rodents.

HABITATS AND DISTRIBUTION

The public perception of South America as a land of tropical forest is over simplified. Habitats range from the lowland rain forests of the Amazon Basin to the harsh high altitude environments of the southern Andes. Climatic differences among major ecosystems can be pronounced (Lacher 1982; Lacher and Mares 1986). Tropical forest systems themselves are very heterogeneous (Hueck 1972) and major differences in flora and fauna exist among the Atlantic Coastal Forest of Brazil, the Amazon, and the Pacific coastal forests of Colombia and Ecuador. The vast savanna-like Cerrado of Brazil is one of the largest and most threatened of the habitats of South America and remains poorly studied as attention is diverted to the Amazon. Important arid ecosystems occur throughout the continent. Thorn scrub habitats range from coastal Venezuela, through the Caatinga of northeastern Brazil, to the Chaco thorn scrub of Paraguay, Bolivia, and Argentina, and true deserts occur in southwestern Argentina and coastal Chile (Mares 1985). Arid high elevation habitats occur along the Andean cordillera from Venezuela to Tierra del Fuego (Mares 1985). Temperate habitats like chaparral (Chile) and deciduous and coniferous forest (southern Chile and Argentina) are present, though limited in area.

Sigmodontines occur in all but the most extreme high altitude habitats of South America. Some genera, such as *Andinomys* and *Auliscomys*, are found in puna grasslands above 4500 meters in the Andes. *Phyllotis* can be found above 5000 meters in Argentina in areas almost completely devoid of vegetation (Olrog and Lucero 1981). Arid areas as well as mesic tropical habitats harbor diverse sigmodontine communities. Because of the great diversity of these rodents and habitats in South America, it would be difficult to present a continent-wide overview of their reproductive biology. In addition, very few populations of sigmodontines have been studied for a long enough period in the wild to collect information on patterns of repro-

TABLE 19.2. Geographical location and climate of the habitats presented in the case studies (Hueck 1972; Lacher 1982).

Region	Vegetation	Temperature	Precipiation
Atlantic Coastal Forest	Lowland and Montane rain forest	Annual mean 24°C Low seasonality	1000–4500 mm Moderate seasonality
Panama	Lowland rain forest	Annual mean 27°C No seasonality	2500 mm Some seasonlity
Llanos	Wet savanna	Annual mean 27°C Low seasonlity	2000 mm Highly seasonal
Cerrado	Upland savanna, Scrub savanna	Annual mean 26°C Low seasonlity	1100–2000 mm Highly seasonal
Caatinga	Dry scrub woodland	Annual mean 27°C Hot year-round	200–1000 mm Highly variable
Chilean Matorral	Dry decidous forest	Annual mean 13–15°C Seasonal, mild winters	1000–4000 mm Very seasonal
Pampas	Open grassland	Annual mean 15°C Seasonal, mild winters	750 mm Seasonal
Monte	Dry desert	Annual mean 15°C Hot summers	200 Dry, spring rains
Chilean/Argentine forest	*Nothofagus* forest	Annual mean 8–14°C Cold winters	1000–3000 mm Dry summers

duction. Several case studies of reproductive patterns in the field in ecosystems with different vegetation and climates can reveal general trends.

Cricetid Reproduction in South America

The latitudinal and altitudinal gradients present in South America rival or exceed those found in North America. Studies conducted in geographic areas that reflect this habitat diversity were selected for the case studies (Table 19.2; Fig. 19.1).

WET TROPICAL FOREST

Few long-term studies of sigmodontines inhabiting wet tropical forests have been published that report annual reproductive patterns. These forests are wet and warm year-round, although there is seasonality in precipitation. Fleming (1971) studied the small mammal community of lowland tropical forests in Panama. The one common sigmodontine in this study, *Oryzomys capito*, reproduced year-round, just like the Mexican *Peromyscus*. *Oryzomys capito* is a typical murid rodent in that females mature rapidly and begin to breed at less than two months of age. Producing an average of 6.12 litters per year (0.51 per month), they clearly are opportunistic breeders. Fonseca and Kierulff (1989) found a similar pattern in a 17-month study in the Atlantic rain forest of Brazil. The Atlantic forest

FIGURE 19.1. Map of South America showing the locations of the ecosystems represented in the case studies. Ecosystems are: (1) Lowland rain forest, Panama; (2) Atlantic coastal rain forest, Brazil; (3) Llanos, Venezuela; (4) Cerrado savannas, central Brazil; (5) Caatinga scrub, northeastern Brazil; (6) Matorral woodlands, central Chile; (7) Pampas, eastern Argentina; (8) Monte Desert, northwestern Argentina; (9) mesic *Nothofagus* temperate forest, southern Chile and Argentina.

site of this study is both drier and more seasonal than Panama. This community contained nine species of sigmodontines; however, reproductive patterns were presented only for the three most common species. *Oryzomys nigripes* (the *O. eliurus* of Carleton and Musser 1989) reproduced year-round as did *Oryzomys trinitatus* (=*concolor*). The third species for which reproductive data were presented, *Akodon cursor*, was observed breeding in 11 of 17 months but, unlike the *Oryzomys*, showed a distinct breeding peak that began at the end of the wet season and continued through to the end of the dry season. All three species were observed to be largely insectivorous. The reason for the difference in reproductive patterns between the *Oryzomys* and *Akodon* is not known.

MOIST SAVANNA

The small mammals of the seasonally moist savannas of Venezuela (the Llanos) have been studied by O'Connell (1982) and Vivas (1986). The Llanos are seasonally inundated with a prolonged dry season between December and April. All four species of sigmodontines that occur in the Llanos (*Oryzomys bicolor*, *Rhipidomys mastacalis*, *Sigmodon alstoni*, and *Zygodontomys brevicauda*) breed year-round, although Vivas (1986) reports some decline in *Sigmodon* breeding during the dry season. Litter sizes are large, ranging from 3.8 for *Rhipidomys* to 5.0 for *Sigmodon*. Again, opportunism seems to be the rule in these seasonally inundated savannas.

The Cerrado of Brazil is an extensive savanna that covers the high central plateau. The Cerrado, like the Llanos, has highly seasonal precipitation. Winter (June, July and August) can be exceedingly dry, with virtually no rain. The Cerrado is never inundated and winter temperatures can drop to less than 10°C at night during the dry season. Seasonality is therefore somewhat more pronounced than in the Llanos. The Cerrado actually is a mosaic of several vegetation formations. Cerrado "sensu lato" contains a mix of savanna woodland (both closed and open canopy) as well as a variety of types of more open grassland. In addition to Cerrado, there are gallery forests that occur along the margins of streams and rivers. These forests are extensions of either the Atlantic rain forest in the east or the Amazon forest in the north and west.

The small mammals of the Cerrado have been studied by Nitikman and Mares (1987), Lacher et al. (1989), and Mares et al. (1989). The region around the city of Brasilia has one of the most diverse sigmodontine communities in South America. Eighteen species have been observed in the area, including seven species of *Oryzomys*, four species of *Akodon*, and two species of *Calomys*. Virtually all of the species that occur in the Cerrado region, including inhabitants of the gallery forests, show a highly seasonal breeding pattern (*Akodon cursor*, *A. reinhardti*, *A. cerradensis*, *A. lindbergi*, *Bolomys lasiurus*, *Calomys callosus*, *C. tener*, *Holochilus brasiliensis*, *Oryzomys bicolor*, *O. capito*, *O. chacoensis*, *O. concolor*, *O. fornesi*, *O. nigripes*, *O. subflavus*, and *Oxymycterus roberti*). Sigmodontines initiate breeding in the late dry season (July to August) and cease breeding in the middle of the rainy season (January to February). The only exceptions are *Rhipidomys mastacalis*, an arboreal inhabitant of the gallery forest, and *Nectomys squamipes*, a semi-aquatic rodent that inhabits stream bank habitats in the gallery forest. These appear to be the only two species capable of breeding year-round in the Cerrado. *Oryzomys bicolor* is also arboreal, yet is a seasonal breeder.

Tropical Semiarid Scrub

The Caatinga is a dry region located in northeastern Brazil. The region has low annual precipitation and is hot year-round. In addition, annual precipitation is extremely variable, with periodic droughts where a year or more may pass without significant precipitation (Streilein 1982a). Small mammals were studied by Karimi et al. (1976) and Streilein (1982b). All species of sigmodontines that occur in the Caatinga (*Bolomys lasiurus*, *Calomys callosus*, and *Wiedomys pyrrhorhinos*) appear to be opportunistic breeders, which generally will result in a seasonal pattern. Data are sparse, however, and this region is ideal for a long-term study of the reproductive response of sigmodontine rodents to a seasonal yet highly unpredictable precipitation regime.

Temperate Scrub Forest

The central coast of Chile is covered by Matorral woodland, a habitat similar to the Chaparral and more arid scrub woodlands of California. The region has a Mediterranean climate, with spring rains (June through September) and mild winters. This habitat is one of the best studied of South America. Small mammal research relevant to this chapter was done by Fulk (1975) and Meserve and Glanz (1978). Four species of sigmodontines were studied: *Akodon olivaceus*, *A. longipilis*, *Oryzomys longicaudatus*, and *Phyllotis darwini*. Reproduction was highly seasonal, occurring during the period of spring rain (July through November). The more northerly regions are subjected to heavy coastal fogs, which deposit moisture through condensation. In these regions the breeding season extends throughout the year. The situation in Chile clearly illustrates the opportunistic nature of murid reproduction.

Temperate Grasslands

Two major studies have been carried out in the temperate pampas of South America. Barlow (1969) did a study on the small mammals of the pampas of Uruguay and Dalby (1975) studied the small mammals of a similar environment in Argentina. The pampas have a mild temperate climate with strong seasonality in temperature. Seasonality in precipitation is far less pronounced than in the savannas. The communities studied in Uruguay and Argentina differ somewhat in composition, but reproductive patterns are very similar. In Uruguay, *Oryzomys flavescens*, *Akodon azarae*, *Calomys laucha*, and *Bolomys obscurus* all bred seasonally, between October and May (summer); however, two species, *Oxymycterus rutilans* (an insectivore) and *Reithrodon physodes* (an herbivore) bred year-round (O'Connell 1982). The community in Argentina was simpler. *Oryzomys nigripes* and *Akodon azarae* both were seasonal breeders, but the breeding season was shorter (November to March) in the cooler southern climate. The insectivore *Oxymycterus rutilans* bred year-round as it did in Uruguay.

Dry Desert

The Monte Desert of northwestern Argentina is a hot desert much like the Sonoran desert of North America (Mares 1985). It has a depauperate fauna with only two species of sigmodontines, *Phyllotis griseoflavus* and *Eligmodontia typus*. Both species appear to be opportunistic breeders but concentrate reproductive activity between early spring (August) and late fall (April). The rains generally fall between October and February. Mares (pers. comm.) feels that reproduction in the Monte is keyed more to photoperiod than to precipitation.

TEMPERATE MOIST FOREST

The temperate moist forests of southern Chile and Argentina bear a great resemblance to the forests of the Pacific Northwest of North America. The climate is one of cool summers and cold winters with some regions receiving up to 3000 mm of precipitation per year (with accumulations of up to 0.5 m of snow; Pearson and Pearson 1982). Precipitation is seasonal with 75% falling between April and September (Murua and Gonzalez 1986). Reproductive patterns in small mammal communities were studied in Chile (Murua and Gonzalez 1986) and in Argentina (Pearson 1983). The studies had two species in common, *Akodon olivaceus* and *Oryzomys longicaudatus*. In the more maritime climate of Chile, the reproductive season ran from November to April (Murua and Gonzalez 1986). The two species have a similar breeding period (October to May) in the cooler forests of Argentina, but the data of Pearson (1983) suggest that most breeding is concentrated in November and December and ends before April, though data are absent for January through March. Other species studied in Argentina (*Akodon longipilis*, *Auliscomys micropus*, *Chelemys macronyx*, *Geoxus valdivianus*, and *Irenomys tarsalis*) have a similar breeding season.

Summary

Reproductive patterns vary greatly among sigmodontine rodents in South America (Table 19.3). The pattern parallels that presented for North America by Bronson (1989), with reproduction varying from aseasonal in the wet tropics to highly seasonal in dry or cold environments. Intraspecific variation is present for several species, even with the limited data set available. *Oryzomys capito* breeds throughout the year in Panama but is a seasonal breeder in the central Brazilian gallery forests. *Oryzomys bicolor* is a seasonal breeder in the gallery forests of central Brazil yet breeds year-round in the Venezuelan Llanos. More interesting is the pattern observed in *Akodon olivaceus* and *Oryzomys longicaudatus*. These two species are aseasonal breeders in the arid, foggy regions of northern Chile, become seasonal breeders (July through November) in the dry scrub further south, shift the season of breeding further into the summer (November through April) in the extreme southern forests, and then appear to restrict the season of breeding even further on the cooler Argentine side.

Opportunism seems to be the prevailing rule in South America as in North America. Vast regions of the continent and a large proportion of the

TABLE 19.3. General pattern in sigmodontine reproduction in selected South American habitats. Latitude is location of the study site. Recall that seasons occur six months later than in North America. See text for citations.

Ecosystem	Location	Latitude	General Pattern
Wet tropical forest	Panama	9°N	Breed year-round
Wet tropical forest	Atlantic coastal Brazil	20°S	Breed year-round
Wet savanna	Northern Venezuela	8°N	Breed year-round
Wet savanna	Central Brazil	15°S	Seasonal in late dry, early wet season
Tropical dry scrub	Northeastern Brazil	7°S	Highly opportunistic seasonal
Temperate scrub woodland	Central Chile	30°S	Breed in spring rainy season
Temperate grassland	Uruguay and eastern Argentina	33–37°S	Breed late spring to early fall
Dry desert	Northwestern Argentina	25°S	Breed early spring to early fall
Moist temperat forest	Southern Chile	38–41°S	Breed in late spring and summer

fauna still remain to be studied. Whether these patterns remain consistent with further research remains to be seen.

REFERENCES

Anderson S, Jones JK Jr (1984) Orders and Families of Recent Mammals of the World. John Wiley and Sons, New York

Barlow JC (1969) Observations on the biology of rodents in Uruguay. Life Sciences Contributions, Royal Ontario Museum 75:1–59

Baskin JA (1978) *Bensonomys*, *Calomys*, and the origin of the phyllotine group of Neotropical cricetines (Rodentia:Cricetidae). Journal of Mammalogy 59:125–135

Bronson FH (1985) Mammalian reproduction: An ecological perspective. Biology of Reproduction 32:1–26

Bronson FH (1989) Mammalian Reproductive Biology. University of Chicago Press, Chicago

Bronson FH, Perrigo G (1987) Seasonal regulation of reproduction in muroid rodents. American Zoologist 27:929–940

Carleton MD (1980) Phylogenetic relationships in neotomine-peromyscine rodents (Muroidea) and a reappraisal of the dichotomy within New World Cricetinae. Miscellaneous Publications, Museum of Zoology, University of Michigan, 157:1–146

Carleton MD, Musser GG (1989) Systematic studies of oryzomyine rodents (Muridae:Sigmodontinae): A synopsis of *Microryzomys*. Bulletin of the American Museum of Natural History 191:1–83

Dalby PL (1975) Biology of pampa rodents, Balcarce Area, Argentina. Publications of the Museum, Michigan State University, Biological Sciences 5:149–272

Delany MJ, Neal BR (1969) Breeding seasons in rodents in Uganda. Journal of Reproduction and Fertility, Supp. 6:229–235

Fleming TH (1971) Population ecology of three species of neotropical rodents. Miscellaneous Publications, Museum of Zoology, University of Michigan, 143:1–77

Fonseca GAB, Kierulff C (1989) Biology and natural history of Atlantic Forest small mammals. Bulletin of the Florida State Museum 34:99–152

Fulk GW (1975) Population ecology of rodents in the semiarid shrublands of Chile. Occasional Papers of the Museum, Texas Tech University, 33:1–40

Hershkovitz P (1966) Mice, land bridges and Latin American faunal interchange. In Wenzel RL, Tipton VJ (eds.). Ectoparasites of Panama. Field Museum of Natural History, Chicago. pp. 725–747

Hershkovitz P (1972) The recent mammals of the neotropical region: A zoogeographic and ecological review. In Keast A, Erk FC, Glass B (eds.). Evolution, Mammals, and Southern Continents. State University of New York Press, Albany. pp. 311–431

Honacki JH, Kinman KE, Koeppl JW (1982) Mammal Species of the World: A Taxonomic and Geographic Reference. Allen Press and the Association of Systematic Collections, Lawrence, Kansas

Hueck K (1972) As Florestas da América do Sul. Editora Polígono S.A., São Paulo

Karimi Y, De Almeida CR, Petter F (1976) Note sur les rongeurs du Nordest du Bresil. Mammalia 40:257–266

Lacher TE Jr (1982) Behavioral research in South America. In Mares MA, Genoways HH (eds.). Mammalian Biology in South America. Special Publication Series, Pymatuning Laboratory of Ecology, University of Pittsburgh Press, Pittsburgh, Pennsylvania. pp. 209–230

Lacher TE Jr, Mares MA (1986) The structure of neotropical mammal communities: An appraisal of current knowledge. Revista Chilena de Historia Natural 59:121–134

Lacher TE Jr, Mares MA, Alho CJR (1989) The structure of a small mammal community in a central Brazilian savanna. In Redford KH, Eisenberg JF (eds.). Advances in Neotropical Mammalogy. Sandhill Crane Press, Gainesville, Florida. pp. 137–162

Mares MA (1975) South American mammal zoogeography: Evidence from convergent evolution in desert rodents. Proceedings of the National Academy of Sciences, Washington, D.C., 72:1702–1706

Mares MA (1976) Convergent evolution of desert rodents: Multivariate analysis and zoogeographic implications. Paleobiology 2:39–63

Mares MA (1985) Mammal faunas of xeric habitats and the Great American Interchange. In Stehli FG, Webb SD (eds.). The Great American Biotic Interchange. Plenum, New York. pp. 489–520

Mares MA, Braun JK, Gettinger D (1989) Observations on the distribution and ecology of the mammals of the Cerrado grasslands of central Brazil. Annals of Carnegie Museum 58:1–60

Marshall LG, Webb SD, Sepkoski JJ Jr, Raup DM (1982) Mammalian evolution and the Great American Interchange. Science 215:1351–1357

Meserve PL, Glanz WE (1978) Geographical ecology of small mammals in the northern Chilean arid zone. Journal of Biogeography 5:135–148

Murua R, Gonzalez LA (1986) Regulation of numbers in two neotropical rodent

species in southern Chile. Revista Chilena de Historia Natural 59:193–200
Nitikman LZ, Mares MA (1987) Ecology of small mammals in a gallery forest of central Brazil. Annals of Carnegie Museum 56:75–95
O'Connell MA (1982) Population biology of North and South American grassland rodents: A comparative review. In Mares MA, Genoways HH (eds.). Mammalian Biology in South America. Special Publication Series, Pymatuning Laboratory of Ecology, University of Pittsburgh Press, Pittsburgh, Pennsylvania. pp. 167–185
Olrog CC, Lucero MM (1981) Guia de los Mamiferos Argentinos. Ministerio de Cultura y Educacion, Fundacion Miguel Lillo, San Miguel de Tucuman
Pearson OP (1983) Characteristics of a mammalian fauna from forests in Patagonia, southern Argentina. Journal of Mammalogy 64:476–492
Pearson OP, Pearson AK (1982) Ecology and biogeography of the southern rainforests of Argentina. In Mares MA, Genoways HH (eds.). Mammalian Biology in South America. Special Publications Series, Pymatuning Laboratory of Ecology, University of Pittsburgh Press, Pittsburgh, Pennsylvania. pp. 129–142
Reig OA (1981) Teoria del origen y desarollo de la fauna de mamiferos de América del Sur. Monographiae Naturae, Museo Municipal de Ciencias Naturales "Lorenzo Scaglia" 1:1–161
Reig OA (1984) Distribuição geográfica e história evolutiva dos roedores muroideos sulamericanos (Cricetidae:Sigmodontinae). Revista Brasileira de Genética 7:333–365
Reig OA (1987) An assessment of the systematics and evolution of the Akodontini, with the description of new fossil species of *Akodon* (Cricetidae: Sigmodontinae). In Patterson BD, Timm RM (eds.). Studies in Neotropical Mammalogy: Essays in Honor of Philip Hershkovitz. Fieldiana, Zoology, New Series 39:347–399
Simpson GG (1980) Splendid Isolation: The Curious History of South American Mammals. Yale University Press, New Haven
Streilein KE (1982a) Ecology of small mammals in the semiarid Brazilian caatinga. I. Climate and faunal composition. Annals of Carnegie Museum 51:79–107
Streilein KE (1982b) The ecology of small mammals in the semiarid Brazilian caatinga. III. Reproductive biology and population ecology. Annals of Carnegie Museum 51:251–269
Temme M (1981) Reproductive parameters of the Polynesian rat *Rattus exulans* in the northern Marshall Islands. Z. Angeu. Zool. 68:315–338
Vivas AM (1986) Population biology of *Sigmodon alstoni* (Rodentia:Cricetidae) in the Venezuelan Llanos. Revista Chilena de Historia Natural 59:179–191

20

Reproductive Biology of South American Manatees

Miriam Marmontel, Daniel K. Odell, and John E. Reynolds III

Introduction

The order Sirenia includes the only mammals that are both fully aquatic and herbivorous. The members of this order possess adaptations for both their habitat and their diet. For example, the large fusiform body is almost hairless; external appendages such as the hind limbs and external ear pinnae are absent; the pectoral flippers are reduced and paddlelike; and the tail is modified into a large paddle (manatees) or deeply notched flukes (dugongs) to facilitate locomotion through an aquatic medium. Pachyostotic (swollen) and osteosclerotic (very hard and solid) bones play a role in buoyancy regulation. A peculiar mode of tooth replacement (with an indeterminate number of molars in manatees), presence of horny plates in the mouth, ventral deflection of the rostrum (most pronounced in *Dugong*), the size, proportions, and structure of the digestive tract, and hindgut fermentation are all adaptations for herbivory. Fossil evidence suggests that sirenians were derived from primitive terrestrial herbivores early in the Tertiary (Reinhart 1971), and biochemical analysis of proteins points to a common origin with elephants and hyraxes (Kleinschmidt et al. 1986).

The present distribution of the Sirenia consists of shallow tropical and subtropical waters. The exclusively marine family Dugongidae is represented by one extant species (the dugong, *Dugong dugon*), found in the Indian and western Pacific Oceans and adjoining smaller bodies of water. Excessive exploitation by Russian hunters exterminated another recent dugongid, the Steller's sea cow *Hydrodamalis gigas*, in 1768, less than three decades after its discovery in the Bering Sea (Bertram and Bertram 1973).

The family Trichechidae contains one genus, *Trichechus*. The West African manatee *T. senegalensis* occurs in some coastal, estuarine, and freshwater systems of tropical West Africa. *Trichechus manatus* (the West Indian manatee) is further subdivided into two subspecies (Domning and Hayek 1986), (1) *T. m. latirostris* (the Florida manatee, in the southeastern United States) and (2) *T. m. manatus* (the Antillean manatee, from Mexico to South America, through Central America and the West Indies). The West Indian manatee also inhabits coastal, riverine, and estuarine habitats. Temperature, water depth, availability of seagrasses or other vegetation for food, and access to freshwater are important determinants of manatee distribution. In South America, trichechids are represented by two species, (1) the Antillean manatee and (2) the Amazonian manatee (*Trichechus inunguis*). *Trichechus manatus manatus* occurs in coastal waters from Colombia through the state of Bahia (Brazil), including the lower Orinoco basin and Lake Maracaibo (Mondolfi 1974; Domning 1982a; O'Shea et al. 1988). The Amazonian manatee is the only sirenian confined to freshwater, occupying lakes, lagoons, and waterways of the Amazon drainage basin (Thornback and Jenkins 1982). It is the smallest of the manatees and may be distinguished from its coastal relative by several characteristics, including the lack of nails on its pectoral fins, smoother skin, a less downwardly-deflected rostrum, and usually the presence of distinctive pink or white patches on its chest and belly.

Presently, *T. m. latirostris* and *D. dugon* are the best studied sirenians. Particular data from these species must sometimes be applied (with great care!) to the other species for which those same data may be lacking. This is valid to an extent with the current topic, reproductive biology.

Manatees may have evolved in South America's Miocene coastal rivers and lagoons, contemporarily with marine dugongids. The Miocene Andean orogeny promoted the creation of a temporary Amazon lake, where *T. inunguis* may have evolved in isolation from coastal relatives. Orogenic and eustatic Miocene-Pliocene changes promoted increased erosion and silt runoff, favoring the appearance of true grasses. In the Pliocene, the manatees, with smaller and more numerous teeth, were better adapted to the prevalent seagrasses and outcompeted dugongids in the New World. Manatees dispersing to Africa probably gave rise to *T. senegalensis* (Domning 1982b).

Manatee grazing as a form of weed and mosquito control has been investigated in semi-captive conditions (Allsopp 1960; MacLaren 1967), and manatees have been suggested as a commercial source of protein (Bertram and Bertram 1968), but present population levels, their slow reproduction, and endangered status limit either use. The greatest ecological contribution by sirenians is perhaps recycling of limiting nutrients in ecosystems, stimulating primary productivity (Lomolino 1977). More than half of what manatees eat returns to the ecosystem in the form of feces and urine (Best 1984). Increased phytoplankton production resulting from the loading of nitrogen (15% of fecal dry matter; Best 1984) should affect fish production as well (Melack 1976).

The large size, tasty meat, and abundant oil have made sirenians favorite targets of coastal native peoples and explorers as well (Bertram and Bertram 1973; Mondolfi 1974; Peterson 1974, Timm et al. 1986; O'Shea et al. 1988). Hunting, although illegal, continues to occur in many areas, but on a local scale. Manatees, both coastal and freshwater, were commercially hunted for their meat for export to the West Indies during the 16th and 17th centuries in the Caribbean, Guianas, and Brazil (Bertram and Bertram 1973, Domning 1982a). The Amazonian manatee was the subject of large-scale commercial exploitation spanning almost two centuries, which involved animals of all ages and sizes, including pregnant females (Domning 1982a). In Brazil, the main product prior to 1915 was "mixira" (fried meat preserved in its own fat); the period of 1935–53 saw an increased demand for the tough, durable hides, with 7000 to 10,000 Amazonian manatees killed annually. With the advent of synthetic leather substitutes, fresh meat became more important from 1954–73, when legal protection was instituted (Domning 1982a). More recently, sirenians have become threatened by incidental mortality (killed by collisions with boats and entanglement in fishing nets), through habitat modification, and by loss or degradation of seagrass beds and floating meadows. As a result of hunting pressure, habitat loss, and incidental take, the World Conservation Union, formerly the International Union for the Conservation of Nature, considers all species of sirenians vulnerable to extinction (Thornback and Jenkins 1982). A general review of sirenian biology, exploitation, and management appears in Reynolds and Odell (1991).

The uniqueness of the sirenian niche defines certain "K" life history correlates that make all species susceptible to over-exploitation. Sirenians have a long life span, high adult survival rates, low fecundity, and delayed reproductive maturity. An understanding of the reproductive biology and life history of these animals is fundamental to predict trends in the population, understand effects of age-specific mortality and habitat loss/degradation, and implement future conservation programs.

A recently developed technique of age determination in Florida manatees, based on growth-layer groups in the periotic bone (Marmontel et al. 1990), represents an important tool in manatee population studies. Among other aspects, it will allow researchers to analyze age-specific patterns of mortality and fecundity which are fundamental aspects of population dynamics. The same procedures derived for Florida manatees are potentially appropriate for other species of manatees and should provide valuable information on their life histories as well.

Reproductive Biology

ANATOMY

Sirenians present little sexual dimorphism. Asymptotic length of females may be slightly larger than that of males (Hartman 1979; Marsh 1980; Marmontel unpub. ms.). The main external difference between the sexes is the position of the reproductive openings. In females, the genital opening is

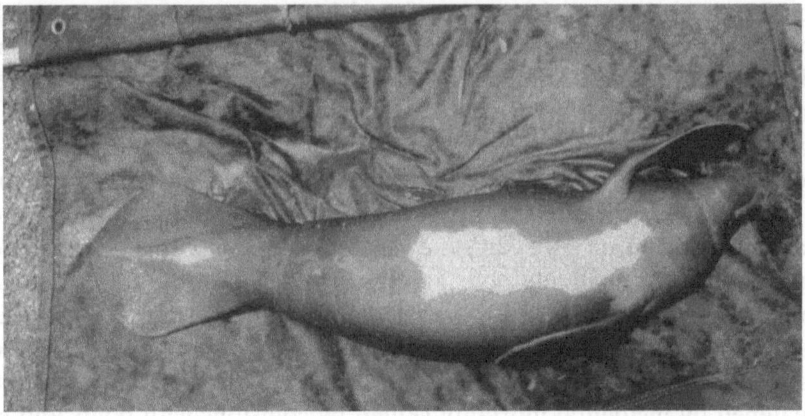

Figure 20.1. Ventral view of a female Amazonian manatee, showing the genital opening at the posterior end of the body, just craniad to the anus. Note the broad, paddle-shaped tail and the characteristic chest patch. (Photo: Daryl P. Domning)

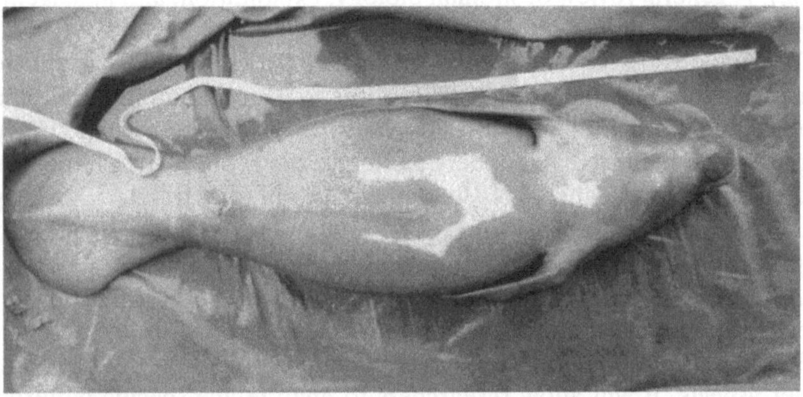

Figure 20.2. Ventral view of a male Amazonian manatee showing the genital opening in close proximity to the umbilicus and distant from the anus. (Photo: Daryl P. Domning)

located immediately anterior to the anus (Fig. 20.1), whereas in males it is slightly posterior to the umbilical scar but well anterior to the anus (Fig. 20.2). The two mammae are axillary, with one located under each flipper. Incisors are lacking in *Trichechus*, but are retained as upper tusks in dugongs, where they are more developed in males (Husar 1978a).

The reproductive tracts of both males and females are located in the caudal portion of the abdominal cavity (Figs. 20.3 to 20.6). The longitudinally plicated uterus is bicornuate, with one cervix and a small uterine

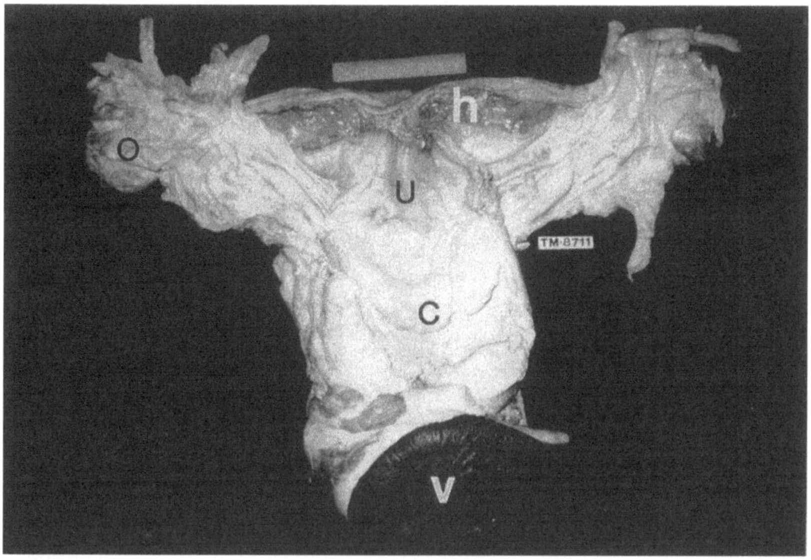

FIGURE 20.3. Dorsal view of reproductive organs of a 329-cm female Florida manatee. v = vulva; c = cervix; u = uterus; h = uterine horn; o = ovary. The uterus is distended due to recent parturition. The animal was lactating. (Photo: Miriam Marmontel)

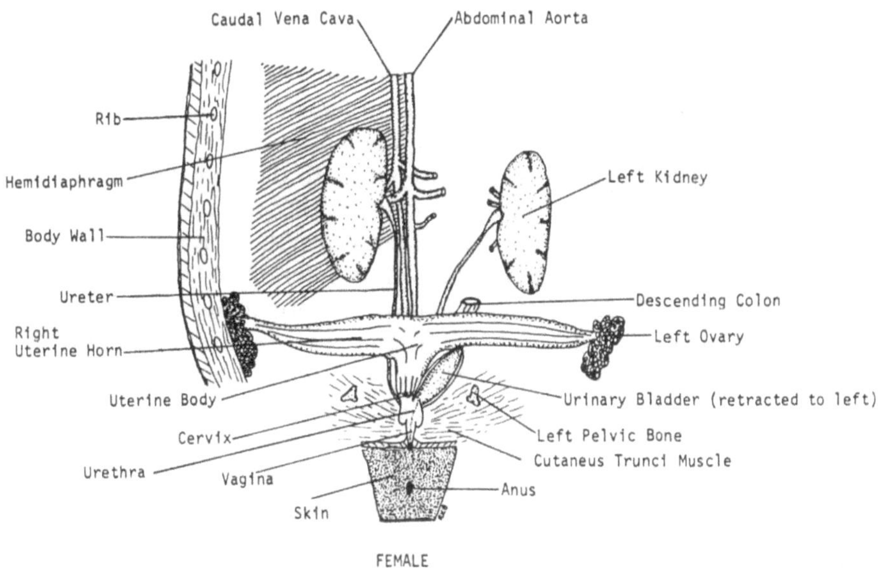

FIGURE 20.4. Diagram of female reproductive tract, in ventral view. (Reproduced with permission from Bonde et al. 1983)

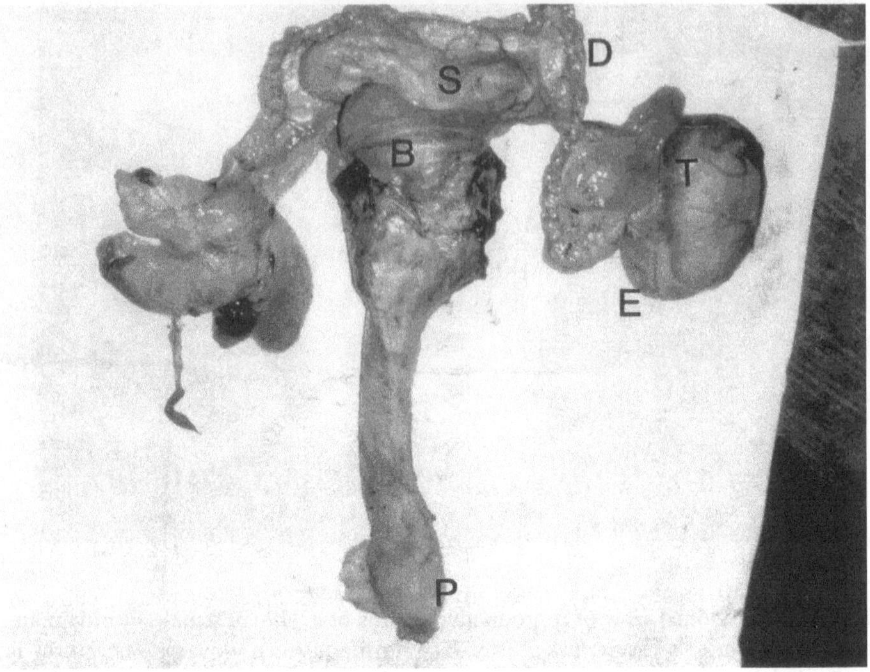

FIGURE 20.5. Dorsal view of reproductive organs from a male 288-cm Florida manatee. P = penis; B = urinary bladder; S = seminal vesicle; D = ductus deferens; E = epididymis; T = testis. (Photo: Sea World of Florida—M. T. Walsh)

body. Placental scars, marking previous fetal attachment sites, provide reproductive rate data covering the life span of female dugongs (Marsh et al. 1984a), but do not persist for life in Florida manatees (Marmontel 1988). The broad, flattened ovaries lie postero-lateral to the kidneys and are enclosed in a peritoneal pouch. Typical parous ovaries present large numbers of large Graafian follicles, corpora lutea, and corpora albicantia.

The ovoid testes are internal, located ventro-laterally to the kidneys, and attached to the peritoneum. The seminal vesicles are bilateral and dorsal to the urinary bladder (Bonde et al. 1983). The prostate is nonglandular and consists of erectile muscular tissue (Caldwell and Caldwell 1985). The penis generally remains internal, but is extruded when the animal becomes stimulated.

Details of the histology of sirenian reproductive organs are given in Marsh et al. (1984a, 1984b) and Marmontel (1988).

PHYSIOLOGY

The age at which Florida manatees reach sexual maturity is not clearly defined (see Table 20.1). Field observations suggest that it may occur

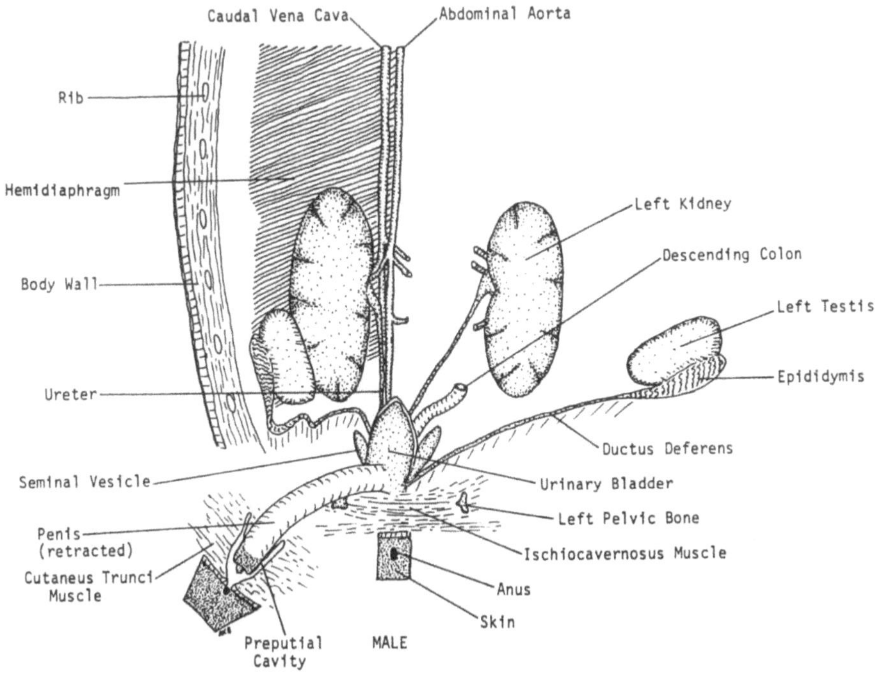

FIGURE 20.6. Diagram of male reproductive tract, in ventral view. (Reproduced with permission from Bonde et al. 1983)

around five years (Hartman 1979; Rathbun and Powell 1982). Analyses of growth curves and observations of captive animals have suggested that the event is delayed until eight to ten years of age (Odell et al. 1981). The smallest female known to have calved measured 260 cm in length (Odell 1977), and a 277-cm female presented two placental scars on its uterine horn, but it is likely that most females mature around 270 cm (Marmontel 1988).

Among mammals, a number of infertile mating episodes is common. Both dugongs and elephants may undergo several sterile estrous periods prior to successful conception (Short 1966; Marsh et al. 1984c). Hartman (1979) observed estrous herds involving cows with calves and, assuming that ovulation does not take place during lactation, suggested that infertile estrous periods may also happen in manatees. Polyestry has been histologically demonstrated in dugongs (Marsh et al. 1984a), but remains to be demonstrated in manatees.

Histological findings suggest that manatees, similar to dugongs, are polyovular (Marsh et al. 1984a; Marmontel 1988). Superovulation is rare in mammals (Weir and Rowlands 1973) and, considering that sirenians are uniparous (Table 20.1), the production of multiple corpora lutea (Fig.

TABLE 20.1. Life history characteristics of manatees and dugongs.

Species	Adult size	Age/length sex maturity	Gestation length	Litter size	Neonatal size	Length calf dependency	Calving interval
Trichechus inunguis	Max. 280 cm (Husar 1977) Up to 300 kg (Best 1983)		~1 year (Best 1984)	1 (Husar 1977; Best 1984; Timm et al. 1986)	85–105 cm 10–15 kg (Best 1982) 60–80 cm (Timm et al. 1986)	1–2 years?	2 years?
Trichechus manatus	270–390 cm max. 1600 kg	8–10 years (Odell et al. 1981); 5 years (Hartman 1979; Rathbun and Powell 1982) 250–270 cm (Hartman 1979), 260cm ♀ (Odell et al. 1981), 275 cm ♀ (Odell 1977), 270 cm ♀ (Marmontel 1988)	12–14 months (Hartman 1979; Rathbun and Powell 1982)	1 (Hartman 1979); Twins reported (Moore 1951; Hartman 1979; Rathbun and Powell 1982)	120–140 cm 30 kg (Odell 1982)	1–2 years (Hartman 1979)	2 (Bengtson 1981) 3 (Reynolds 1981) 3–3.5 (Hartman 1979) 5 (Brownell and Ralls 1981)
Trichechus senegalensis	300–400 cm <500 kg (Husar 1978b)	?	~1 year ?	1 (Beal 1939)	100 cm (Cadenat 1957)	1–2 years?	2+ years?
Dugong dugon	240–305 cm (Heinsohn 1972) 230–908 kg (Mani 1960 in Husar 1978a)	9–10 (♀, ♂) (Marsh et al. 1984c)	13–14 months (Marsh 1988)	1 (Marsh et al. 1984c)	100–130 cm 20–35 kg (Marsh et al. 1984c)	1.5 year	3–7 years (Marsh et al. 1984c)

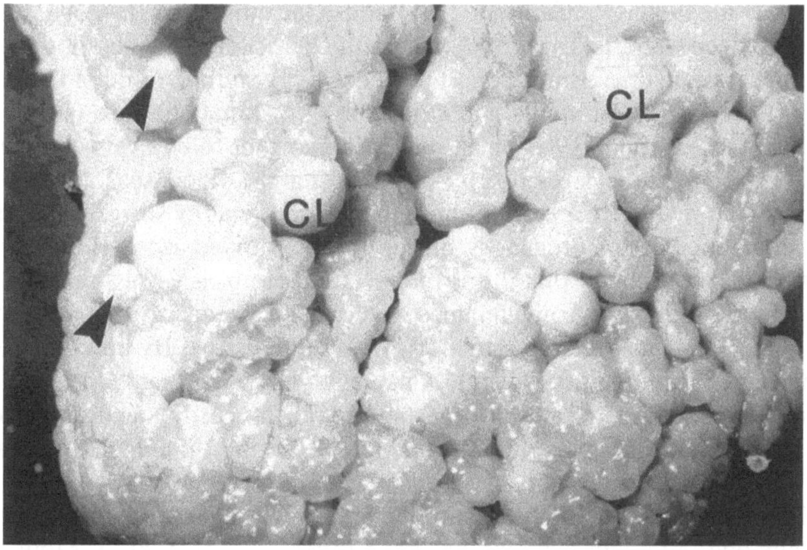

FIGURE 20.7. Ovary of a 329-cm female Florida manatee, showing multiple corpora lutea (CL). Arrows show corpora albicantia. Translucid rounded structures are follicles in various stages of development. (Photo: Miriam Marmontel)

20.7) probably has a hormonal implication. Corpora lutea are endocrine structures that produce progesterone and prepare the body for gestation and parturition. If the sirenian corpus luteum produces very small amounts of progesterone, it may be necessary to accumulate a certain number of them before maximal endometrial proliferation occurs. A minimum of 21 corpora lutea seems necessary to allow pregnancy in Florida manatees (Marmontel 1988).

The gestation period in manatees is approximately one year (Table 20.1), and lactation may last one to two years (Hartman 1979). Therefore, manatee cows typically calve no more often than every two to three years in the wild and may do so every five years (Brownell and Ralls 1981). The minimum calving interval was two years for two radio-tagged females in the St. Johns River (Bengtson 1981). Both field and captive observations attest that cows that lose their calves may cycle shortly afterwards. Hartman (1979) observed behaviors suggesting that a cow came into estrus two weeks after the death of her young calf. In mammals, follicular activity is limited, but not necessarily suspended, during pregnancy because the secretion of progesterone by corpora lutea prevents ovulation and causes atresia. It is likely that when the suppressing effect of the corpora lutea and the placenta is removed after parturition, a new follicular cycle may occur (Perrin and Donovan 1984).

During the dry seasons, manatees in the Orinoco River (Mondolfi 1974;

Timm et al. 1986; O'Shea et al. 1988) and Amazon basin (Best 1983) migrate to deep water lakes (poços), where the lack of living vegetation leads to either fasting or consumption of dead organic material found at the bottom (Best 1983). Fat stores and extremely low metabolic rate (Gallivan and Best 1980; Irvine 1983) must play an important role in carrying manatees through these stressful periods. During the flood season, female Amazonian manatees are able to attain sufficient energy stores to sustain the energetic cost of pregnancy during the next dry season (Best 1984). Best (1982) speculated that if nutritional status controls estrus, in years of poor food availability, some females would avoid the energy demands of a long pregnancy and lactation by not coming into estrus. By breeding seasonally (see section on Behavior), Amazonian manatees match the energetically demanding latter third of gestation and the costly lactation period with the period of highest food availability, allowing the female to regain her physiological condition.

The sex ratio at birth is 1:1 (Rathbun and Powell 1982), and litter size is usually one, although twinning has been reported (Table 20.1). Mondolfi (1974) reported a very young nursing calf (*T. manatus*) captured in the Orinoco River in 1973 that measured 0.8 m and weighed 40 kg; this size compares to newborn *T. manatus* calves in Florida, which range in size from 120 to 140 cm long and weigh about 30 kg (Odell 1982). Newborn Amazonian manatees measure 60 to 105 cm and weigh 10 to 15 kg. The gestation period is 12 to 14 months in all sirenians (Table 20.1).

Manatee milk is high in protein (7.4%) and lipids (12.7%) and low in carbohydrates (0.3%). No lactose has been detected (Bachman and Irvine 1979). Orphaned Amazonian manatee calves have been raised in captivity at INPA (Brazil's National Institute for Amazonian Research) since 1974 on artificial milk diets, consisting of a mixture of milk powder and vegetable oil, supplemented by a multivitamin complex (Colares et al. 1990). Calves consuming this diet have gained approximately 1 to 2 kg per week and grown 1.4 cm per week (Best et al. 1983).

Reproductive senescence, generally rare in mammals, has been suggested for dugongs (Marsh et al. 1984a), but may not occur in *T. m. latirostris* (Marmontel 1988). Two manatees (a pair) that have been in captivity at Miami Seaquarium since 1957 and 1958 are still reproductively active.

Male fertility is more difficult to assess in the field, and relevant information is derived primarily from histological observations. In Florida manatees, testes and epididymides have been examined histologically (Hernandez, Reynolds, and Marsh, unpubl.), using specimens from animals recovered dead through a carcass salvage network. Spermatogenesis occurred in a manatee as small as 237 cm, which was found to belong to "age class 3" (between 3 and 4 years old). Almost 50% of the male manatees measuring 240 to 270 cm long had mature sperm when death was between March and November; only 1 out of 8 animals of this length that were recovered in December, January, or February had mature sperm present. At a body length of approximately 270 cm and longer, manatee testis weight rises rapidly (Odell et al. 1981). Male manatees longer than 270 cm found dead from March through November had mature sperm in the testis or epididymis 95% of the time ($n = 21$); sperm production was, again, lower during winter (December through February), with only two of nine (22%) animals over 270 cm long having mature sperm at that time. A number of manatees had abundant sperm in the epididymis, without having sperm in the seminiferous tubules.

Manatees in Florida suffer stresses in winter. They sometimes die due to the effects of cold weather; to try to avoid cold stress, they migrate to several natural and artificial warm water discharges along the coasts of Florida and southern Georgia. Even in the discharge areas, extremely cold weather can stress the manatees thermally. In the vicinity of some discharges (e.g., some power plants), food is scarce. If the weather is particularly cold, manatees remain in the vicinity of the plants to avoid cold stress, even though this means fasting for some time. James Reid (U.S. Fish and Wildlife Service, pers. comm. 1990) has reported emaciated manatees at certain warm water discharge areas, and manatees dying during extremely cold winters showed fat atrophy, emaciation, and absence of food in the gastrointestinal tract (O'Shea et al. 1985). It is not unlikely that the stresses imposed by migrating, being exposed to cold temperatures, and fasting may, especially in very cold winters, limit spermatogenesis.

In more tropical regions, outside of Florida, the main environmental stress affecting manatee movements is the wet-dry cycle, rather than temperature itself. Possibly stresses during the dry season, especially due to lack of food for Amazonian manatees, could stop spermatogenesis at that time.

BEHAVIOR

A solitary or transient social system predominates among manatees. There is no selection for monogamy and the result is a "promiscuous" mating system. Most of our knowledge on manatee reproductive behavior derives from direct observations of live animals in warm water springs in Florida (Hartman 1979). Except for the relationship between a cow and her calf, most associations appear temporary (Reynolds 1981). Cows and calves maintain contact through vocal communication (Hartman 1979). Florida manatees travel long distances within seasons, and males systematically patrol and search overlapping areas (Bengtson 1981). An estrous female attracts temporary associations of 5 to 17 adult and juvenile courting males (Hartman 1979; Fig. 20.8). The mating herd may escort the female for one week to a month (Hartman 1979) and for distances of up to 160 km (Bengtson 1981). Males mouth and embrace the female and roll upside down in an effort to approach her abdomen (Hartman 1979). Generally the female flees from the pursuit, sometimes even stranding herself in an effort to

FIGURE 20.8. Florida manatee mating herd at the warm water discharge of a power plant. (Photo: U.S. Fish and Wildlife Service—Sirenia Project)

discourage male approach (Hartman 1979). When the female becomes receptive, she copulates quickly and may choose more than one male. It is possible that males identify estrous females through chemoreception. Although a Jacobsen's organ has never been found, the manatee tongue contains taste buds (Barrett 1979); manatees rub body parts on available objects and animals often mouth each other's genital and anal areas (Hartman 1979). External labial swelling is observable in estrous female Florida manatees.

In Amazonia, fishermen refer to "cavalgação" as the period when groups of male manatees are said to pursue one or more females (Pereira 1944). Pereira (1944) believed there was no certain reproductive season in Amazonian manatees, with females giving birth both in winter and summer. He did notice more intense sexual activity during high water level, however, and associated it with water depth and temperature difference from summer.

Pereira's (1944) ideas about aseasonal breeding are not considered accurate today, although his thoughts regarding the relationship between water level and extent of mating are valid. The Amazon basin is subject to an annual cycle of high and low water levels in response to seasonal rains. Rising river levels stimulate new growth and greater productivity of aquatic and semi-aquatic plants. Apparently in response to the changes in nutrient content and food availability, *T. inunguis* became a seasonal breeder (Best

FIGURE 20.9. Calf nursing from a Florida female manatee, in Crystal River, Florida. (Photo: U.S. Fish and Wildlife Service—Sirenia Project)

1982). Calves in Brazilian Amazonia, at least, are born between December and July, but the peak of parturition occurs during the months when the water level is rising and plant productivity is highest (February to May). In Ecuador pregnant females are noticed during the second half of the year and young are said to be born in January and June (Timm et al. 1986). In Florida, where a parturition peak occurs in spring and summer (Hartman 1979), seasonal breeding may have the additional advantage of limiting calves from being exposed to potentially dangerous winter temperatures (Hartman 1979). Both wild and captive Florida manatee calves may start nibbling on plants a few weeks after birth (Moore 1957; Hartman 1979; Odell 1982), so synchronizing calving with the optimal nutritional conditions supplements milk intake with abundant new shoots. In the Brazilian northeast, cows accompanied by young calves have been observed from November to March (Borobia and Lodi 1990; K. Silva pers. comm. 1990).

Pregnant female manatees seek shallow waters and/or sheltered areas to give birth (Hartman 1979; Bengtson 1981). Calves are born both tail and head first. Adoption of orphaned or abandoned calves has been observed in the field (Hartman 1979). Suckling takes place underwater, with the female in a horizontal position and the calf lying at an angle to the cow's body (Moore 1956, 1957; Fig. 20.9).

Implications for Conservation

Despite legal protection in Florida and all South and Central American countries (Domning 1982b; Thornback and Jenkins 1982; Lefebvre et al. 1989) *T. manatus* and *T. inunguis* are still subject to poaching for subsistence and local marketing (Colares et al. 1990; K. Silva pers. comm. 1990). Manatees are occasionally caught incidentally in commercial fishing gear (drift and gill nets; Thornback and Jenkins 1982; O'Shea et al. 1988; Oliveira et al. 1990), and the problem is likely to increase as commercial fisheries develop and expand in all areas. Boat-related fatalities in South America (the major human-related cause of manatee mortality in Florida) have recently been recorded in Venezuela and Colombia (O'Shea et al. 1988; M. Prieto pers. comm. 1990).

Sirenian populations grow at a very slow rate, on the order of 5% annually (Marsh et al. 1984c; Packard 1985). Even though population estimates are nonexistent for South American manatee populations, the numbers are likely to be below carrying capacity in Amazonia and very reduced along the coast (Domning 1982a; K. Silva pers. comm. 1990). The low reproductive rates shown by manatees preclude a rapid recovery and exploited populations could take years to recover, especially if mortality rates were to increase.

With continued human population growth in coastal environments and as the Amazon basin becomes altered by human activities, manatees will also suffer from habitat destruction (Caldwell and Caldwell 1985). Threats include extensive land reclamation, construction of dams and dikes, logging of mangroves, dredging projects, toxicants in the water due to mining operations, draining of wetlands for agricultural uses (Mondolfi 1974; O'Shea et al. 1988; K. Silva pers. comm. 1990), potential oil spills, and increased barge traffic associated with oil exploration, industrial wastes, and sewage discharge (Mondolfi 1974). Large-scale deforestation will increase erosion in the Amazon basin (Timm et al. 1986); a potential reduction in the annual rainfall might extend the dry season and increase natural mortality through starvation (Best 1982). Habitat destruction/alteration that reduces human protein sources could increase illegal harvest of manatees; so could military operations during which soldiers harvest manatees for food. Habitat modifications that lead to reduced food, reduced freshwater availability, or reduced warm water availability could impact manatee reproduction by impairing spermatogenesis, if not oogenesis, for all species in all areas.

Summary and Conclusions

As suggested by Reynolds and Gluckman (1988), no single solution will preserve manatees. A combination of habitat protection, acquisition of critical habitat and subsequent creation of manatee sanctuaries or reserves, law enforcement, scientific research, and public education is needed in all countries where sirenians exist. All species' ranges cross political boundaries, requiring joint efforts for effective conservation.

Although they are long-lived, manatees have a low reproductive capacity. Age of sexual maturation is delayed, litter size is small, and calving intervals are long. In the absence of male care, there is high maternal investment in each calf, reflected in the long periods of gestation, suckling, and close association between mother and young. In many parts of their range, manatees have been reduced by hunting and incidental take. If manatee numbers are to be maintained, intensive efforts must be made to protect the animals as well as the habitats critical to their survival.

Acknowledgments

We thank Dr. Thomas J. O'Shea of U.S. Fish and Wildlife Service for review of the manuscript. This is a contribution from the Program for Studies in Tropical Conservation. This is Sea World of Florida Technical Contribution No. 9005-F.

References

Allsopp WHL (1960) The manatee: Ecology and use for weed control. Nature 188:762

Bachman KC, Irvine AB (1979) Composition of milk from the Florida manatee, *Trichechus manatus latirostris*. Comp. Biochem. Physiol. 62A:873–878

Barrett S (1979) Taste receptors in the West Indian manatee, *Trichechus manatus*. Abstr. 3rd Biennial Conf. Bio. Mar. Mamm., Seattle, Washington, 7–11 October 1979

Beal WP (1939) The manatee as a food animal. Nigerian Field 8:124–126

Bengtson JL (1981) Ecology of manatees (*Trichechus manatus*) in the St. Johns River, Florida. Ph.D. Dissertation, University of Minnesota, Minneapolis, 126 pp

Bertram GCL, Bertram CKR (1968) The Sirenia as aquatic meat-producing herbivores. Symp. Zool. Soc. Lond. 21:385–391

Bertram GCL, Bertram CKR (1973) The modem Sirenia: Their distribution and status. Biol. J. Linn. Soc. 4:297–338

Best RC (1982) Seasonal breeding in the Amazonian manatee, *Trichechus inunguis* (Mammalia:Sirenia). Biotropica 14(1):76–78

Best RC (1983) Apparent dry-season fasting in Amazonian manatees (Mammalia:Sirenia). Biotropica 15(10):61–64

Best RC (1984) The aquatic mammals and reptiles of the Amazon. In Sioli H (ed.). The Amazon—Limnology and Landscape Ecology of a Mighty Tropical River and Its Basin. Dr. W. Junk Publ., Dordrecht Monographieae Biologicae, Vol. 56, pp. 371–412

Best RC, Ribeiro GA, Yamakoshi M, da Silva VMF (1983) Artificial feeding for unweaned Amazonian manatees. Intl. Zoo. Yrbk. 22:263–267

Bonde RT, O'Shea TJ, Beck CA (1983) Manual of procedures for the salvage and necropsy of carcasses of the West Indian manatee (*Trichechus manatus*). National Technical Information Service Publication PB83-255273, Springfield, Virginia

Borobia H, Lodi L (1990) O peixe-boi marinho (*Trichechus manatus*) no nordeste brasileiro: occurência e conservação. Abstract, IV Workshop on South American Aquatic Mammals, Valdívia, Chile, 12–15 November

Brownell RL Jr, Ralls K (eds.) (1981) The West Indian Manatee in Florida. Proceedings of a workshop held in Orlando, Florida, 27–29 March 1978. Florida Department of Natural Resources, Tallahassee, Florida

Cadenat J (1957) Observations de cétaces, siréniens, cheloniens et sauriens en 1955–1956. Bull. Inst. F. Afr. Noire 19A:1358–1383

Caldwell DK, Caldwell MC (1985) Manatees—*Trichechus manatus* Linnaeus, 1758; *Trichechus senegalensis* Link, 1795 and *Trichechus inunguis* (Natterer, 1883). In Ridgway SH, Harrison RJ (eds.). Handbook of Marine Mammals, Vol. 3: The Sirenians and Baleen Whales. Academic Press, London. pp. 33–66

Colares FAP, Colares IG, Rosas FCW, Colares EP (1990) Amazonian manatees (*Trichechus inunguis*): A 15 year long-term study. Proc. Am. Assoc. Zoo Vets. pp. 43–47

Domning DP (1982a) Commercial exploitation of manatees *Trichechus* in Brazil c. 1785–1973. Biol. Cons. 22:101–126

Domning DP (1982b) Evolution of manatees: A speculative history. J. Paleontol. 56(3):599–619

Domning DP, Hayek L-AC (1986) Interspecific and intraspecific morphological variation in manatees (Sirenia:*Trichechus*). Mar. Mamm. Sci. 2(2):87–144

Gallivan G, Best R (1980) Metabolism and respiration of the Amazonian manatee (*Trichechus inunguis*). Physiol. Zool. 53(3):245–253

Hartman DS (1979) Ecology and behavior of the manatee (*Trichechus manatus*) in Florida. The American Society of Mammalogists, Special Publication No. 5

Heinsohn GE (1972) A study of dugongs (*Dugong dugong*) in Northern Queensland, Australia. Biol. Conserv. 4(3):205–213

Husar SL (1977) *Trichechus inunguis*. Mammalian Species 72:1–4

Husar SL (1978a) *Dugong dugon*. Mammalian Species 88:1–7

Husar SL (1978b) *Trichechus senegalensis*. Mammalian Species 89:1–3

Irvine AB (1983) Manatee metabolism and its influence on distribution in Florida. Biol. Conserv. 25:315–334

Kleinschmidt T, Czelusniak J, Goodman M, Braunitzer G (1986) Paenungulata: A comparison of the hemoglobin sequences from elephant, hyrax and manatee. Mol. Biol. Evol. 3(5):427–435

Lefebvre LW, O'Shea TJ, Rathbun GB, Best RC (1989) Distribution, status, and biogeography of the West Indian manatee. In Woods CA (ed.). Biogeography of the West Indies; Past, Present, and Future. Sandhill Crane Press, Gainesville, Florida. pp. 567–610

Lomolino MV (1977) The ecological role of the Florida manatee (*Trichechus manatus latirostris*) in water hyacinth-dominated ecosystems. M.Sc. Thesis, University of Florida, Gainesville

MacLaren JP (1967) Manatees as a naturalistic biological mosquito control method. Mosquito News 27(3):387–393

Mani SB (1960) Occurrence of the sea cow, *Halicore dugong* (Erxl.) off the Saurashtra Coast. J. Bombay Nat. Hist. Soc. 57:216–217

Marmontel M (1988) The reproductive anatomy of the female manatee *Trichechus manatus latirostris* (Linnaeus 1758) based on gross and histologic observations. M.Sc. Thesis, University of Miami, Coral Gables, Florida

Marmontel M, O'Shea TJ, Humphrey SR (1990) An evaluation of bone growth-layer counts as an age-determination technique in Florida manatees. National Technical Information Service PB-91-103564, Springfield, Virginia

Marsh H (1980) Age determination of the dugong (*Dugong dugon* (Müller)) and its biological implications. Int. Whal. Comm. Rep. (Spec. Issue 3):181–201

Marsh H (1988) An ecological basis for dugong conservation in Australia. In Augee ML (ed.). Marine Mammals of Australia—Field biology and captive management. Royal Zool. Soc. New South Wales, Mosman, NSW, Australia, pp. 9–21

Marsh H, Heinsohn GE, Channells PW (1984a) Changes in the ovaries and uterus of the dugong, *Dugong dugon* (Sirenia:Dugongidae), with age and reproductive activity. Aust. J. Zool. 32:743–765

Marsh H, Heinsohn GE, Glover TD (1984b) Changes in the male reproductive organs of the dugong, *Dugong dugon* (Sirenia:Dugongidae) with age and reproductive activity. Aust. J. Zool. 32:721–741

Marsh H, Heinsohn GE, Marsh LM (1984c) Breeding cycle, life history and population dynamics of the dugong, *Dugong dugon* (Sirenia:Dugongidae). Aust. J. Zool. 32:767–788

Melack JM (1976) Primary productivity and fish yields in tropical lakes. Trans. Am. Fish. Soc. 105:575–580

Mondolfi E (1974) Taxonomy, distribution and status of the manatee in Venezuela. Mem. Soc. Ci. Nat. La Salle 34(97):5–23

Moore JC (1951) The range of the Florida manatee. Quart. J. Fla. Acad. Sci. 14(1):1–19

Moore JC (1956) Observations of manatees in aggregations. Am. Mus. Novit. 1811:1–24

Moore JC (1957) Newborn young of a captive manatee. J. Mammal. 38:137–138

Odell DK (1977) Age determination and biology of the manatee. U.S. Fish and Wildlife Service, Natl. Fish Wildl. Lab. Rep., Contract 14-16-0008-930

Odell DK (1982) West Indian manatee, *Trichechus manatus*. In Chapman JA, Feldhamer GA (eds.). Wild Mammals of North America: Biology, Management and Economics. The Johns Hopkins Univ. Press, Baltimore. pp. 828–837

Odell DK, Forrester DJ, Asper ED (1981) A preliminary analysis of organ weights and sexual maturity in the West Indian manatee (*Trichechus manatus*). In Brownell RL Jr, Ralls K (eds.). The West Indian Manatee in Florida. Florida Department of Natural Resources, Tallahassee. pp. 52–65

Oliveira EM, Silva KG, Soavinski RJ, Pinto R (1990) Mortalidade do peixe-boi marinho *Trichechus manatus* na costa do nordeste do Brasil. Abstract, IV Workshop on South American Aquatic Mammals, Valdívia, Chile, 12–15 November

O'Shea, TJ, Beck CA, Bonde RK, Kochman HI, Odell DK (1985) An analysis of manatee mortality patterns in Florida, 1976–81. J. Wildl. Manage. 49(1):1–11

O'Shea TJ, Correa-Viana M, Ludlow ME, Robinson JG (1988) Distribution, status, and traditional significance of the West Indian manatee *Trichechus manatus* in Venezuela. Biol. Conserv. 46:281–301

Packard JM (1985) Preliminary assessment of uncertainty involved in modeling manatee population dynamics. Manatee Population Research Rept. No. 9. Tech.

Rep. No. 8–9. Fla. Coop. Fish and Wildl. Res. Unit, University Florida, Gainesville, Florida

Pereira MN (1944) O peixe-boi da Amazônia. Bol. Minist. Agric. (Rio de Janeiro) 33:21–95

Perrin WP, Donovan GP (1984) Report of the workshop. In Perrin WP, Brownell RL Jr, DeMaster DP (eds.). Reproduction in Whales, Dolphins and Porpoises. Intl. Whaling Comm. Spec. Issue No. 6, Cambridge, England. pp. 1–24

Peterson SA (1974) Man's relationship with the Florida manatee *Trichechus manatus latirostris* (Harlan): A historical perspective. MS Thesis, University of Michigan, Ann Arbor

Rathbun G, Powell J (1982) Reproduction in the manatee (*Trichechus manatus*). Abstr. 62nd Ann. Meeting, Am. Soc. Mammal., Snowbird, Utah

Reinhart RH (1971) Fossil Sirenia of Florida. The Plaster Jacket (Florida State Museum, University of Florida) 15:1–11

Reynolds JE III (1981) Aspects of the social behavior and herd structure of a semi-isolated colony of West Indian manatees, *Trichechus manatus*. Mammalia 45(4):431–451

Reynolds JE III, Gluckman CJ (1988) Protection of West Indian manatees (*Trichechus manatus*) in Florida. National Technical Information Service PB-222922, Springfield, Virginia

Reynolds JE III, Odell DK (1991) Manatees and Dugongs. Facts on File Press, New York 192 pp

Short RV (1966) Oestrous behavior, ovulation and the formation of the corpus luteum in the African elephant, *Loxodonta africana*. East Afr. Wildl. J. 4:56–68

Thornback J, Jenkins M (1982) The IUCN Red Data Book Part 1: Threatened Mammalian Taxa of the Americas and Australasian Zoogeographic Region (excluding Cetacea). Int. Union Cons. Nature Nat. Res., Gland, Switzerland

Timm RM, Albuja VL, Clauson BL (1986) Ecology, distribution, harvest and conservation of the Amazonian manatee *Trichechus inunguis* in Ecuador. Biotropica 18(2):150–156

Weir BH, Rowlands IW (1973) Reproductive strategies of mammals. Ann. Rev. Ecol. Syst. 4:139–163

21
Reproduction in Collared Peccaries

Robert L. Lochmiller and Eric C. Hellgren

Introduction

The collared peccary *Tayassu tajacu*, a gregarious mammal in the order Artiodactyla, belongs to the family Tayassuidae with the white-lipped peccary *T. pecari* and the recently discovered Chacoan peccary *Catagonus wagneria*. Distribution of the three extant members of this unique family of ungulates overlaps throughout a considerable portion of their South American range. Collared peccaries are found in diverse habitats from the southwestern United States to the northern regions of Argentina. Collared peccaries comprise an important source of protein for subsistence among South American Indians and are harvested routinely by sport and commercial interests (Bodmer et al. 1988; Vickers 1988).

Detailed information on the reproductive biology of peccaries is essential to understanding the consequences of habitat destruction, intensive harvesting, and resource limitations on the dynamics of wild populations. Currently, only the reproductive biology of the collared peccary is well understood, having been studied intensively in both captive and wild environments in the southwestern United States. Basic descriptions of the reproductive biology of the other two peccary species are available in Roots (1966), McDonald and Lasley (1978), and Mayer and Brandt (1982). Additional information on sexual maturation, fecundity rates, conception rates, neonatal survival, and basic biology for these two species will be necessary before biologists interested in their management can adequately model herd and population dynamics in response to environmental stresses. In this paper, we review basic morphology and physiology of reproduction in the most widely distributed member of the Tayassuidae, the collared peccary, and also examine the consequences of resource limitations (primarily nutritional) on various aspects of their reproductive biology.

Female Reproductive Biology

FUNCTIONAL ANATOMY

Detailed descriptions of the gross and microscopic anatomy of the reproductive organs of female collared peccaries are provided in several studies (Wislocki 1931; Low 1970; Lochmiller et al. 1986a). The general morphology of their reproductive tract (Table 21.1) is similar to domestic swine (*Sus scrofa*). The peccary uterus is bicornuate with cornua (10 to 25 cm) and oviducts (9 to 12 cm) slightly shorter and more coiled than the porcine sow. The gravid uterus of peccaries undergoes substantial changes in size, increasing from 20 to 150 g during the first 100 days of gestation (Lochmiller et al. 1986a). Microscopically, the morphology of the uterus is remarkably similar to the porcine sow (Wislocki 1931).

Characteristics of the peccary placenta have been compared with Suidae and Hippopotamidae by MacDonald (1983) and MacDonald and Bosma (1985). As with other Suina, the peccary has a diffuse epitheliochorial placenta. The chorion is folded, with rows of shallow villi on its surface and interrupted by round to irregularly shaped areolae (Wislocki 1931; MacDonald and Bosma 1985). Like Suids, peccary umbilical cords contain two arteries and one vein (MacDonald and Bosma 1985). Fetuses tend to be equally divided between the uterine horns in pregnant females.

The ovaries of the collared peccary are relatively large (Table 21.1) and symmetrical in shape, but often appear bilobed in the presence of large functional corpora lutea. Typically, a large number of follicles will develop into primary follicles during an estrous cycle, but only 10 to 40 normally

TABLE 21.1. Gross morphology of the adult female and male reproductive tract in collared peccaries (from Lochmiller et al. 1986a).

Reproductive organ	Weight (g)	Volume (ml)
Female		
Uterus	48.6	45.9
Cervix	11.7	11.2
Paired ovaries	2.03	1.91
Vagina	13.3	12.6
Total tract	75.6	71.7
Mammary glands	5.7	—
Male		
Paired testes	36.4	35.3
Paired epididymides	6.1	5.8
Bulbo-urethral glands	33.6	32.8
Seminal vesicle glands	17.8	17.4
Mammary glands[a]	5.4	—

[a] Total weight of paired inguinal and abdominal mammae (functional glands) in nonlactating females and males.

reach a size greater than 1.5 mm in diameter and only one to four will actually ovulate at 4 to 8 mm (Low 1970; Lochmiller et al. 1986b). Detailed histological descriptions of changes in luteal tissue throughout gestation can be found in Sowls et al. (1976).

Sexual Maturity

Sowls (1966) reported that female peccaries in captivity breed as early as 33 weeks of age; however, young females begin cycling as early as 30 weeks of age (Low 1970). Physiologically, most females on nutritionally adequate diets will be fully capable of breeding by 12 months of age. Considering the likelihood of periods of nutritional stress in the wild, peccaries normally do not breed until they are at least two years of age. The youngest documented age at conception in a wild population is 16 months (Low 1970).

Estrous Cycle

Collared peccaries possess a polyestrous-type reproductive pattern where the duration of the estrous cycle varies greatly among individuals and between cycles within individuals. Using vasectomized adult males, Sowls (1966), Low (1970), and Coscarelli (1985) determined that the average length of the estrous cycle for individual females ranged from 22.0 to 26.0 days, and the duration of estrus (heat) ranges from one to five days (Sowls 1966; Low 1970; Coscarelli 1985). Information from Lochmiller et al. (1984) suggested that a short estrus of one day is relatively infertile and that litter size may be influenced by copulation frequency (Table 21.2).

Research on nutritional stress in the adult female collared peccary has revealed the estrous cycle and estrus to be extremely sensitive to prior nutritional history (Lochmiller et al. 1986b). Females fed a diet similar

Table 21.2. Average duration of estrus in a captive colony of adult female collared peccaries from south Texas. Observe the apparent relationship between copulation frequency and incidence of pregnancy and litter size (calculated from Lochmiller et al. 1984).

Litter size	No. adults bred	No. of days copulating		Duration of estrus (days)	
		Mean	Range	Mean	Range
0[a]	4	1.0	1 to 1	1.0	1 to 1
1	3	1.3	1 to 2	1.3	1 to 2
2	12	3.8	2 to 6	3.2	2 to 5
3	1	3.0	—	5.0	—

[a] Females that failed to conceive a litter.

TABLE 21.3. General behavioral and biological characteristics of the estrous cycle of adult female collared peccaries fed either a nutritionally adequate diet (N = 7) or a diet restricted in energy and protein (N = 6) for 10 weeks prior to the peak breeding season. Adult females were mated to captive males for 35 consecutive days beginning in the 10th week of treatment (adapted from Lochmiller et al. 1986b).

	No. of females	
Reproductive parameter	Adequate diet	Inadequate diet
In anestrous condition	0	3
Cycling	7	3
Exhibiting ovalatory estrus	7	2
Exhibiting anovulatory estrus	0	1
Conceiving and implanting	6	1
Length of estrus (days)		
Mean	3.5	1.3
Range	2–4	1–2

nutritionally to the prickly pear cactus (*Opuntia* sp., a primary component of peccary diets during periods of below average rainfall) for 10 weeks (resulting in an 18% reduction in body weight) showed a high incidence of anestrous females, a substantially shortened estrus, and a low incidence of conception (Table 21.3). The short duration of estrus among malnourished peccaries might have contributed to fertilization failures and low pregnancy rate in the above study. Reproductive responses reported from experimental trials of malnutrition partially explain the frequent herd recruitment failures that were observed accompanying periods of below average rainfall in the southwestern United States (Low 1970; Bissonette 1982; Hellgren et al. 1984).

OVULATION AND CONCEPTION

Coscarelli (1985) documented the actual events of ovulation, including the time of ovulation relative to the onset of estrus, in captive female peccaries using laparoscopy. He reported that mature follicles appear ruby to bright red and translucent, measure 6 to 12 mm, and are prominent on the ovary surface just prior to ovulation. Coscarelli (1985) observed that ovulation of the most mature follicles (usually two) occurred 12 to 24 hours after the onset of estrus and all ovulations occur at the apex of the follicle.

Low (1970) observed average ovulation rates of 2.2 ova/estrus in south Texas and 2.0 ova/estrus in west Texas; rates did not differ with age. Although Low (1970) did not observe any strong relationship between range condition and ovulation rate, experimental evidence indicates that nutritional stress can significantly depress the number of ova shed during estrus (Lochmiller et al. 1986b).

Table 21.4. Documented litter sizes and incidence of twinning among captive and wild populations of collared peccaries.

Location	N	Average litter size[a]	% Twins	Authority
Captive	32	1.78 (born)	66	Lochmiller et al. 1984
Captive	78	2.00 (born)	68	Sowls 1984
Captive	29	2.17 (born)	79	Sowls 1966
Wild	7	1.28 (in utero)	—	Jennings and Harris 1953
Wild	20	2.10 (in utero)	50	Sowls 1966
Wild	34	1.73 (in utero)	68	Low 1970
Wild	45	1.44 (young)	38	Knipe 1957
Wild	26	1.81 (young)	27	Neal 1959
Wild	82	1.50 (young)	50	Ellisor and Harwell 1979
Chacoan peccary	11	2.72 (in utero)	—	Mayer and Brandt 1982

[a] Litter size estimates based on field observations of young (young), in utero fetal counts (in utero), or counts of newborns in captivity (born).

Gestation and Fetal Development

Although uncommon, migration of ova from the side of ovulation to the opposite cornua for implantation occurs in peccaries. Litter size varies from one to four with a mode of two (Table 21.4), although a litter of five has been documented (Halloran 1945). Intrauterine mortality is not great and probably varies from 3 to 10% (Smith and Sowls 1975; Mayer and Brandt 1982; Lochmiller et al. 1986b). Field observations of litter size based on the number of neonatal young observed with females within herds are nearly always below estimates derived from in utero fetal counts, indicating that substantial neonatal mortality can occur in the wild. Gestation averages 145 days (range 140 to 149 days; Sowls 1966; Low 1970; Lochmiller et al. 1984). Lochmiller et al. (1987) found no evidence that moderate restrictions in protein and energy intake during gestation influence litter size, number of young born alive, and length of gestation.

Birth weight of peccaries varies inversely with litter size so that the largest young are produced in singleton litters. Mean birth weights have been reported by Neal (1959; 454 g), Smith and Sowls (1975; 639 g for males, 608 g for females), and Lochmiller et al. (1984; 693 g). Size discrepancies among litter mates from larger litters (>3) can be dramatic (Lochmiller et al. 1984; Hellgren unpubl. data).

The endocrinology of gestation has been examined in considerable detail by Sowls et al. (1976) and Hellgren et al. (1985). Moderate nutritional (caloric and protein) stress does not adversely influence the production of reproductive steroids during gestation (Hellgren et al. 1985). This suggests that recruitment failures which typically occur in wild peccary populations when food conditions are poor (Bissonette 1982) may be the result of preconception or postpartum events.

Postpartum Estrus and Lactation

The collared peccary is similar to domestic swine in their ability to express an ovulatory postpartum heat. Low (1970) reported that 41% of female wild peccaries collected from south Texas during a period of good range conditions had evidence of consecutive pregnancies while only 9% had similarly re-mated during drought conditions. The ability of the collared peccary to override lactational anestrus in the presence of good range nutritional conditions provides a powerful mechanism for recruitment into the population.

The duration of lactation is probably 6 to 8 weeks (Sowls 1984), although Byers (1983) showed that nursing activity did not decrease until after 12 to 16 weeks postpartum. The gross and fatty acid composition of peccary and porcine milks are fairly similar, with some differences in mineral composition (Sowls et al. 1961; Brown et al. 1963; Lochmiller et al. 1985a). Although experimental evidence that nutrition influences lactation is not available, mortality rates for nursing offspring are high on poor quality rangelands, often approaching 100% during drought (Bissonette 1982).

Male Reproductive Biology

Functional Anatomy

Anatomy of the reproductive tract in male collared peccaries is very similar to that of the domestic boar. Adult testes measure approximately 45 mm × 25 mm (Low 1970; Hellgren 1984), weigh 15 to 25 g, and have an estimated testicular volume of 30 to 50 mm^3 (Hellgren et al. 1989). Peccaries have large, well-developed accessory glands, notably the seminal vesicles and bulbo-urethral glands (Table 21.1).

Sexual Maturity

The testes are scrotal from at least one month prior to birth and remain scrotal throughout life. In wild peccaries, they grow from 0.1 g at birth to an adult weight by 18 and 24 months in south and west Texas animals, respectively (Low 1970). At birth, only spermatogonia and Sertoli cells are present in the seminiferous tubules. By three to four months of age, spermatocytes begin to form and a few spermatozoa appear in the tubules by 9 to 10 months; sperm become common in the tubules by 11 months of age (Low 1970). Age of first successful fertilization has been documented in pen-reared peccaries at 10.7 months in Arizona (Sowls 1966). Tubule diameter and sperm counts increase to a maximum at 30 months, where it is maintained for at least nine years (Low 1970).

Reproductive Physiology

Semen characteristics are only available from electroejaculates obtained from captive animals (Hellgren et al. 1989). Electroejaculation-obtained samples average 1.1 ml liquid volume, 0.9 ml gel volume, and 2.0 ml total volume. These samples are likely underestimates of actual peccary ejaculate volume; two gel fractions collected from bred female peccaries had volumes of 45 and 80 ml, respectively (Hellgren et al. 1989). Sperm concentrations (spermatozoa/ml) and total sperm counts in electroejaculates obtained from eight adult male peccaries over 14 months averaged 371×10^6 and 596×10^6 sperm/ml, respectively, with little seasonal variation (Hellgren et al. 1989).

Environmental Influences on Reproduction

The question of seasonal reproductive patterns in the male has not been clearly elucidated. Hellgren et al. (1989) showed a low-amplitude circannual rhythm in serum testosterone, testicular volume, and scrotal circumference in captive peccaries, with maximal levels between October and March, coincident with the peak breeding season in the northern part of their range. Nevertheless, evidence of parturition in all months suggests that the male peccary, at least in the northern end of its range, is fertile throughout the year. No information on male reproductive biology is available from more southern parts of the collared peccary's range to assess seasonality in male reproductive function, especially in relation to wet and dry seasons in tropical habitats.

Nutrition may also play a role in male reproductive biology. Severe dietary restriction (25% of ad libitum) of captive male peccaries for nine weeks produced no apparent effects on spermatogenic function, but did depress scrotal circumference and serum testosterone concentrations (Lochmiller et al. 1985b).

Breeding Seasons and Behavior

Reproductive Seasons

Although the collared peccary is basically a polyestrous breeder capable of breeding year-round, peak periods of reproductive activity are evident in populations throughout their range. Evidence indicates that breeding activity is susceptible to changing nutritional conditions as influenced (primarily) by rainfall in xeric environments from the southwestern United States (Low 1970; Bissonette 1982) to Brazil and Paraguay (Mayer and Brandt 1982; Sowls 1984). Peak parturition periods typically coincide with the early summer rainy season, which undoubtedly has a strong influence on neonatal and juvenile survival and development.

TABLE 22.5. Progesterone concentrations in 50–100 μl aliquots of dorsal scent gland fluid recovered from female collared peccaries in various reproductive states.

Animal	Progesterone (ng/ml)	Reprodcutive status
R16	1.69	anestrus
R20	2.03	follicular phase of cycle
R7	13.85	pregnant, one fetus
R21	17.61	pregnant, two fetuses

COURTSHIP

Several authors have provided detailed descriptions of the courtship behavior of collared peccaries in both captive and wild environments (Schweinsburg 1969; Bissonette 1982; Sowls 1984). Basic "greeting" behaviors (scent gland rubbing, nuzzling, nose-nose contact) usually precede behaviors directly related to courtship, such as nose-vulva contact, "inhibited" biting of ears and neck, and nose-penis contact (Sowls 1984). Involvement of the prominent dorsal scent gland of peccaries in reproduction is unclear. The scent produced by this gland may function as a semi-ochemical means of communicating reproductive status, as suggested by the presence of steroid hormones (Table 21.5).

Although rarely witnessed in the wild, we have observed several instances of aggression between alpha and subordinate males in the presence of a receptive female in captivity. In general, aggressive encounters such as these are considered highly unusual in wild peccary herds during the breeding season (Sowls 1984). However, dominant alpha males appear to breed with a higher percentage of estrous females than do subordinate males in wild herds (Bissonette 1982).

Future Reproductive Research Needs

Although considerable progress has been made towards understanding the basic physiology of reproduction in the collared peccary, a substantial void remains in the area of reproductive ecology. This is particularly apparent in the southern portions of their range where virtually nothing is known about the biology of this species, as well as the white-lipped and Chacoan peccaries. Research has clearly documented that reproduction is sensitive to changes in nutritional status. Future research activities should examine how natural (rainfall) and man-induced (livestock grazing, deforestation) perturbations impact on the nutritional resources within the habitat of collared peccaries and document their impact on reproduction and recruitment in the population.

Acknowledgments

This work was supported by the National Science Foundation (BSR-8657043) and National Rifle Association.

References

Bissonette JA (1982) Ecology and social behavior of the collared peccary in Big Bend National Park. U.S. National Park Service Scientific Monograph Series No. 16. 85 pp

Bodmer RE, Fang TG, Ibanez LM (1988) Ungulate management and conservation in the Peruvian Amazon. Biological Conservation 45:303–310

Brown WH, Stull JW, Sowls LK (1963) Chemical composition of the milk fat of the collared peccary. Journal of Mammalogy 44:112–113

Byers JA (1983) Social interactions of juvenile collared peccaries, *Tayassu tajacu* (Mammalia:Artiodactyla). Journal of Zoology (London) 201:83–96

Coscarelli KP (1985) Ovulation in the collared peccary as documented by laparoscopy. M.S. Thesis, Texas A & M University, College Station. 53 pp

Ellisor JE, Harwell F (1979) Ecology and management of javalina in south Texas. Federal Aid Report Serial #16, Texas Parks and Wildlife, Austin. 25 pp

Halloran A (1945) Five fetuses for *Pecari angulatus* from Arizona. Journal of Mammalogy 26:434

Hellgren EC (1984) The effect of diet and season on selected parameters of the reproductive physiology of the collared peccary (*Tayassu tajacu*). Unpubl. M.S. Thesis, Texas A & M University, College Station

Hellgren EC, Lochmiller RL, Amoss MS, Grant WE (1985) Serum progesterone, estradiol-17β, and glucocorticoids in the collared peccary during gestation and lactation as influenced by dietary protein and energy. General and Comparative Endocrinology 59:358–368

Hellgren EC, Lochmiller RL, Amoss MS, Seager SWJ, Magyar SJ, Coscarelli KP, Grant WE (1989) Seasonal variation in serum testosterone, testicular measurements, and semen characteristics in the collared peccary (*Tayassu tajacu*). Journal of Reproduction and Fertility 85:677–686

Hellgren EC, Lochmiller RL, Grant WE (1984) Demographic, morphologic and reproductive status of a herd of collared peccaries (*Tayassu tajacu*) in south Texas. American Midland Naturalist 112:402–407

Jennings WS, Harris (1953) The collared peccary in Texas. Federal Aid Report Series #12, Texas Game and Fish Commission. 31 pp

Knipe T (1957) The javalina in Arizona. Arizona Game and Fish Wildlife Bulletin No. 2. 96 pp

Lochmiller RL, Hellgren EC, Grant WE (1984) Selected aspects of collared peccary (*Dicotyles tajacu*) reproductive biology in a captive Texas herd. Zoo Biology 3:145–149

Lochmiller RL, Hellgren EC, Grant WE (1986a) Absolute and allometric relationships between internal morphology and body mass in the adult collared peccary, *Tayassu tajacu* (Tayassuidae). Growth 50:296–316

Lochmiller RL, Hellgren EC, Grant WE (1986b) Reproductive responses to nutritional stress in adult female collared peccaries. Journal of Wildlife Management 50:295–300

Lochmiller RL, Hellgren EC, Grant WE (1987) Influence of moderate nutritional stress during gestation on reproduction of collared peccaries (*Tayassu tajacu*). Journal of Zoology (London) 211:321–328

Lochmiller RL, Hellgren EC, Grant WE, Greene LW, Dill CW (1985a) Description of collared peccary (*Tayassu tajacu*) milk composition. Zoo Biology 4:375–379

Lochmiller RL, Hellgren EC, Varner LW, Greene LW, Amoss MS, Seager SWJ, Grant WE (1985b) Physiological responses of the adult male collared peccary to severe nutritional restriction. Comparative Biochemistry and Physiology 82A:49–58

Low WA (1970) The influence of aridity on reproduction of the collared peccary (*Dicotyles tajacu* [Linn.]) in Texas. Ph.D. Dissertation, Univ. British Columbia, Vancouver. 170 pp

MacDonald AA (1983) A comparative histological study of placentae from pigs, peccaries, and hippopotami. Acts Morphological Neer-Scandanavian 21:180

MacDonald AA, Bosma AA (1985) Notes on placentation in the suina. Placenta 6:83–91

McDonald LE, Lasley BL (1978) Infertility in a captive group of white-lipped peccaries. Journal of Zoo Animal Medicine 9:90–95

Mayer JJ, Brandt PN (1982) Identity, distribution, and natural history of the peccaries, *Tayassuidae* In Mares MA, Genoways HH (eds.). Mammalian Biology in South America, Vol. 6. Spec. Publ. Ser., Pymatuning Lab. Ecol., Univ. Pittsburgh, Pittsburgh Pennsylvania. pp. 433–455

Neal BJ (1959) A contribution on the life history of the collared peccary in Arizona. American Midland Naturalist 61:177–190

Roots CG (1966) Notes on the breeding of the white-lipped peccaries *Tayassu albirostus* at Dudly Zoo. International Zoo Yearbook 6:198–199

Schweinsburg RE (1969) Social behavior of the collared peccary (*Pecari tajacu*) in the Tucson Mountains. Ph.D. Thesis, University of Arizona, Tucson. 129 pp

Smith NS, Sowls LK (1975) Fetal development of the collared peccary. Journal of Mammalogy 56:619–625

Sowls LK (1966) Reproduction in the collared peccary (*Tayassu tajacu*). In Rowlands IW (ed.). Comparative Biology of Reproduction in Mammals. Academic Press, London. pp. 155–172

Sowls LK (1984) The Peccaries. The University of Arizona Press, Tucson. 251 pp

Sowls LK, Smith NS, Holtan DW, Moss GE, Estergreen L (1976) Hormone levels and corpora lutea cell characteristics during gestation in the collared peccary. Biology of Reproduction 14:572–578

Sowls LK, Smith VP, Jenness R, Sloan RE, Regeher E (1961) Chemical composition and physical properties of the milk of the collared peccary. Journal of Mammalogy 42:245–251

Vickers WT (1988) Game depletion hypothesis of Amazonian adaptation: Data from a native community. Science 239:1521–1522

Wislocki GB (1931) Notes on the female reproductive tract (ovaries, uterus and placenta) of the collared peccary (*Pecari angulatus bangsi* Goldman). Journal of Mammalogy 12:143–149

Index

MIX
Papier aus verantwortungsvollen Quellen
Paper from responsible sources
FSC® C105338

If you have any concerns about our products,
you can contact us on
ProductSafety@springernature.com

In case Publisher is established outside the EU,
the EU authorized representative is:
Springer Nature Customer Service Center GmbH
Europaplatz 3, 69115 Heidelberg, Germany

Printed by Libri Plureos GmbH
in Hamburg, Germany